SATURN V

THE COMPLETE MANUFACTURING AND TEST RECORDS
Plus Supplemental Material

Alan Lawrie
&
Robert Godwin

An Apogee Books Publication

Dedication

This book is dedicated to Clive B Taylor, my first boss at British Aerospace, and the man responsible for getting me into the space propulsion business. Thank you Clive.

Alan Lawrie

All rights reserved under article two of the Berne Copyright Convention (1971).
We acknowledge the financial support of the Government of Canada through the Book Publishing Industry Development Program for our publishing activities.

Published by Apogee Books, Box 62034, Burlington, Ontario, Canada, L7R 4K2.
http://www.apogeebooks.com

Printed and bound in Canada
Saturn - The Complete Manufacturing & Test Records & Supplemental Material
by Alan Lawrie & Robert Godwin

DVD by Robert Godwin

ISBN 1-894959-19-4

©2005 Alan Lawrie & Robert Godwin

Unless otherwise stated all numbered photos are from NASA (either via NARA or from the NASA web), whilst un-numbered photos have been supplied by Alan Lawrie

Acknowledgements

I am indebted to the many people who have assisted me with the research, production and support needed for me to be able to write this book.

First and foremost my wife Olwyn Georgina who supported and encouraged me throughout the project and who "lived" Saturn V for two years. Thank you for helping with the proof reading and for your wise words, and also thanks to my mother for your encouragement over the years.

Cathy Carr at Cranfield University library gave me access to their archives of NASA STAR catalogs that list every NASA published report. Pauline Roe and Sandra Horsell at the EADS Astrium library obtained many obscure microfiche and paper copy reports from the British Library, many of which have not surfaced anywhere else. Thank you also for your library facilities and Pauline for your words of encouragement. The British Library should be congratulated for holding these obscure Saturn reports, so obscure that they didn't want them back.

Bob Jaques at the NASA Marshall Space Flight Center located many documents and took the time to show me the test stands and arrange a NASA photographer to be there. Mike Wright runs the history office at MSFC and set up the useful on-line document database. Jonathan Baggs continues to help in that office.

At the NASA Stennis Space Center Carol Castle of the History Office assisted me and gave me a tour of the test stands. Bob Bell at the Michoud Assembly Facility gave me a great tour in a golf cart, loaded me with freebees and was very helpful. Thank you to Margaret Pepersack in the Michoud library for helping setting up the visit. It was a pleasure meeting you after all the e-mails Margaret.

At the National Archives and Records Administration at East Point, Georgia, I was given wonderful support by Charlie Reeves and Arlene Royer. They didn't complain as I requested box after box from the vaults. Arlene continued to assist in the provision of long-lost negatives of Saturn photographs.

Anne Coleman maintains several wonderful collections of space-related archives in the Salmon Library at the University of Alabama at Huntsville. Foremost is the unique Saturn collection. Anne hosted me wonderfully in the archive for a day, providing copying facilities and home made snacks to keep me going. Thank you Anne for answering my many requests for document copies and for keeping such a well-ordered archive. I recommend UAH as the first port-of-call for any serious Saturn student. Anne also introduced me to Dave Christensen who was there when the Saturn History project was being set up at UAH in the late 60s.

Rocketdyne, in Canoga Park, California, turned out to be the unexpected gem. After trying unsuccessfully for some time to gain access I must thank Terry Murphy, Division Director, for opening the door at the last minute, and Cindy Saruwatari for coordinating arrangements. Paul Coffman, a J-2 engine veteran at Rocketdyne, gave Olwyn and me a tour of the Leadership and Learning Center at Canoga Park and has provided timely answers to all of my questions however obscure. If it wasn't for Paul this book would have had many gaps – thank you so much.

Bill Larsen at the NASA JSC history office provided first hand knowledge about the movement of Saturn stages to JSC. Shelly Henley Kelly at the University of Houston Clear Lake, helped in locating an unpublished draft of a MSFC chronology and also gave advice about book writing as she has already achieved that feat.

Thank you to Phil Broad for allowing me to use some of the S-IVB photographs posted on his web site.

Thank you to Tori Nichols, a Masters student in propulsion engineering at Cranfield University, who very kindly spent many hours researching and organizing the on-line photographs which were used in this book, drafting the SACTO battleship chapter, compiling the S-IC manufacturing chart and proof reading various sections of the book.

I thank my various space propulsion bosses over the years, Clive Taylor, Dave Gale and Ali Hussain, who have encouraged me and continue to employ me to work in an area that I enjoy.

Thanks to the British Interplanetary Society for publishing an extract of this book in the June 2004 issue of Spaceflight magazine. If you are interested in space history and also current events I recommend a subscription to Spaceflight. Thanks to my long time friend and fellow propulsion engineer Geoff Statham, despite the fact that you wouldn't lend me that book of rare Russian space photos from the BIS library in 1983!

Finally to Rob Godwin and the team at Apogee, firstly for producing such a great set of space books, but then for having the faith in me to complete this project and publish the book.

Thank you one and all. (See pictures on page 324)

Alan Lawrie.

I would like to thank Ed Buckbee and Dennis Wingo for helping provide materials for this book and Gayle Frere for assistance with the DVD.

Robert Godwin

Introduction

The genesis for this book occurred in the fall of 2002. Up until that time I had always had an interest in space history and had spent over 20 years in the satellite propulsion industry in England. That autumn NASA released an on-line version of the Saturn Illustrated Chronology that covers in words and pictures the development of the Saturn family of rockets up until April 1968. At about the same time Spacecraft Films released a DVD collection covering the Saturn I and IB rockets. I studied these two publications and realized that I didn't know quite as much as I thought I did about the Saturn rocket.

With my appetite whetted I searched for more information. Surprisingly, for a subject that had such a major impact on the middle part of the last century, very little has been published. After all, the Saturn rockets were the biggest ever produced and got man to the moon for the first time. Where was all the background information I was looking for? Without doubt the best book so far on this subject is the official NASA history, Stages to Saturn by Roger Bilstein, published in 1980. This book was over ten years in the making and provides a good summary of the overall project.

However, I was looking for specifics. Details of tests performed, engines used, problems encountered, logistics of transportation. Almost none of this information exists in the open literature.

I started on a two year quest to track down the information that had not appeared anywhere else. What exactly were the serial numbers of the F-1 engines on the Apollo 11 Saturn V? When were the engines tested? When was the stage tested? For how long? What problems did they have in manufacturing the stage?

It became apparent that I would have to gather the information in piecemeal form from many different sources. There simply isn't a central library that has everything. Part of the problem is that much of the information was only published in internal Contractor reports. Some of these were entered into the NASA CR system and can be obtained (but not that easily I have to say.) However, most reports are gone forever. The companies that produced the reports have changed hands several times since the 1960s and there is no central library. Even Bilstein noted that many documents were not available, and that was when he was researching in the 1970s!

The official position is that the original reports have been sent to the National Archives. The archive responsible for Saturn documentation is near Atlanta. I spent time there and it is hit and miss whether you find what you want as there are not the resources to catalog the contents of hundreds and hundreds of boxes that have sat there for 30 years.

I was not deterred and I raided the British Library of its Saturn collection, which turned out not only to be a gold mine, but was unique as I never came across these NASA reports anywhere else.

After that it was clear that I needed to visit the US. In the wake of 9/11 it was difficult obtaining clearance but I managed to visit the History Offices at Marshall and Stennis, and toured the facilities and test stands at Marshall, Stennis and Michoud. I searched the Saturn Collection at the University of Alabama, examined the Saturn hardware at the Alabama Space and Rocket Center in Huntsville, drove out to the Alabama welcome center where a Saturn IB rocket is on display and finally searched the vaults of the National Archives.

At the National Archives they have the original negatives for the Saturn photos but no index. At Stennis I found the index, put the two together, and what you see in this book are Saturn photos published for the first time in the best possible clarity.

A second trip took me to Rocketdyne at Canoga Park. Previously I had seen Saturn hardware at NASM in Washington, JSC in Houston and at the Kennedy Space Center in Florida.

Several of the reports I obtained were done so with the assistance of the Freedom of Information Act, but generally people were as helpful as they could be. The main problem was that it is now 40 years since these activities and many of the documents have been lost forever. The people that worked on the Saturn program are generally retired and many, sadly, are no longer with us. It has become a quest – gather this information before it is lost once and for all.

Along the way I have met some wonderful characters who have offered me help and encouragement and I am grateful to all of you.
Translating all of this data into a logical book took seven months.

I struggled with deciding on the scope of the book. It could have been all Saturn rockets, but in the end Saturn V was quite enough. Saturn I and IB will have to wait for Volume 2! I decided to stop at the point that each stage was delivered to KSC as the processing after the stages were mounted together and the details of the launch itself could fill yet another book. The only places where I cover activities at KSC are when a stage was returned to the manufacturer or was never launched.

The units used in this book are imperial (English) as the Saturn program used these units and the US continues to. Although I am fully metric, for me the Saturn V will always be 363 feet tall and have a thrust of 7.5 million pounds.

I decided to broadly arrange the chapters with each one covering the story of a Saturn stage. I have covered not only the flight stages but also all of the ground test stages. It would have been remiss not to cover the engines, the manufacturing and testing facilities, and the transportation logistics. So there are chapters on all of these. In addition I have compiled the factual data into a number of useful spreadsheets.

I am confident that 95% of the information in this book has not appeared in published form anywhere. If there are mistakes, some will be mine, but there were discrepancies between various NASA and contractor reports on such things as dates and firing times for which I had to make a best estimate.

In the event the only things I wanted to locate, but could not, were the time of day that some of the S-II and S-IVB stages were fired at. I found all the dates. In addition some of the details of the later S-IVB Battleship firings at MSFC were rather elusive.

I hope that you find the book a useful reference on aspects of the Saturn V program not available anywhere else. Its been two years of my life (done in my spare time) and I'm satisfied that I have brought this information out of the grave it was sliding into.

One last note. I was touring the Michoud Assembly Facility the day that President Bush announced the Moon/Mars initiative. Hopefully that will renew interest in the rocket designed and built 40 years ago – the Saturn V.

Happy reading.

Alan Lawrie - March 2005 - Hitchin, England.

Supplemental Information

In the following pages are some of the rarer documents provided by NASA regarding the Saturn V. In conjunction with Alan, I chose the material which we thought best represented an overview of this remarkable machine. The first document is the Saturn V News Reference. This was originally distributed as a ring binder to members of the press corps and it is now highly desirable, selling for as much as $1500. It is a very insightful overview of the vehicle. The next document is an extremely rare booklet from 1965 called "Saturn V Payload Planner's Guide." This comes from the era when no one could have possibly foreseen the early demise of the vehicle. It looks in detail at the sort of payloads which the booster was capable of sending aloft. One can only wish that we still had such a vehicle today.

Robert Godwin March 2005 - Burlington, Canada

Contents

The Saturn V News Reference

THE SATURN V	13
Description	13
Typical Lunar Landing Mission	14
Earlier Saturns	15
How Saturn V Design Was Reached	16
Program Highlights	18
FIRST STAGE	21
First Stage Fabrication and Assembly	21
Post-manufacturing Checkout	26
First Stage Systems	27
Visual Instrumentation	35
First Stage Flight	37
F-1 ENGINE	39
Thrust Chamber Assembly	39
Turbopump	41
Gas Generator System	42
Propellant Feed Control System	43
Pressurization System	44
Engine Interface Panel	45
Electrical System	45
Hydraulic Control System	45
Flight Instrumentation System	46
Engine Operation	47
Engine Cutoff	48
SECOND STAGE	50
Structure	50
Propellant System	56
Ullage Motors	59
Thermal Control System	60
Flight Control System	60
Measurement System	60
Electrical System/ Ordnance System	61
Ground Support	62
THIRD STAGE	64
Stage Fabrication and Assembly	64
Third Stage Systems	67
J-2 ENGINE	78
J-2 Engine Description	78
Thrust Chamber and Gimbal System	78
Propellant Feed System	79
Gas Generator and Exhaust System	82
Control System	82
Start Tank Assembly System	83
Flight Instrumentation System	83
Engine Operation	83
INSTRUMENT UNIT	86
Instrument Unit Description	86
Instrument Unit Fabrication	86
Instrument Unit Systems	86
FACILITIES	93
Boeing Facilities	93
North American Space Division Facilities	94
Douglas Facilities	95
IBM Facilities	96
North American Rocketdyne Facilities	97
Huntsville Facilities	98
Mississippi Test Facility	100
Kennedy Space Center	101
TESTING	106
Introduction	106
Qualification Testing	107
Reliability Testing	107
Development Testing	108
Acceptance Testing	108
Automatic Checkout	108
Flight Testing	108
Test Documentation	108
VEHICLE ASSEMBLY AND LAUNCH	109
Assembly and Checkout	109
Testing at the Launch Site	110
PROGRAM MANAGEMENT	111
NASA Organization	111
Marshall Center Project Man. Organization	111
Management Personnel	113
FLIGHT HISTORY	**119**
APPENDIX - GLOSSARY	**122**
APPENDIX A- SUBCONTRACTORS	123
APPENDIX B- INDEX	129

The Complete Manufacturing & Test Records

Facilities and transportation	135
F-1 engine	141
J-2 engine	145
S-IC-S stage	149
S-IC-F stage	150
S-IC-D stage	152
S-IC-T stage	155
S-IC-1 stage	162
S-IC-2 stage	163
S-IC-3 stage	165
S-IC-4 stage	167
S-IC-5 stage	168

S-IC-6 stage	172
S-IC-7 stage	174
S-IC-8 stage	176
S-IC-9 stage	179
S-IC-10 stage	180
S-IC-11 stage	182
S-IC-12 stage	186
S-IC-13 stage	188
S-IC-14 stage	189
S-IC-15 stage	192
S-II-S stage	194
S-II-TS-A stage	196
S-II-TS-B stage	197
S-II-TS-C stage	199
S-II common bulkhead test article	200
S-II high force test article	201
S-II-EMM test article	202
S-II Battleship stage	203
S-II-F stage	205
S-II-D stage	208
S-II-T stage	208
S-II simulated spacer	211
S-II-1 stage	212
S-II-2 stage	214
S-II-3 stage	216
S-II-4 stage	219
S-II-5 stage	223
S-II-6 stage	226
S-II-7 stage	229
S-II-8 stage	231
S-II-9 stage	234
S-II-10 stage	236
S-II-11 stage	239
S-II-12 stage	241
S-II-13 stage	243
S-II-14 stage	244
S-II-15 stage	246
S-IVB-500FS stage	248
S-IVB-500ST stage	249
S-IVB common bulkhead test article	249
S-IVB-S stage	249
S-IVB MSFC Battleship	250
S-IVB SACTO Battleship	252
S-IVB-F stage	255
S-IVB-D stage	257
S-IVB-T stage	259
S-IVB-501 stage	259
S-IVB-502 stage	260
S-IVB-503 stage	262
S-IVB-503N stage	264
S-IVB-504N stage	266
S-IVB-505N stage	267
S-IVB-506N stage	270
S-IVB-507 stage	272
S-IVB-508 stage	274
S-IVB-509 stage	275
S-IVB-510 stage	277
S-IVB-511 stage	278
S-IVB-512 stage	280
S-IVB-513 stage	281
S-IVB-514 stage	282
S-IVB-515 stage	282
S-IVB-212 stage	284
REFERENCES	286
Tables	
Stage allocation chart	288
Engine allocation chart	288
S-IC-T tests	288
S-IC flight stage tests	289
S-II flight stage firings tests	289
S-IVB flight stage firings tests	289

Saturn V Payload Planners Guide

Introduction	293
Saturn V Configurations	293
First Stage	293
Second Stage	293
Third Stage	293
Instrument Unit	294
SaturnV Capability	294
Payload Consideration	294
Planning & Schedules	294
Launch Vehicle Accomodations	296
Prime Payloads Above the S-IVB stage	298
Payload Thermal Environment	301
Payload Acoustics & Vibration Environment	303
S-IVB Stage Subsystem Information	304
Auxiliary Propulsion System	304
Electrical Power System	305
Orbital & Deep Space Tracking Data And Control Stations	306
Launch Support Facilities	306
Data Reduction & Evaluation	307
Checklist for Auxiliary Payload/ Vehicle Interface Requirements	308
Saturn V Configuration	310
LOR/Apollo Configuration	310
Man Rating, Reliability & Quality Control	315
Saturn V Performance	316
LOR/Apollo Mission Profile	316
General Three Stage Mission	316
Saturn V Growth Potential	317
High Energy Mission Vehicles	318
INDEX	326
Figures	
Saturn V Three Stage LOR Configuration	294
Saturn V Payload Potential	295
Payload Planning & Implementation Flow	296
S-IVB Integration Schedule for Aux. Payload	297
Typical KSC Saturn Preparation schedule	297
Saturn S-IVB Delivery Schedule	297
Proposed Concepts for S-IVB Payload Volume	299
S-IVB Forward Skirt for Config. Volumes	299
S-IVB Alternate Config. for Aux. Payloads	300
S-IVB Fwd Skirt thermal ConditionedPanels	300
Auxiliary Payload Adapter	300
Prime Payload Fairing Configurations	301
Prime Payload Adapter	302
Prime Payload Using S-IVB Propulsion	302
Acoustic Noise-Time History	303
Design Specs for acoustic noise	303
Design Specs for Random Vibration	303
Design Specs for sinusoidal Vibration	303
S-IVB Auxiliary Propulsion System	304
S-IVB Auxiliary Propulsion System Module	306
Saturn V on Pad 39 KSC	308
Saturn V Launch Complex	309
Operational Saturn V Configurations	310
Saturn Launch Vehicles	312
Saturn V First Stage/ S-IC Inboard Profile	312
Saturn V Second Stage/ S-II Inboard Profile	313
Saturn V Third Stage/ S-IVB Inboard Profile	314
Instrument Unit	315
Saturn V nominal LOR/Apollo Mission Seq.	316
Saturn V Payload Capabilities	319
Apogee Altitude Vs Payload	319
Payload Vs. Velocity	319
Altitude Vs Range	319
Dynamic Pressure Vs Flight Time	319
Mach Number Vs Flight Time	320
Inertial Flight Path Angle Vs Flight Time	320
Axial Acceleration Vs Flight Time	320
Inertial Velocity Vs Flight Time	320
Altitude Vs Flight Time	320
Saturn V/Centaur statistics	321
High Energy Mission Configurations	322
Acknowledgements	324

SATURN V

news
REFERENCE

AUGUST 1967

National Aeronautics and Space Administration
George C. Marshall Space Flight Center
John F. Kennedy Space Center

The Boeing Company
Launch Systems Branch

Douglas Aircraft Company
Missile & Space Systems Division

International Business Machines Corporation
Federal Systems Division

Rocketdyne Division
North American Aviation, Inc

Space Division
North American Aviation, Inc

FOREWORD

This volume has been prepared by the five Saturn V major contractors: The Boeing Company; Douglas Aircraft Company; Space Division of North American Aviation, Inc.; Rocketdyne Division of North American Aviation, Inc.; and International Business Machines Corporation in cooperation with the National Aeronautics and Space Administration.

It is designed to serve as an aid to newsmen in present and future coverage of the Saturn V in its role in the Apollo program and as a general purpose large launch vehicle. Every effort has been made to present a comprehensive overall view of the vehicle and its capabilities, supported by detailed information on the individual stages and all major systems and subsystems.

Weights and measurements cited throughout the book apply to the AS-501 vehicle, the first flight version of the Apollo/Saturn V. All photographs and illustrations in the book are available for general publication. The first letter in each photo number is a code identifying the organization holding that negative: B for Boeing; R for Rocketdyne Division of North American; D for Douglas; IBM for IBM; S for Space Division of North American; H for NASA, Huntsville, Ala.; and K for NASA, Kennedy Space Center, Fla.

Addresses are :

The Boeing Company P. O. Box 29100
New Orleans, La. 70129
Attention: William W. Clarke

Douglas Aircraft Company
Missile & Space Systems Division
Space Systems Center
5301 Bolsa Avenue
Huntington Beach, Calif. 92647
Attention: Larry Vitsky

International Business Machines Corporation
Federal Systems Division
150 Sparkman Drive
Huntsville, Ala. 35807
Attention: James F. Harroun

Rocketdyne Division
North American Aviation, Inc.
6633 Canoga Avenue
Canoga Park, Calif. 91304
Attention: R. K. Moore

National Aeronautics and Space Administration
George C. Marshall Space Flight Center
Public Affairs Office
Huntsville, Ala. 35812
Attention: Joe Jones

National Aeronautics and Space Administration
Public Affairs Office
Kennedy Space Center, Fla. 32931
Attention: Jack King

Space Division
North American Aviation, Inc.
Seal Beach, Calif. 90241
Attention: Richard E. Barton

SATURN V FACT SHEET

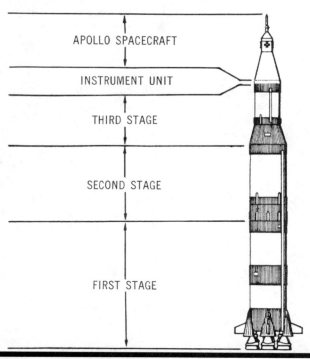

PHYSICAL CHARACTERISTICS

	DIAMETER	HEIGHT	WEIGHT
OVERALL VEHICLE	33 ft.	364 ft.*	6,100,000 lb. (total liftoff)
FIRST STAGE	33 ft.	138 ft.	300,000 lb. (dry)
SECOND STAGE	33 ft.	81 ft. 7 in.	95,000 lb. (dry)**
THIRD STAGE	21 ft. 8 in.	58 ft. 7 in.	34,000 lb. (dry)**
INSTRUMENT UNIT	21 ft. 8 in.	3 ft.	4,500 Lb.
APOLLO SPACECRAFT		80 ft.	95,000 Lb.

*SINCE INDIVIDUAL STAGE DIMENSIONS OVERLAP IN SOME CASES, OVERALL VEHCILE LENGTH IS NOT THE SUM OF INDIVIDUAL STAGE LENGTHS
**INCLUDES AFT INTERSTAGE WEIGHT

PROPULSION SYSTEMS
FIRST STAGE - Five bipropellant F-I engines developing 7,500,000 lb. thrust
RP-1 Fuel - 203,000 gal. (1,359,000 lb.), LOX-331,000 gal. (3,133,000 lb.)
SECOND STAGE - Five bipropellant J-2 engines developing more than 1,000,000 lb. thrust
LH_2 - 260,000 gal. (153,000 lb.), LOX-83,000 gal. (789,000 lb.)
THIRD STAGE -One bipropellant J-2 engine developing up to 225,000 lb. thrust
LH_2 - 63,000 gal. (37,000 lb.), LOX-20,000 gal. (191,000 lb.)

CAPABILITY
FIRST STAGE - Operates about 2.5 minutes to reach an altitude of about 200,000 feet (38 miles) at burnout
SECOND STAGE - Operates about 6 minutes from an altitude of about 200,000 feet to an altitude of 606,000 feet (114.5 miles)
THIRD STAGE - Operates about 2.75 minutes to an altitude of about 608,000 feet (115 miles) before second firing and 5.2 minutes to translunar injection
PAYLOAD - 250,000 Lb. into a 115 statute-mile orbit

THE SATURN V

INTRODUCTION

When the United States made the decision in 1961 to undertake a manned lunar landing effort as the focal point of a broad new space exploration program, there was no rocket in the country even approaching the needed capability. There was a sort of "test bed" in the making, a multi-engine vehicle now known as Saturn I. It had never flown. And it was much too small to offer any real hope of sending a trio to the moon, except possibly through as many as a half dozen separate launchings from earth and the perfection of rendezvous and docking techniques, which had never been tried. That was the situation that brought about the announcement on Jan. 10, 1962, that the National Aeronautics and Space Administration would develop a new rocket, much larger than any previously attempted. It would be based on the F-1 rocket engine, the development of which had been underway since 1958, and the hydrogen-fueled J-2 engine, upon which work had begun in 1960. The Saturn V, then, is the first large vehicle in the U. S. space program to be conceived and developed for a specific purpose. The lunar landing task dictated the make-up of the vehicle, but it was not developed solely for that mission. As President Kennedy pointed out when he issued his space challenge to the Congress on May 25, 1961, the overall objective is for "this Nation to take a clearly leading role in space achievement which in many ways may hold the key to our future on earth." He said of the lunar landing project: "No single space project in this period will be more exciting, or more impressive to mankind, or more important for the long-range exploration of space; and none will be so difficult or expensive to accomplish... The Saturn V program is the biggest rocket effort undertaken in this country. Its total cost, including the production of 15 vehicles between now and early 1970, will be above $7 billion. NASA formally assigned the task of developing the Saturn V to the Marshall Space Flight Center on Jan. 25, 1962. Launch responsibility was committed to the Kennedy Space Center. (The Manned Spacecraft Center, the third center in manned space flight, is responsible for spacecraft development, crew training, and inflight control.)

DESCRIPTION

Marshall Center rocket designers conceived the Saturn V in 1961 and early 1962. They decided that a three-stage vehicle would best serve the immediate needs for a lunar landing mission and would serve well as a general purpose space exploration vehicle. One of the more important decisions made early in the program called for the fullest possible use of components and techniques proven in the Saturn I program. As a result, the Saturn V third stage (S-IVB) was patterned after the Saturn I second stage (S-IV). And the Saturn V instrument unit is an outgrowth of the one used on Saturn I. In these areas, maximum use of designs and facilities already available was incorporated to save time and costs. Many other components were necessary, including altogether new first and second stages (S-IC and S-11). The F-1 and J-2 engines were already under development, although much work remained to be done. The guidance system was to be an improvement on that of the Saturn I. Saturn V, including the Apollo spacecraft, is 364 feet tall. Fully loaded, the vehicle will weigh some 6.1 million pounds. The 300,000-pound first stage is 33 feet in diameter and 138 feet long. It is powered by five F-1 engines generating 7.5 million pounds thrust. The booster will burn 203,000 gallons of RP-1 (refined kerosene) and 331,000 gallons of liquid oxygen (LOX) in 2.5 minutes. Saturn V's second stage is powered by five J-2 engines that generate a total thrust of a million pounds. The 33-foot diameter stage weighs 95,000 pounds empty and more than a million pounds loaded. It burns some 260,000 gallons of liquid hydrogen and 83,000 gallons of liquid oxygen during a typical 6-minute flight. Third stage of the vehicle is 21 feet and 8 inches in diameter and 58 feet and 7 inches long. An inter-stage adapter connects the larger diameter second stage to the smaller upper stage. Empty weight of the stage is 34,000 pounds and the fueled weight is 262,000 pounds. A single J-2 engine developing up to 225,000 pounds of thrust powers the stage. Typical burn time is 2.75 minutes for the first burn and 5.2 minutes to a translunar injection. The vehicle instrument unit sits atop the third stage. The unit, which weighs some 4,500 pounds, contains the electronic gear that controls engine ignition and cutoff, steering, and all other commands necessary for the Saturn

V mission. Diameter of the instrument unit is 21 feet and 8 inches, and height is 3 feet. Directly above the instrument unit in the Apollo configuration is the Apollo spacecraft. It consists of the lunar module, the service module, the command module, and the launch escape system. Total height of the package is about 80 feet.

TYPICAL LUNAR LANDING MISSION

The jumping-off place for a trip to the moon is NASA's Launch Complex 39 at the Kennedy Space Center. After the propellants are loaded, the three astronauts will enter the spacecraft and check out their equipment. While the astronauts tick off the last minutes of the countdown in the command module, a large crew in the launch control center handles the complicated launch operations. For the last two minutes, the countdown is fully automatic. At the end of countdown, the five F-1 engines in the first stage ignite, producing 7.5 million pounds of thrust. The holddown arms release the vehicle, and three astronauts begin their ride to the moon. Turbopumps, working together with the strength of 30 diesel locomotives, force 15 tons of fuel per second into the engines. Steadily increasing acceleration pushes the astronauts back into their couches as the rocket generates 4-1/2 times the force of earth gravity. After 2.5 minutes, the first stage has burned its 4,492,000 pounds of propellants and is discarded at about 38 miles altitude. The second stage's five J-2 engines are ignited. Speed at this moment is 5,330 miles per hour. The second stage's five J-2 engines burn for about 6 minutes, pushing the Apollo spacecraft to an altitude of nearly 115 miles and near orbital velocity of 15,300 miles per hour. After burnout the second stage drops away and retrorockets slow it for its fall into the Atlantic Ocean west of Africa. The single J-2 engine in the third stage now ignites and burns for 2.75 minutes. This brief burn boosts the spacecraft to orbital velocity, about 17,500 miles an hour. The spacecraft, with the third stage still attached, goes into orbit about 12 minutes after liftoff. Propellants in the third stage are not depleted when the engine is shut down. This stage stays with the spacecraft in earth orbit, for its engine will be needed again. Throughout the launch phase of the mission, telemetry systems are transmitting continuously, tracking systems are locked on, and voice communications are used to keep in touch with the astronauts. All stage separations and engine thrust terminations are reported to the Mission Control Center at Houston. The astronauts are now in a weightless condition as they circle the earth in a "parking orbit" until the timing is right for the next step to the moon. The first attempt at a lunar landing is

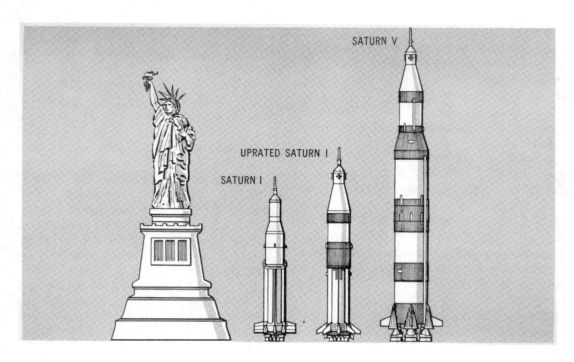

planned as an "open-ended" mission with detailed plans at every stage for mission termination if necessary. A comprehensive set of alternate flight plans will be laid out and fully rehearsed for use if such a decision should prove necessary. For example, a decision might be made in the earth parking orbit not to continue with the mission. At every stage of the mission, right up to touchdown on the moon, this termination decision can be made and an earth flight plan initiated. During the one to three times the spacecraft circles the earth, the astronauts make a complete check of the third stage and the spacecraft. When the precise moment comes for injection into a trans-lunar trajectory, the third stage J-2 engine is re-ignited. Burning slightly over 5 minutes, it accelerates the spacecraft from its earth orbital speed of 17,500 miles an hour to about 24,500 miles an hour in a trajectory which would carry the astronauts around the moon. Without further thrust, the spacecraft would return to earth for re-entry. If everything is operating on schedule, the astronauts will turn their spacecraft around and dock with the lunar landing module. After the docking maneuver has been completed, the lunar module will be pulled out of the forward end of the third stage, which will be abandoned. Abandonment completes the Saturn V's work on the lunar mission.

EARLIER SATURNS

Saturn I

Studies which led to the Saturn family of rockets were started by the Wernher von Braun organization in April of 1957. The aim of the program was to create a 1.5 million-pound-thrust booster by clustering previously developed and tested engines. On Aug. 15, 1958, the Advanced Research Projects Agency (ARPA) formally initiated what was to become the Saturn project. The agency, a separately organized research and development arm of the Department of Defense, authorized the Army Ballistic Missile Agency to conduct a research and development program at Redstone Arsenal for a 1.5 million-pound-thrust vehicle booster. A number of available rocket engines were to be clustered and tested by a full-scale static firing by the end of 1959. The program objectives were expanded by ARPA in October of 1958 to include a multi-stage carrier vehicle capable of performing advanced space missions.

Concurrent with the development of a multi-stage vehicle, static test facilities at Redstone Arsenal and launch complex facilities at Cape Canaveral - now Cape Kennedy were being constructed. The proposed large vehicle project was officially renamed Saturn on Feb. 3, 1959, by ARPA memorandum. The space agency assumed technical direction of the Saturn project in late 1959. The project was transferred officially on Mar. 16, 1960, and the Army development group at Huntsville was transferred to NASA and became the nucleus of the new Marshall Space Flight Center. The first static firing of a Saturn I booster was conducted April 29, 1960.

Test Vehicle-The first assembled Apollo Saturn V vehicle approaches the launch pad at Kennedy Space Center. It was used to verify launch facilities, train launch crews, and develop test and checkout procedures at KSC. It was rolled out on May 25, 1966.

The NASA Saturn Vehicle Evaluation Committee (Silverstein Committee) on Dec. 15, 1959, recommended a long-range development program for a Saturn vehicle with upper stage engines burning liquid hydrogen and liquid oxygen. The initial vehicle, identified as Saturn C-1 and now as Saturn I, was to be a stepping stone to a larger vehicle. A building-block concept was proposed

Saturn V Launch Vehicle at Sunset

that would yield a variety of Saturn configurations, each using previously proven developments as far as possible. Early in 1960 the Saturn program was given the highest national priority, and a 10-vehicle research and development program was approved. The two-stage Saturn I vehicle with the Apollo spacecraft was about 188 feet tall and weighed some 1,125,000 pounds at liftoff. While plans for the lunar mission were progressing, the Saturn I project made history. On Oct. 27, 1961, the first Saturn I booster was flight tested successfully from Cape Kennedy. The first flight booster with dummy upper stages was called SA-1. This vehicle was followed by successful flights of SA-2 on April 25, 1962, SA-3 on Nov. 16, 1962, and SA-4 on Mar. 28, 1963. The SA-5 vehicle, combining the first stage (S-1) with the second stage (S-IV), was successfully launched on Jan. 29, 1964, with both stages functioning perfectly to place a 37,700-pound payload into earth orbit. SA-6, launched on May 28, 1964, and SA-7, launched on Sept. 18,1964, each placed "unmanned" boilerplate configurations of Apollo spacecraft into earth orbit. SA-9, launched on Feb. 19, 1965, was the first Saturn I vehicle to launch a Pegasus meteoroid technology satellite into earth orbit. The SA-8 and SA-10 Saturn I vehicles were successfully launched on May 25, 1965, and July 30, 1965, respectively, also placing a Pegasus satellite into earth orbit to complete the test and launch pro- gram with an unprecedented 100 per cent record of success.

Uprated Saturn I (Saturn IB)

The space agency, using the building-block approach, conceived the Uprated Saturn I as the quickest, most reliable, and most economical means of providing a vehicle with greater payload than the Saturn I. This vehicle was planned for orbital missions with the Apollo spacecraft before the Saturn V vehicle would be available. The Uprated Saturn I is based on a blending of existing elements of Saturn I and Saturn V. A redesigned Saturn I booster (designated the S-IB stage), and an S-IVB upper stage and instrument unit from the Saturn V are used on this launch vehicle. Maximum use of designs and facilities available from the earlier approved Saturn programs saved both time and costs. The Saturn I first stage was redesigned in several areas by NASA and the Chrysler Corporation, the stage contractor, for the expanded role as the Up-rated Saturn I booster. Basically, it retained the same shape and size, but required some modification for mating with the upper stage, which has a greater diameter and weight than the Saturn I upper stage. Stage weight was cut by more than 20,000 pounds to increase payload capacity. The Rocketdyne H-1 engine was uprated to 200,000 pounds of thrust, compared with 188,000 pounds of thrust for each engine in the final Saturn I configuration. The engines will be improved again to 205,000 pounds beginning with the SA-206. For the Uprated Saturn I, a guidance computer used in the early Saturn I was replaced by another IBM computer of completely new design which incorporates the added flexibility and extreme reliability necessary to carry out the intended Uprated Saturn I missions. The Uprated Saturn I, topped by the Apollo spacecraft, stands approximately 224 feet tall, and is about 21.7 feet in diameter. Total empty weight is about 85 tons, and liftoff weight fully fueled is approximately 650 tons. Several uprated Saturn I vehicles have been launched since the original SA 201 launch on Feb. 26, 1966.

HOW SATURN V DESIGN WAS REACHED

While a major effort of this country's space commitment was to explore the moon, the broader target was to build a capability—people, launch vehicles, propulsion, spacecraft, production, testing, and launching sites to explore a vast new frontier and develop a long-range space faring capability that would establish continuing national preeminence. The questions facing national space planners in 1961 and 1962 were complex. Although the use of a Saturn I for a manned lunar landing was theoretically possible, it would have been extremely difficult. About six Saturn I launches would have been required, their payloads being assembled in earth orbit to form a moon ship. No space rendezvous and docking had taken place at that time. During the first half of 1962, two paramount decisions were announced: to develop a new general purpose launch vehicle in the middle range of several under consideration, and to conduct the manned lunar landing by use of a lunar orbit rendezvous (LOR) technique. The Saturn V, as the chosen vehicle was named, was given the go-ahead in January, 1962. It was to be composed of three propulsive stages and a small instrument unit to contain guidance and control. It could perform earth orbital missions through the use of the first two stages, while all three would be required for lunar and planetary expeditions. The ground stage was to be powered by five F-1 engines, each developing 1.5 million pounds of thrust, and the stage would have five times the power of the Saturn I booster then under development. The upper stages would use the J 2 hydrogen/ oxygen engine, five in the second stage and one in the third. Each would develop up to 225,000 pounds of thrust. Such a rocket would be capable of placing 120 tons into earth orbit or dispatching 45 tons to the moon. (The numbers have been uprated now to about 125 and 47½.) During its assembly, checkout, and launch, the Saturn V would use a new mobile launch concept. It would be assembled in a huge Vehicle Assembly Building, and then transported in an upright position to a launch pad several miles away. Propulsion development decisions preceded those for the vehicles. The need for a building-block rocket engine in the million-pound-thrust class was apparent even as ARPA was ordering work to begin on the first stage cluster of engines for the Saturn I. In January, 1959, NASA contracted with North American Aviation's Rocketdyne Division for development of the F-1. Late in 1959, the Silverstein Committee recommended the development of a new high-thrust hydrogen engine to meet upper stage requirements. In June, 1960, Rocketdyne was selected to develop the J-2 engine after evaluation of competitive proposals by NASA. Three proposed Apollo modes which were considered in detail were: the direct flight mode, using a very large launch vehicle called "Nova"; the earth orbital rendezvous (EOR) mode, requiring separate Saturn launches of a tanker and a manned spacecraft; and the lunar orbital rendezvous mode, requiring a single launch of the manned spacecraft and the lunar module. Selected was the LOR mode, in which the injected spacecraft weight would be reduced from 150,000 pounds to approximately 80,000 pounds by eliminating the requirement for the propulsion needed to soft-land the entire spacecraft on the lunar surface. A small lunar excursion module, or LEM. now referred to as the lunar module, would be detached after deboost into lunar orbit. The lunar module would carry two of the three-man Apollo crew to a soft landing on the moon and would subsequently be launched from the moon to rendezvous with the third crew member in the "mother ship." The entire crew would then return to earth aboard the command module. NASA concluded that LOR offered the greatest assurance of successful accomplishment of the Apollo objectives at the earliest practical date. Members of NASA's Manned Space Flight Management Council recommended LOR unanimously in 1962 because it:

1. Provided a higher probability of mission success with essentially equal mission safety;
2. Promised mission success some months earlier than did other modes;
3. It would cost 10 to 15 per cent less than the other modes; and
4. Required the least amount of technical development beyond existing commitments while advancing significantly the national technology.

As a part of the Saturn V decision, it was determined that elements of the existing Saturn I vehicle and the planned Saturn V would be combined to form a new mid-range vehicle, the uprated Saturn I (Saturn IB). The Uprated Saturn I would have a payload capability 50 per cent greater than the Saturn I and would make possible the testing of the Apollo spacecraft in earth orbit about one year earlier than would be possible with the Saturn V. By the end of 1962, all elements of the new program were under way, with the

Marshall Space Flight Center directing the work for NASA. The Boeing Company; Space Division of North American Aviation, Inc.; and Douglas Aircraft Company were acting as prime contractors for the Saturn V first, second, and third stages, respectively. Engines were being developed by the Rocketdyne Division of North American. MSFC designed the instrument unit and awarded a production contract to International Business Machines Corp. (Chrysler Corp. had been selected to produce the first stage of the Uprated Saturn I.) A large network of production, assembly, testing, and launch facilities was also being prepared by the end of 1962. Aside from the provision of various facilities at contractor plants and the augmentation of the Marshall Space Flight Center resources, three new government operations were established: the launch complex in Florida operated by the NASA- Kennedy Space Center and two new elements of MSFC -Michoud Assembly Facility in New Orleans, La., for the production of boosters, and Mississippi Test Facility, Bay St. Louis, Miss., for captive firing of stages. Four years after its establishment, the Saturn V program was progressing on schedule, pointing toward the launch of the first vehicle in 1967 and fulfillment of the manned lunar landing before the end of the decade.

PROGRAM HIGHLIGHTS

Following are highlights of the Saturn V development program:

1961

Aug. 24 NASA announced the selection of the 88,000-acre site at Merritt Island, Fla., adjacent to Kennedy Space Center, then Cape Canaveral, for the assembly, checkout, and launch of the Saturn V.

Sept. 7 NASA selected the government-owned Michoud plant, New Orleans, as production site for Saturn boosters. It became a part of the Marshall Space Flight Center.

Sept. 11 NASA selected North American Aviation, Inc., to develop and build the second stage for an advanced Saturn launch vehicle (as yet undefined) for manned and unmanned missions. One month later the Marshall Center directed NAA to design the second stage using five J-2 engines. A preliminary contract was signed in February, 1962.

Oct. 6 NASA selected the Picayune-Bay St. Louis, Miss., area for its Mississippi Test Facility-an arm of the Marshall Center-for use in static testing of rocket stages and engines.

Dec. 15 The Boeing Company was selected as prime contractor for the first stage of the advanced Saturn vehicle-not yet fully defined. A preliminary contract was signed in February, 1962, with the work to be conducted at the Michoud Assembly Facility.

Dec. 21 NASA selected the Douglas Aircraft Company to negotiate a contract to develop the third stage (S-IVB) of the advanced Saturn, based on the Saturn I's S-IV stage. A supplemental contract for production of 11 third stages was signed in August, 1962.

1962

Jan. 10 Announcement was made that the advanced Saturn vehicle would have a first stage powered by five F-1 engines, a second stage powered by five J-2 engines, and for lunar missions a third stage with one J-2 engine.

Jan. 25 NASA formally assigned development of the three-stage Saturn C-5 (Saturn V became the name in February, 1963) to MSFC.

April 11 NASA Headquarters gave the Apollo/Saturn I/Saturn V highest national priority.

May 26 Rocketdyne Division of NAA conducted the first full-thrust, long-duration F-1 engine test.

July 11 It was announced that the Saturn IB (Uprated Saturn I) would be developed and that the lunar orbit rendezvous method of accomplishing a lunar landing had been selected.

Dec The U. S. Army Corps of Engineers awarded a contract for the design of the Vehicle Assembly Building (VAB) at the Florida launch complex.

1963

Feb. 27 The first contract for the Mississippi Test Facility (MTF) Saturn V test facilities was awarded.

May The J-2 engine was successfully fired for the first time in a simulated space altitude of 60,000 feet.

Oct. 31 The Marshall Center received the first production model of the F-1 engine.

Nov. 12 NASA contracted for the first Saturn V launch pad at the Kennedy Space Center.

1964

March IBM was awarded an instrument unit contract for the digital computer and data adapter by the Marshall Center. IBM became the prime IU contractor in May.

Oct. 9 The Edwards AFB test facility was accepted as the F-1 test complex, amounting to a cost of $34 million.

Dec. 1 The first mainstage shakedown firing of the third stage battleship was accomplished, lasting 10 seconds.

Dec. 23 First full-duration firing of the third stage battleship occurred.

1965

April 16 All five engines of the S-IC-T, first stage test vehicle, were fired at the Marshall Center for 6.5 seconds.

April 24 The first cluster ignition test of the second stage battleship was successfully completed.

Aug. 5 The first full-duration firing of the first stage was conducted successfully at the Marshall Center.

Aug. 8 Third stage flight readiness test of 452 seconds, fully automated, was accomplished at Sacramento.

Aug. 13 The IU was qualified structurally and man rated for Saturn V use by withstanding a 140 per cent load limit.

Aug. 17 The third stage battleship was tested in Saturn V configuration for full duration (start-stop-restart).

Dec. 16 The S-IC-T static firings were completed at the Marshall Center with a total of 15 firings—three of full duration.

1966

Feb. 17 & 25 The S-IC-1 underwent static firing at the Marshall Center and required no more static firings.

Mar. 30 The S-IU-500F was mated to the three stages of the Saturn V facilities vehicle at the Kennedy Space Center's VAB.

May 20 First full-duration firing of the second stage flight stage was conducted at MTF.

May 25 The Apollo/Saturn V facilities vehicle, AS-500-F, was transported to Pad A at Launch Complex 39, KSC, on the crawler.

May 26 Full-duration acceptance firing of the S-IVB-501, the first flight version of the third stage for Saturn V, was accomplished.

Sept The F-1 and J 2 engines were qualified for manned flights.

Dec. 1 Initial static firing of the first flight version of the second stage occurred at MTF.

Nov. 15 The first flight version of the first stage was static fired at MSFC.

First Stage Separation During an Apollo/Saturn V Shot

FIRST STAGE FACT SHEET

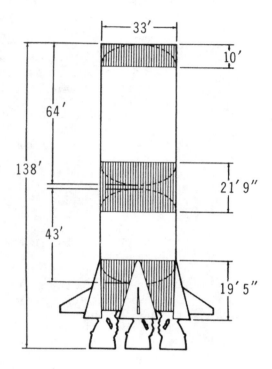

WEIGHT: 300,000 lb. (dry)
DIAMETER: 33 ft.
HEIGHT: 138 ft.
BURN TIME: About 2.5 min.
VELOCITY: 6,000 miles per hour at burnout (approx.)
ALTITUDE AT BURNOUT: About 38 miles
MAJOR STRUCTURAL COMPONENTS
   THRUST STRUCTURE
   FUEL TANK
   INTERTANK
   LOX TANK
   FORWARD SKIRT
MAJOR SYSTEMS
PROPULSION : Five bipropellant F-1 engines
4,792,000 lb. (loaded)
  Total thrust: 7.5 million Lb.
  Propellant: RP-1 - 203,000 gal. or 1,359,000 Lb.
  LOX - 331,000 gal or 3,133,000 Lb.
  Pressure: Control 1.27 cubic feet of gaseous nitrogen at 3,250 psig
  Fuel pressurization - 124 cubic feet or 636 Lb. of gaseous helium at 3,100 psig
  LOX pressurization - gaseous oxygen converted from 6,340 pounds of LOX by the engines
HYDRAULIC: Power primarily for engine start and for gimbaling four outboard engines
ELECTRICAL: Two 28 VDC batteries, basic power for all electrical functions
INSTRUMENTATION : Handles approx. 900 measurements
TRACKING: ODOP Transponder

FIRST STAGE

FIRST STAGE DESCRIPTION

The Saturn V first stage (SIC) is a vertical grouping of five cylindrical major components and a cluster of five F-l rocket engines. Upward from the engines are the thrust structure, fuel tank, inter-tank structure, LOX tank, and forward skirt. The total stage measures 138 feet in height and 33 feet in diameter without its fins. It weighs 6,100,000 pounds at liftoff and delivers 7.5 million pounds of thrust.

FIRST STAGE FABRICATION AND ASSEMBLY

Design, assembly, and test of the first stage booster are the prime tasks being performed by The Boeing Company at the Marshall Space Flight Center, Huntsville, Ala., the Michoud Assembly Facility, New Orleans, La., and the Mississippi Test Facility in southwestern Mississippi. Launch operations support is provided by the Boeing Atlantic Test Center, Kennedy Space Center, Fla. Contractor suppliers lend support for much of the first stage fabrication. Several ground test stages were completed before manufacture of a series of flight stages was begun. Huntsville and Michoud installations shared responsibility for assembly of four ground test stages and the first two flight stages. All other flight stages are being assembled at Michoud.

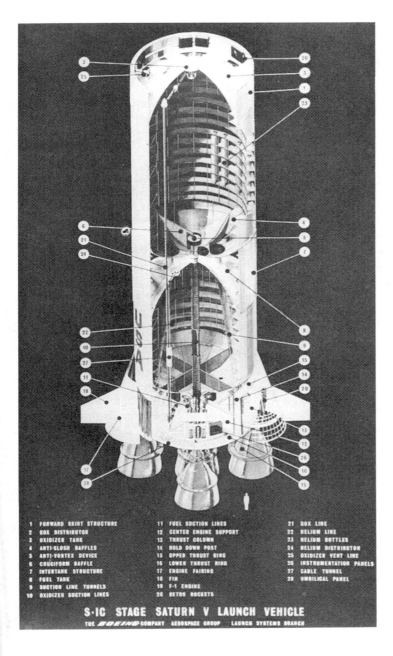

First Stage Cutaway

1. FORWARD SKIRT STRUCTURE
2. GOX DISTRIBUTOR
3. OXIDIZER TANK
4. ANTI-SLOSH BAFFLES
5. ANTI-VORTEX DEVICE
6. CRUCIFORM BAFFLE
7. INTERTANK STRUCTURE
8. FUEL TANK
9. SUCTION LINE TUNNELS
10. OXIDIZER SUCTION LINES
11. FUEL SUCTION LINES
12. CENTER ENGINE SUPPORT
13. THRUST COLUMN
14. HOLD DOWN POST
15. UPPER THRUST RING
16. LOWER THRUST RING
17. ENGINE FAIRING
18. FIN
19. F-1 ENGINE
20. RETRO ROCKETS
21. GOX LINE
22. HELIUM LINE
23. HELIUM BOTTLES
24. HELIUM DISTRIBUTOR
25. OXIDIZER VENT LINE
26. INSTRUMENTATION PANELS
27. CABLE TUNNEL
28. UMBILICAL PANEL

S-IC STAGE SATURN V LAUNCH VEHICLE
THE *BOEING* COMPANY AEROSPACE GROUP LAUNCH SYSTEMS BRANCH

Saturn V

Assembled First Stage B-7031-10

Base Assembly Workmen cover the thrust structure shell with aluminum skin. B-10648-5

Thrust Structure-The 24-ton base of the booster is being taken to the Vertical Assembly Building for mating with other first stage components. B-6018-7

Thrust Structure

The thrust structure is the heaviest of first stage components, weighing 24 tons. It is 33 feet in diameter and about 20 feet tall with these major components: the lower thrust ring assembly, the center engine support assembly, four holddown posts, engine thrust posts, an upper thrust ring assembly, intermediate rings, and skin panel assemblies. The upper ring provides stability for the corrugated skins around the structure. Four F-1 engines are mounted circumferentially upon the thrust posts and the fifth upon the center engine support assembly. The center engine remains rigid while the others gimbal or swivel, allowing the stage to be guided. A base heat shield protects internal parts from engine heat, and four holddown posts restrain the vehicle while the engines build up power for liftoff. The thrust structure supports the entire vehicle weight and distributes the forces of the engines.

Fuel Tank

The fuel tank holds 203,000 gallons of kerosene and encloses a system of five LOX tunnels. The tank, weighing more than 12 tons dry, is capable of releasing 1,350 gallons of kerosene per second to the engines through 10 fuel-suction lines. The LOX tunnels carry liquid oxygen from the LOX tank, through the fuel tank, and to the engines. Bound by eight aluminum skin panels, the fusion-welded fuel tank assembly is 33 feet in diameter and 44 feet tall. Ends are enclosed by ellipsoidal bulkheads. The bulkheads consist of eight pie-shaped gores mated with a polar cap to form a dome shape. Connecting links between the skin rings and bulk-heads are circular bands known as the Y-rings. The Y-rings are used on both propellant tanks and link them to other segments of the booster at final assembly.

Fuel Tank-Kerosene is fed to the engines at 1,300 gallons per second from this 203,000 gallon tank. Here the finished tank is being lowered onto its transporter. B-6469-20

Inside View - The fuel tank contains horizontal baffles, which are designed to prevent sloshing of fuel. B-5622-4

LOX Tank - the completed 331,000-gallon tank is being carried to the hydrostatic testing facility where it will be tested for leaks. B-8219-3

Fuel Tank Assembly - Workmen weld the base of the 27-inch-high Y-ring to the cylindrical segment of the fuel tank. This ring joins the tank sides to the dome and to the intertank structure. B-4780-6

LOX Tunnel - Five 42-foot tunnels bring liquid oxygen from the LOX tank through the fuel tank and to the engines. Here a tunnel is being fitted into the fuel tank. B-5773-6

LOX Tank

The 331,000-gallon liquid oxygen tank is the largest component of the first stage booster, standing more than 64 feet in height. Its content is 297 degrees below zero Fahrenheit and provides the oxidizer to support combustion of the kerosene. Mixing of the two propellants is in a proportion to ensure complete combustion. Each second during flight, the engines consume more than 2,000 gallons of liquid oxygen. The LOX tank's construction is similar to that of the fuel tank with the LOX tunnels beginning at the tank base, running through the intertank and fuel tank and to the engines. Dry weight of the LOX tank exceeds 19 tons.

Intertank

The intertank is not a tank in itself but serves as a 6-1/2-ton link between fuel and LOX tanks. Its composition is 18 corrugated skin panels supported by five frame ring assemblies. The lower bulkhead of the LOX tank dips into the intertank while the upper bulkhead of the fuel tank extends upward into the intertank. Around the edges of the intertank are attached 216 fittings, which fasten the tank together with the Y-rings of the fuel and LOX tanks. The intertank structure also contains a personnel access door.

Umbilical Openings

An umbilical opening in the intertank provides for electrical and instrumentation requirements, emergency LOX drain, line pressurization, electrical conduit, and provisions for venting internal pressure. The thrust structure contains three of four other umbilical openings on the booster. The fourth is located in the forward skirt. The thrust structure umbilicals carry the fuel line, liquid oxygen drain, ground supply fluid lines, and all control functions essential in case of a vehicle abort.

Forward Skirt

The forward skirt tops the first stage and provides a connecting link for the first and second stages of the Saturn V. Weighing 2-1/2 tons, the structure consists of 12 skin panels attached to three circumferential support rings. It contains a small personnel access door; an umbilical opening for telemetry cables, an environmental air duct, and minor pneumatic lines; and an umbilical disconnect door.

Fins and Fairings

Four fairings attach to the thrust structure and partially surround the outboard engines at the foot of the booster. They house the eight retrorockets and the actuator support structures. Fairings are shaped like cone halves and are constructed of aluminum. Their purpose is to smooth the air flow over the engines. The fins are airfoil attachments to the fairings. Fins are rigid and add to the vehicle's flight stability. A titanium skin covers the fin for greatest protection against temperatures as high as 2,000 degrees Fahrenheit. Each of the eight retrorockets generates about 86,600 pounds of thrust for two-thirds of a second and, upon firing, blows off the tips of the fairings. (Retrorocket thrust varies with propellant temperature.)

A Completed Intertank B-3291-6

Forward Skirt - The structural link between the LOX tank and the engine shroud of the second stage is shown being lowered for dimensional inspection. B-2835-34

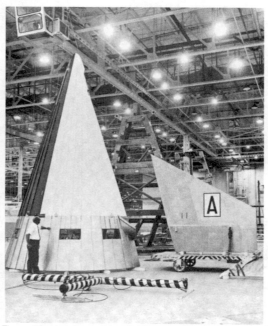

Fin and Fairing Assembly - Fairings are fitted over each of the outboard engines to smooth the air flow. Fins are attached to the fairings. B 6733-3

Michoud Manufacturing Area - In the foreground of this Michoud plant view, fairings are being assembled. B-9580-8

Tube and Valve Cleaning Vat - Each stage component is treated in a cleaning solution before final assembly. B-9940-7

Vertical Assembly

When all major components of the first stage are assembled in NASA's Michoud Assembly Facility, they are routed to the Vertical Assembly Building to be assembled. Manipulated by an overhead crane, the components are placed in final assembly position in the single-story building rising the equivalent of 18 stories. First the thrust

Vertical Assembly-Booster sections are mated in the Vertical Assembly Building. At top left the thrust structure is shown. Fuel tank, intertank assembly, LOX tank, and forward skirt are added in successive pictures. B-8565-5, B-7594-6, B-7594-1

structure is placed on four heavy pylons 20 feet above floor level. Meanwhile, two of the segments-the fuel and LOX tanks which are brought to the Vertical Assembly Building in segments -are being completed on two tank assembly bays. Then, in building-block fashion, the thrust structure is joined by the fuel tank, intertank, LOX tank, and forward skirt. When the forward skirt is secured, the first stage stands 138 feet high. Vertical assembly completed, the 180-ton-capacity overhead crane lifts the booster by a forward handling ring attached to the forward skirt and returns it to horizontal position on its 435,000-pound transporter.

As assembly jobs approach completion, installation of internal systems and engines is made in preparation for systems test and checkout. Mechanical, hydraulic, and pneumatic systems tests are conducted to leak-check and functionally check the propellant systems and the engine complex. Checks then are performed to demonstrate the proper operation of the electrical and instrumentation systems. All systems are operated and checked individually and then checked as an integrated system in the automatic all-systems checkout.

The Stage Test Building with four giant test cells provides the facility. Inside the building are four control rooms, four computer rooms, and two telemetry rooms. These rooms house equipment that demonstrates the acceptability of the integrated systems of the booster. This includes telemeter calibration, continuity checks, and discrete-function monitoring. RF (radio frequency) also is evaluated.

Moving - A completed first stage is readied for post-manufacturing checkout. B-7733-10

Monitoring - Technicians check booster performance during a simulated flight from a stage test control room. B-9964-2

Engines - One of the first stage's F-1 engines is mounted. Together the five will consume 4,492,000 pounds of propellants in 2.5 minutes. B-10872-4

POST MANUFACTURING CHECKOUT

Before a booster leaves Michoud for test firing, its electrical and mechanical systems are tested extensively by Boeing technicians and engineers.

After the operation of the test and checkout equipment is verified, all electrical, pneumatic, and hydraulic connections are made to the stage, resistance checks are run, and the stage undergoes physical examination. The environmental control system is connected and checked for proper operation, and the stage's electrical circuits are physically checked for resistance. Stage electrical power is applied in sequential steps and the distribution monitored. The stage instrumentation transmission system is checked out on both coaxial hardwire and RF links. The electrical systems checkout includes checks of the power distribution circuits, heater power subsystems, destruct system, sequencing subsystem, separation subsystem, and emergency detection system. The range safety systems undergo a complete end- to-end checkout including transmittal of RF commands to the range safety command receiver and monitoring the arm, cutoff, and destruct signals generated by the system. Instrumentation system testing during stage check- out includes: identification of data channels, gain adjustment of signal conditioners, and checks of measurement systems, telemetry systems, and operational RF systems.

First Stage in Test Cell B-9908-1

Pressure and leak checks are conducted on fuel and LOX tanks and associated lines, engines, fuel and LOX delivery systems, fuel and LOX pressurization systems, and the control pressure system. Checks are made of the calibration pressure switch simulation, fill and drain operation, and prevalve operation on both fuel and LOX systems. Propulsion system checks include checks of firing command preparation and execution, engine shut-down prior to "launch commit," malfunction cutoff, and normal propulsion sequences. Most of the above-mentioned tests are run for a second time prior to static testing and again during post-static checkout.

FIRST STAGE SYSTEMS

Fuel System

The first stage fuel system supplies RP-1 fuel to the F-1 engines. The system consists of a fuel tank, fuel feed lines, pressurization system, fill and drain components, fuel conditioning system, and associated hardware to meet the propulsion system requirements.

FUEL TANK
The fuel tank, previously described, holds 203,000 gallons of kerosene and is capable of providing 1,350 gallons of fuel per second to the engines through 10 fuel-suction lines.

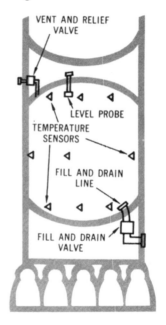

Fuel Fill and Drain

FUEL FILL AND DRAIN SYSTEM
The fuel tank is filled through a 6-inch duct at the bottom of the tank. Fill rate is 200 gallons per minute until the tank is 10 per cent full. After reaching the 10 per cent mark, filling is increased to 2,000 gallons per minute until the tank is full. Normal non-emergency drain takes place through the same duct. A ball-type valve in the fill and drain line provides fuel shutoff. The fuel fill and drain system consists of a fill and drain line, a fill and drain valve, a fuel loading level probe, and nine temperature sensors. During fuel fill, the temperature sensors provide continuous fuel

temperature information used to compute fuel density. When the fuel level in the fuel tank rises to about 102 per cent of flight requirements, the fuel loading probe indicates an overload. After adjusting fuel to meet requirements, the fill and drain valve is closed. The fuel tank can be drained under pressure by closing the fuel tank vent and relief valve, supplying a pressurizing gas to the tank through the fuel tank pre-pressurization system, and opening the fuel fill and drain valve.

FUEL FEED SYSTEM
Ten fuel suction lines (two per engine) supply fuel from the fuel tank to the five F-1 engines. The suction line outlets attach directly to the F-1 engine fuel pump inlets. Each suction line has a pneumatically controlled fuel prevalve which normally remains open. This prevalve serves as an emergency backup to the main engine fuel shutoff valves to terminate fuel flow to the engines.

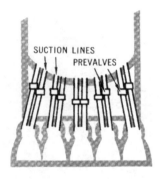

Fuel Feed

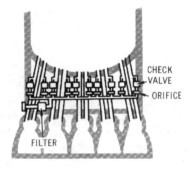

Fuel Conditioning

FUEL-CONDITIONING (BUBBLING) SYSTEM
The fuel-conditioning system bubbles gaseous nitrogen through the fuel feed lines and fuel tank to prevent fuel temperature stratification prior to launch. A wire mesh filter in the nitrogen supply line prevents discharge of contaminants into the conditioning system.

A check valve in the outlet of each fuel-conditioning line prevents fuel from entering the nitrogen lines when the fuel-conditioning system is not operating. An orifice located near each fuel-conditioning check valve provides the proper nitrogen flow into each fuel duct.

FUEL LEVEL SENSING AND ENGINE CUTOFF SYSTEMS
A cutoff sensor mounted on the bottom of the fuel tank provides signal voltages to shut off fuel after a predetermined level of depletion is reached. The fuel is measured during flight by four fuel slosh probes and a single liquid level measuring probe. Fuel levels are detected electronically and reported through the stage telemetry system. Telemetry signals are transmitted to ground support either by radio frequency or, before launch, by coaxial cable. The cutoff sensor, mounted in the lower fuel tank bulkhead, initiates engine cutoff as fuel level falls below two sensing points on the probe. Engine cutoff will normally be initiated by sensors in the LOX system. The cutoff capability is provided as a backup system should fuel be depleted before LOX.

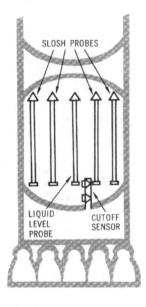

Fuel Level Sensing and Engine Cutoff

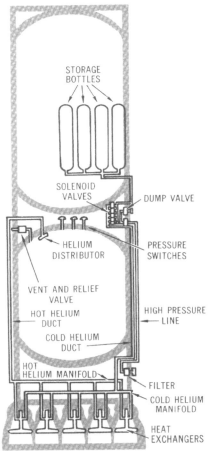

Fuel Pressurization

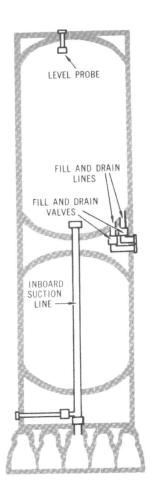

Lox Fill and Drain

FUEL PRESSURIZATION SYSTEM

The fuel pressurization system maintains enough pressure in the fuel tank to provide proper suction at the fuel turbopumps to start and operate engines. The system consists of a helium supply, a helium flow controller, helium fill and drain components, a pre-pressurization subsystem, a fuel tank vent and relief valve, and associated ducts. Four 31-cubic-foot, high pressure storage bottles in the LOX tank store the helium required for in-flight pressurization of the fuel tank ullage. A high pressure line is used for filling the bottles and routing the helium to the flow controller. A solenoid dump valve is installed for emergencies. The helium flow controller uses five solenoid valves mounted parallel in a manifold to control helium flow to the fuel tank ullage. The cold helium duct routes helium from the flow controller to the cold helium manifold. From there, it is distributed to the heat exchangers on the five F-1 engines. The hot helium manifold receives the heated, expanded helium from the engine heat exchangers and routes it to the hot helium duct which then carries it through the helium distributor and on to the fuel tank ullage. Three absolute pressure switches, mounted atop the fuel tank, monitor and control fuel tank pre-pressurization before engine ignition, fuel tank pressurization during flight, and overpressure.

Design strength of the four helium bottles at atmospheric temperatures and prior to LOX loading is about 1,660 pounds per square inch gage (psig). After LOX loading, when the bottles are cold, pressure is increased to about 3,100 psig. A filter in the helium fill line prevents contaminants from entering the flight pressurization system.

LOX System

The liquid oxygen (LOX) system supplies LOX to the five F-1 engines. The system consists of a LOX tank, fill and drain components, LOX suction lines, pressurization subsystem, and associated hardware.

LOX TANK
In addition to the components of the LOX tank previously described, the tank contains internal ring baffles which line the tank walls to provide wall support and prevent excessive sloshing of LOX. A cruciform baffle in the lower tank head limits LOX swirling. Four LOX liquid level probes continuously monitor LOX level in the tank. The probes are made up of a series of continuous capacitive level sensors separated by discrete level sensors.

LOX FILL AND DRAIN SYSTEM
LOX is forced under pressure through two 6-inch LOX fill and drain lines into the tank at a slow fill rate of 1,500 gallons per minute until the tank is 6.5 per cent full. The slow fill rate avoids splash damage to LOX tank components. After a visual leak check, a fill rate of 10,000 gallons per minute takes place until the tank is 95 per cent full. After this, the rate is reduced to 1,500 gallons per minute until the LOX loading level probe senses a full tank and terminates LOX fill. In addition to the two 6-inch lines used for LOX fill and drain, a third line is available for filling the tank through the inboard suction line. LOX boils continuously to maintain the temperature of -297 degrees Fahrenheit at sea level pressure. It is replenished between the periods of loading and pre-pressurization through the fill and drain line. Before LOX drain can be performed, the helium cylinders in the LOX tank must have their pressure decreased from about 3,100 psig to about 1,660 psig. Fill and drain valves are opened to complete drainage of the LOX tank although total evacuation of LOX from the tank requires draining the engines or waiting for boil-off of residual LOX. LOX drain can be speeded with the aid of a pressurizing gas, usually nitrogen.

LOX DELIVERY SYSTEM
LOX is delivered to the engines by five 17-inch suction lines which pass through the fuel tank in five LOX tunnels. LOX suction ducts make up the lines from the LOX tank to the prevalves in the thrust structure. The ducts are equipped with gimbals and sliding joints to counteract vibration and swelling or contraction caused by temperature. Inside the tunnels, air acts as the insulation between the LOX-wetted lines and the fuel-wetted tunnels. LOX level engine cutoff sensors in the suction lines assure safe engine shutdown and leave a minimum amount of unused LOX in the system. In case of emergency, LOX prevalves in each suction line can stop the flow of LOX to the engines.

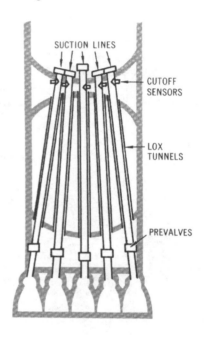

LOX Dellivery

LOX CONDITIONING SYSTEM
LOX cannot exceed -297 degrees Fahrenheit or it will result in gaseous oxygen (GOX). If heat is increased, the result is boiling and not temperature increase since evaporation is a cooling process. Depth in a body of LOX can increase due to the increase in hydrostatic pressure. The greatest chance for overheating in the LOX system is in the transmission surface of the suction lines. Also, the suction lines are too slender for maintenance of self-contained convection currents. This situation is unacceptable since intense boiling can lead to LOX geysering, which in turn can damage the LOX tank structurally. In addition, too high a LOX temperature near the engine inlets can cause a cavity in the LOX pumps and interfere with normal engine starting. Emergency bubbling or thermal pumping is used to correct this situation.

The bubbling technique sends helium into all five suction lines to cool the LOX rapidly. Ground support supplies helium through an umbilical coupling, and filter valves and orifices control the flow of helium into the suction lines. Thermal pumping is a term used to define pumping relatively cold LOX from the LOX tank into the suction lines.

engine heat exchanger where hot gases exhausted from each engine turbine transform LOX into GOX. The GOX flows from each heat exchanger into the GOX line manifold through the flow control valve, up the GOX line, and into the LOX tank through the GOX distributor. The GOX flow is approximately 40 pounds per second to maintain a LOX tank ullage pressure of 18 to 23 psia.

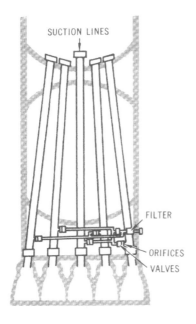

LOX Conditioning

LOX PRESSURIZATION SYSTEM

Pressurizing gases used in the LOX tank are helium, gaseous oxygen, and nitrogen. These gases are used in pre-pressurization, flight pressurization and storage pressurization. Pre-pressurization is necessary 45 seconds prior to engine ignition to give sufficient tank ullage pressure for engine start and thrust buildup. Helium, used as the pressurizing gas to reduce flight weight, is supplied by ground support through the helium ground connection. It proceeds up the gaseous oxygen line into the LOX tank through the GOX distributor. The flow of helium is monitored by the pressure duct and stopped at 26 pounds per square inch absolute (psia) maximum and is resumed when the pressure drops to 24.2 psia during engine start. Ground-supplied helium is available until liftoff. GOX is added to the LOX tank for pressurization during flight. Each engine contributes to GOX pressurization. A portion of LOX-6,340 pounds - passing through the engine is diverted from the LOX dome into the

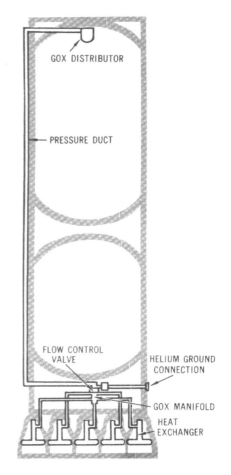

LOX Pressurization

While the booster is being stored or transferred from one location to another, a slight positive nitrogen pressure is maintained for cleanliness and low humidity conditions. The external nitrogen pressure source is removed during flight operations.

Fluid Power System

An unusual but convenient type of fluid power or hydraulic system is in use on the Saturn V first stage. It incorporates the same types of fuels-RP-

1 and RJ-1 (kerosene) - that are used in the stage fuel system. Ordinarily a different and weaker type of fluid is used for hydraulics. This system eliminates the use of a separate pumping system.

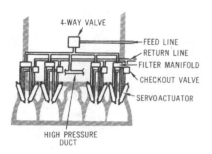

Fluid Power System

The fluid power system provides ground and flight fluid power for valve actuation and thrust vectoring. It gives power primarily to the engine start system and the engine gimbaling system. Its source is the fuel system. RJ-1 is provided from the ground before liftoff, and RP-1 is supplied from the fuel tank during flight. The ground supply of RJ-1 is routed to all five engines at 1,500 psig and eventually back to the ground supply. After ignition, RP-1 is routed from the high pressure fuel duct to the servoactuators for hydraulic power to position the engines. The center engine, which has no thrust vectoring system, directs its hydraulic fluid through the feed line and 4-way hydraulic control valve to supply pressure to the closing ports of the gas generator, main fuel valves, and main LOX valves. The fuel passes through orifices and then is ducted through the ground checkout valve and back to ground supply through the return line. The four outboard engines direct RJ-1 through the servoactuators to the ground checkout valve where it is returned through a coupling to ground supply.

Electrical System

The electrical power and distribution system of the first stage provides power for controlling and measuring functions of the vehicle. The system operates during static firing, launch preparation and check-out, launch, and flight. The electrical system consists of two batteries, a main power distributor, a sequence and control distributor, propulsion distributor, timer distributor, measuring distributors, thrust OK distributor, and measuring power distributor. Two independent 28-volt DC power systems are installed on the stage.

System No. 1, the main power battery, energizes the stage controls. The battery has a 640-ampere-minute rating, weighs about 22 pounds, and is used to control various solenoids. Battery No. 2, the instrumentation battery, energizes the flight measurement system and gives power to redundant systems for greater mission reliability. It has a 1,250-ampere-minute rating and weighs approximately 55 pounds. The range safety system can be operated by either battery. Preflight power is supplied from ground equipment through umbilical connections. The supply for each system is 28 volts. Ground sources supply power for heaters, ignitors, and valve operators that are not operated during flight. The distributors subdivide the electrical circuits and serve as junction boxes. Both electrical systems share the same distributors. The main power distributor houses relays, the power transfer switch, and electrical distribution buses. The relays control circuits that must be time-programmed. The motor-operated, multi-contact, power transfer switch transfers the stage load from the ground supply to the stage batteries. The transfer is tried several times during countdown to verify operation. Power is distributed by the main buses. The switch selector, actuated by the instrument unit (IU), commands the sequence and control distributor, which in turn amplifies the signals received. The sequence and control distributor then energizes the various circuit relays required to implement the flight program. The switch selector is an assembly of redundant low power relays and transistor switches, which control the sequence and control distributor. It is activated by a coded signal from the instrument unit computer. The propulsion distributor contains the monitor and control circuits for the propulsion system. The thrust OK distributor contains the circuits that shut down the engines when developed thrust is inadequate. Two of the three thrust OK switches must operate or the engine will be shut down. The timer distributor houses the circuits to delay the operation of relay valves and other electro-mechanical devices. The programmed delays are essential for optimum performance and safety. The measuring power distributor contains electrical buses, and the measuring distributors route data from measuring racks, serve as measurement signal junction boxes, and switch data between the hardwire and telemetry.

Instrumentation System

The first stage instrumentation system measures and reports information on stage systems and components and provides data on internal and external environments. It keeps abreast of approximately 900 measurements on the stage, such as measurements of valve positions, propellant levels, temperatures, voltages, and pressures. The measurements are telemetered by coaxial cable to ground support equipment and by radio frequency transmission to ground stations. The instrumentation system consists of a measurement system, a telemetry system, and the Offset Doppler tracking system. A remote automatic calibration system provides remote rapid checkout of the measurements and telemetry systems.

MEASUREMENT
The measurement system reports environmental situations and how the first stage reacts to them. Making use of transducers, signal conditioners, measuring rack assemblies, measuring distributors, and the onboard portion of the remote automatic calibration system, this system involves many phases of stage operation. Included are measurements of acceleration, acoustics, current, flow, flight angles, valve position, pressure, RPM's, stress, temperature, vibration, and separation.

TELEMETRY
Telemetry is a method of remote monitoring of flight information accomplished by means of a radio link. The first stage telemetry system is composed of six radio frequency links. Most of the components of the telemetry systems are located in the thrust structure; RF assemblies and a tape recorder are located in the forward skirt. The telemeter transmits data through two common antenna systems. Links F1, F2, and F3 are identical systems which transmit narrow-band, frequency-type data such as that generated by strain gages, temperature gages, and pressure gages. The system can handle 234 measurements on a time-sharing basis and 14 measurements transmitted continuously. Data may be sampled either 120 times per second or 12 times per second. Links S1 and S2 transmit wide-band, frequency-type data generated by vibration sensors. Each link provides 15 continuous channels or a maximum of 75 multiplexed channels depending on the specific measuring program. Telemeter P1 transmits either pulse code modulated or digital type data. Five multiplexers, four analogs, and one digital supply data to the PCM assembly. This provides the most accurate data and is used for ground checkout as well. A telemetering calibrator is used to improve the accuracy of the telemetry systems. The calibrator supplies known voltages to the telemeters periodically during the stage operation. Their reception at tracking stations provides a valid reference for data reduction. The effects of ullage and retrorocket firing attenuation can seriously degrade the telemetry transmission during stage separation; therefore, a tape recorder installed in the forward skirt records data for delayed transmission. The commands for tape recorder operation originate in the digital computer located in the instrument unit.

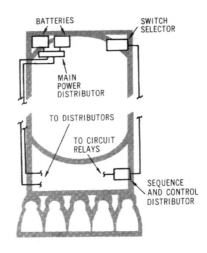

Electrical System

ODOP SYSTEM (Offset Doppler Tracking System)
The ODOP system is an elliptical tracking system that measures the rate of motion at which the vehicle is moving away from or toward a tracking station. The total Doppler shift in the frequency of a continuous wave, ultrahigh frequency signal transmitted from the ground to the first stage is measured. The signal is received by the transponder at the stage, modified, and then retransmitted back to the ground. Retransmitted signals are received simultaneously by three tracking stations. Separate antennas on the stage are used for receiving and retransmitting the signals.

SEPARATION SYSTEM

A redundant initiation system actuates the separation of the first stage from the second stage. A command signal for arming and another for firing the initiation systems are programmed by the instrument unit computer. After LOX depletion, the computer signals operate relays in the switch selector and sequence and control distributor to control the exploding bridgewire firing units. When armed, the firing units store a high voltage electrical charge. When fired, the electrical charge actuates the ordnance. Two firing units are installed on the first stage for the eight retrorockets, and two are installed on the second stage for the separation ordnance.

Range Safety System

The function of the range safety system is to provide ground command with the capability of flight termination by shutting off the engines, blowing open the stage propellant tanks, and dispersing the fuel in event of a flight malfunction. The system is redundant, consisting of two identical, independent systems, each made up of electronic and ordnance subsystems. Flight termination by way of the range safety system goes into effect upon receipt of the proper radio frequency commands from the ground. A frequency-modulated RF signal transmitted from the ground range safety transmitter is received by the antennas and transmitted by way of a hybrid ring to the range safety command receiver. There, the signal is conditioned, demodulated, and decoded. The resulting signal simultaneously causes arming of the exploding bridgewire firing unit and shut-down of the stage engines. A second command signal transmitted by the ground range safety transmitter ignites the explosive train (detonating fuses and shaped charges) to blow open the stage propellant tanks.

Control Pressure System

The control pressure system supplies pressurized gaseous nitrogen for the pneumatic actuation of propellant system valves and purging of various F-1 engine systems. The complete integrated system is made up of an onboard control pressure system, a ground control pressure system, and an onboard purge pressure system. The object in each system is to deliver an actuating or purge medium to an interfacing stage system.

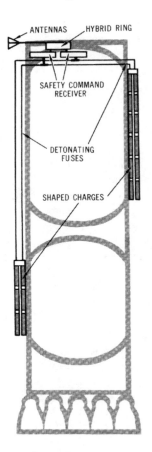

Range Safety System

ONBOARD CONTROL PRESSURE SYSTEM

The onboard control pressure system consists of a high-pressure nitrogen storage bottle, an umbilical coupling and tubing assembly for filling the storage bottle, a manifold assembly, and control valves at the terminal ends of various nitrogen distribution lines. In some cases, two valves are paired with other associated equipment and block-mounted to form a control assembly. The nitrogen onboard storage bottle has 2,200-cubic inch capacity and is made of titanium alloy. It is designed for a maximum proof pressure of 5,000 psig. It is filled and discharged through a port in the single boss. During flight launch preparation, the bottle is filled from a ground supply first to a pressurization of 1,600 psig well in advance of final countdown. This weight pressure is adequate for any prelaunch operational use. The second step occurs in the last hour of the launch countdown and brings the storage bottle pressure up to its normal capacity of 3,250 plus or minus 50 psig. The manifold assembly serves as a gaseous

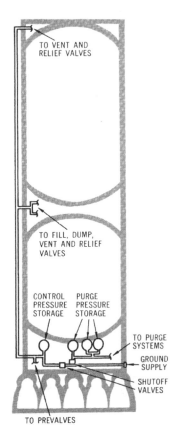

Control Pressure System

bottle, an umbilical coupling and tubing for filling the bottles, and a manifold assembly and tubing for receiving and delivering the gas to the engine and calorimeter purge systems. These purge systems expel propellant leakage and are necessary from the time of loading throughout flight.

Environmental Control System

The environmental control system protects stage equipment from temperature extremes in both the forward skirt and thrust structure areas and provides a nitrogen purge during prefiring and firing operations. Temperature-controlled air is provided by a ground air conditioning unit from approximately 14 hours before launch to approximately 6 hours before launch. At this time, gaseous nitrogen from an auxiliary nitrogen supply unit is introduced into the system and used to purge and condition the forward skirt and thrust structure areas until umbilical disconnect at launch. A distribution manifold vents air and gaseous nitrogen through orifices into the thrust structure to maintain proper temperature. Air and nitrogen are supplied from the ground. The system also distributes air and gaseous nitrogen to instrumentation canisters mounted in the forward skirt. Temperatures in the canisters are held to meet requirements of electrical equipment. From the canisters, the conditioning gas is vented into the forward skirt compartment.

nitrogen central receiving and distributing center as well as a mounting block for filters, shutoff solenoid valves, a pressure regulator, a relief valve, and pressure transducers. Ported manifolds provide tubing assembly connections to the storage bottle, umbilical coupling, and various tubing assembly distribution lines to control valves throughout the stage.

GROUND CONTROL PRESSURE SYSTEM

The ground control pressure system provides a direct ground pressure supply for some of the first stage pneumatically actuated valves. The valves are involved with propellant fill and drain and emergency engine shutdown system operations. Direct ground control assures a backup system in case of emergency and conserves the onboard nitrogen supply.

ONBOARD PURGE PRESSURE SYSTEM

The onboard purge pressure system consists of three high-pressure nitrogen storage bottles identical to the onboard control pressure storage

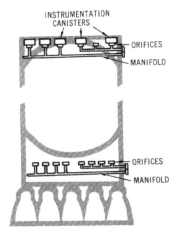

Environmental Control System

Visual Instrumentation

Visual instrumentation, presently planned to be installed on two flight stages, is designed to monitor critical stage functions prior to and during static test and flight conditions.

FILM CAMERAS

The first stage film cameras provide photographic coverage of the LOX tank interior during launch, flight, and separation. The stage carries four film cameras. The two LOX-viewing cameras will provide color motion pictures to show the following: behavior of the liquid oxygen, possible wave or slosh motions, and cascading or waterfall effects of the liquid from the internal tank structure. The capsules, which contain the cameras, are ejected automatically about 25 seconds after separation and are recovered after descent into the water. First stage flight versions of the camera consist of the LOX tank-viewing configuration plus two direct-viewing stage separation capsules. The installation is in the forward skirt area. The tank-viewing optical lenses and the two strobe flash light assemblies are mounted in the LOX tank manhole covers. Connecting the remotely located camera capsules and the flash head are the optical assemblies. Consisting of coupling lens attached to the ejection tube, a 9-foot length of fiber optics, and the objective lens mounted in the flash-head assembly. The equipment required to complete the system, such as batteries, power supplies, timer, and synchronizing circuitry, is contained in the environmentally controlled equipment racks or boxes mounted around the interior of the forward skirt structure. The combined timer and synchronizing unit serves two functions. The digital pulse timer supplies real time correlation pulses which are printed on one edge of the film. The timer also supplies event marker pulses to the opposite edge of the film to record selected significant events such as liftoff, engine shutdown, and stage separation. The synchronizing unit times the intermittent illumination provided by the strobe lamps to coincide with the open portion of the rotating shutter as it passes the motion picture film gate. The capsule assembly consists of the heavy nose section and quartz window, which protect the capsule during re-entry heating and impact on the water. The body of the capsule, including the camera, is sealed and water-tight. A paraloon and drag skirt aid its descent and flotation. A radio beacon and flashing light are mounted on the capsule to aid in recovery.

TELEVISION SYSTEM

The television system on the first stage will transmit four views of engine operation and other engine area functions in the interval from fueling

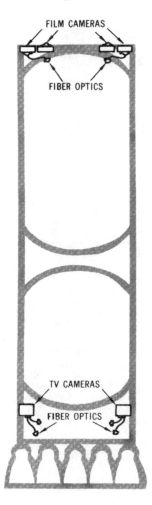

Visual Instrumentation

to first stage separation. The system utilizes two split fiber optics viewing systems and two cameras. Extremes in radiant heat, acoustics, and vibration prohibit the installation of the cameras in the engine area; therefore, fiber optics bundles are used to transmit the images to the cameras located in the thrust structure. Quartz windows are used to protect the lens. Both nitrogen purging and a wiping action are used to prevent soot buildup on the protective window.

Image enhancement improves the fiber-optical systems by reducing the effects of voids between fibers and broken fibers. An optically flat disc with parallel surfaces rotates behind each objective lens.

The drive motor rotates in synchronism with the master drive motor. A DC to AC inverter energizes the synchronous drive motors. A camera control unit houses amplifiers, fly back, sweep, and other circuits required for the video system. Each vidicon output (30 frames/second) is amplified and sampled every other frame (15 frames a second) by the video register. A 2.5 watt FM transmitter feeds the 7- element yagi antenna array covered by a radome.

FIRST STAGE FLIGHT

The first stage is loaded with RP-1 fuel and LOX at approximately 12 and 4 hours respectively, before launch. With all systems in a ready condition, the stage is ignited by sending a start signal to the five F-1 rocket engines. The engine main LOX valves open first allowing LOX to begin to enter the main thrust chamber. Next the engines' gas generators and turbopumps are started. Each engine's turbopump assembly will develop approximately 60,000 horsepower. Combustion is initiated by injecting a hypergolic solution into the engine's main thrust chamber to react with the LOX already present. The main fuel valves then open, and fuel enters the combustion chamber to sustain the reaction previously initiated by the LOX and hypergolic solution. Engine thrust then rapidly builds up to full level. The five engines are started in a 1-2-2 sequence, the center engine first and opposing out-board pairs at 300-millisecond stagger times. The stage is held down while the engines build up full thrust. After full thrust is reached and all engines and stage systems are functioning properly, the stage is released. This is accomplished by a "soft release" mechanism. First, the restraining hold-down arms are released. Immediately thereafter the vehicle begins to ascend but with a restraining force caused by tapered metal pins being pulled through holes. This "soft release" lasts for about 500 milliseconds.

The vehicle rises vertically to an altitude of approximately 430 feet to clear the launch umbilical tower and then begins a pitch and roll maneuver to attain the correct flight azimuth. As the vehicle continues its flight, its path is controlled by gimbaling the outboard F-1 engines consistent with a preprogrammed flight path and commanded by the instrument unit.

At approximately 69 seconds into the flight the vehicle experiences a condition of maximum dynamic pressure. At this time, the restraining drag force is approximately equal to 460,000 pounds.

At 135.5 seconds into the flight most of the LOX and fuel will be consumed, and a signal is sent from the instrument unit to shut down the center engine. The outboard engines continue to burn until either LOX or fuel depletion is sensed. LOX depletion is signaled when a "dry" indication is received from at least two of the four LOX cutoff sensors; one sensor is located near the top of each outboard LOX suction duct. Fuel depletion is signaled by a "dry" indication from a redundant fuel cutoff sensor bolted directly to the fuel tank lower bulkhead. The LOX depletion cutoff is the main cutoff system with fuel cutoff as the backup.

Six hundred milliseconds after the outboard engines receive a cutoff signal, a signal is given to fire the first stage retrorockets. Eight retrorockets are provided and each produces an average effective thrust of 88,600 pounds for 0.666 seconds. The first stage separates from the second stage at an altitude of about 205,000 feet. It then ascends to a peak altitude near 366,000 feet before beginning its descent. While falling, the stage assumes a semi-stable engines down position and impacts into the Atlantic Ocean at approximately 350 miles down range of Cape Kennedy.

F-1 ENGINE FACT SHEET

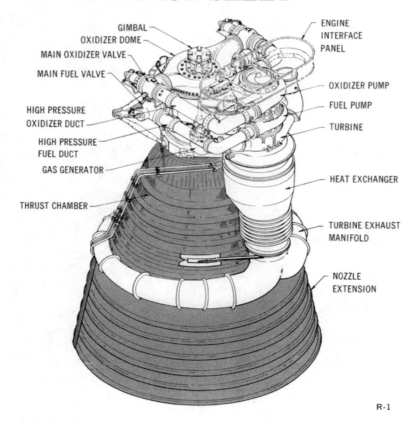

LENGTH	19 ft.
WIDTH	12 ft. 4 in.
THRUST (sea level)	1,500,000 lb.
SPECIFIC IMPULSE (mini mum)	260 sec.
RATED RUN DURATION	150 sec.
FLOWRATE: Oxidizer	3,945 lb. sec. (24,811 gpm)
Fuel	1,738 lb. sec. (15,471 gpm)
MIXTURE RATIO	2.27:1 oxidizer to fuel
CHAMBER PRESSURE	965 psia
WEIGHT FLIGHT CONFIGURATION	18,500 lb. maximum
EXPANSION AREA RATIO	16:1 with nozzle extension
	10:1 without nozzle extension
COMBUSTION TEMPERATURE: Thrust Chamber	5,970°F
Gas Generator	1,465°F
MAXIMUM NOZZLE EXIT DIAMETER	11 ft. 7 in.

NOTE: F-1 engine will be uprated to 1,522,000 lb. thrust for Vehicle 504 and all subsequent operational vehicles.

F-1 ENGINE

ENGINE DESCRIPTION

The F-1 engine is a single-start, 1,500,000-pound fixed-thrust, bipropellant rocket system. The engine uses liquid oxygen as the oxidizer and RP-1 (kerosene) as fuel. The engine is bell-shaped with an area expansion ratio—the ratio of the area of the throat to the base—of 16:1. RP-1 and LOX are combined and burned in the engine's thrust chamber assembly. The burning gases are expelled through an expansion nozzle to produce thrust. The five-engine cluster used on the first stage of the Saturn V produces 7,500,000 pounds of thrust. All of the engines are identical with one exception. The four outboard engines gimbal; the center engine does not. The major engine systems are the thrust chamber assembly, the propellant feed system, the turbo-pump, the gas generator system, the propellant tank pressurization system, the electrical system, the hydraulic control system, and the flight instrumentation system.

Assembly- Thrust chambers of the F-1 rocket engine-the most powerful engine under development by the United States-are assembled in this manufacturing line. R-2

THRUST CHAMBER ASSEMBLY

The thrust chamber assembly consists of a gimbal bearing, an oxidizer dome, an injector, a thrust chamber body, a thrust chamber nozzle extension, and thermal insulation. The thrust chamber assembly receives propellants under pressure supplied by the turbopump, mixes and burns them, and imparts a high velocity to the expelled combustion gases to produce thrust. The thrust chamber assembly also serves as a mount or support for all engine hardware.

Gimbal Bearing

The gimbal bearing secures the thrust chamber assembly to the vehicle thrust frame and is mounted on the oxidizer dome. The gimbal is a spherical, universal joint consisting of a socket-type hearing with a bonded Teflon-fiberglass insert which provides a low-friction bearing surface. It permits a maximum pivotal movement of 6 degrees in each direction of both the X and Z axes (roughly analogous to pitch and yaw) to facilitate thrust vector control. The gimbal transmits engine thrust to the vehicle and provides capability for positioning and thrust alignment.

Oxidizer Dome

The oxidizer dome serves as a manifold for distributing oxidizer to the thrust chamber injector, provides a mounting surface for the gimbal hearing, and transmits engine thrust forces to the vehicle structure. Oxidizer at a volume flowrate of 24,811 gpm enters the dome through two inlets positioned 180 degrees apart (to maintain even distribution of the propellant).

Thrust Chamber Injector

The thrust chamber injector directs fuel and oxidizer into the thrust chamber in a pattern which ensures efficient and satisfactory combustion. The injector is multi-orificed with copper fuel rings and copper oxidizer rings forming the face (combustion side) of the injector and containing the injection orifice pattern. Assembled to the face are radial and circumferential copper baffles which extend downward and compartmentalize the injector face. The baffles and rings, together with a segregated igniter fuel system, are installed in a stainless steel body. Oxidizer enters the injector from the oxidizer dome. Fuel enters the injector from the thrust chamber fuel inlet manifold, and in order to facilitate the engine start phase and to reduce pressure losses, part of the flow is introduced directly into the thrust chamber. The remaining fuel (controlled by orifices) flows through alternate tubes which run the length of the thrust chamber body to the nozzle exit. There, it enters a return manifold and flows back to the injector through the remaining tubes.

Thrust Chamber Body

The thrust chamber body provides a combustion chamber for burning propellants under pressure and an expansion nozzle for expelling gases produced by the burned propellants at the high velocity required to produce the desired thrust. The thrust chamber is tubular-walled and regeneratively fuel-cooled, and the nozzle is bell-shaped. There are four sets of outrigger struts attached to the exterior of the thrust chamber; two sets of the struts are turbo-pump mounts and the other two are attach points for the vehicle contractor's gimbal actuators. The thrust chamber incorporates a turbine exhaust manifold at the nozzle exit and a fuel inlet manifold at the injector end which directs fuel to the fuel down tubes. Brackets and studs welded to the reinforcing "hatbands" surrounding the thrust chamber provide attach points for thermal insulation blankets. Fuel enters the fuel inlet manifold through two diametrically opposed inlets. From the manifold, 70 per cent of the fuel is diverted through 89 alternate CRES "down" tubes the length of the chamber. A manifold at the nozzle exit returns the fuel to the injector through the remaining 89 return tubes. The fuel flowing through the chamber tubes provides regenerative cooling of the chamber walls during engine operation. The thrust chamber tubes are bifurcated; that is, they are comprised of a primary tube from the fuel manifold to the 3:1 expansion ratio area. At that point, two secondary tubes are spliced into each primary tube. This is necessary to maintain a desired cross-sectional area in each of the tubes through the large-diameter belled nozzle section. The turbine exhaust manifold, which is fabricated from preformed sheet metal shells and which forms a torus around the aft end of the thrust chamber body, receives turbine exhaust gases from the heat exchanger. Upon entering the manifold, the gases are distributed uniformly. As the gases are expelled from the manifold, flow vanes in the exit slots provide uniform static pressure distribution in the nozzle extension. Radial expansion joints compensate for thermal growth of the manifold.

Thrust Chamber Nozzle Extension

The thrust chamber nozzle extension increases the expansion ratio of the thrust chamber from 10:1 to 16:1. It is a detachable unit that is bolted to the exit end ring of the thrust chamber. The interior of the nozzle extension is protected from the engine exhaust gas environment (5800 Fahrenheit) by film cooling, using the turbine exhaust gases (1200 Fahrenheit) as the coolant. The gases enter the extension between a continuous outer wall and a shingled inner wall, pass out through injection slots between the shingles, and flow over the surfaces of the shingles forming a boundary layer between the inner wall of the nozzle extension and the hotter exhaust gases exiting from the main engine combustion chamber. The nozzle extension is made of high strength stainless steel.

Hypergol Cartridge

The hypergol cartridge supplies the fluid to produce initial combustion in the thrust chamber. The cartridge, which is cylindrical and has a burst diaphragm welded to either end, contains a hypergolic fluid consisting of 85 per cent triethylborane and 15 percent triethylaluminum. As long as the fluid is in the hermetically sealed cartridge, it is stable, but it will ignite spontaneously upon contact with oxygen in any form. During the start phase of operation, increasing fuel pressure in the igniter fuel system ruptures the burst diaphragms. The hypergolic fluid and the fuel enter the thrust chamber through a segregated igniter fuel system in the injector and contact the oxidizer. Spontaneous combustion occurs and thrust chamber ignition is established.

Pyrotechnic Igniter

Pyrotechnic igniters, actuated by an electric spark, provide the ignition source for the propellants in the gas generator and re-ignite the fuel-rich turbine exhaust gases as they exit from the nozzle extension.

Thermal Insulation

The thermal insulation protects the F-1 engine from the extreme temperature environment (2550 Fahrenheit maximum) created by radiation from the exhaust plume and backflow during clustered-engine flight operation. Two types of thermal insulators are used on the engine—foil-batt on complex surfaces and asbestos blankets on large, simple surfaces. They are made of lightweight material and are equipped with various mounting provisions, such as grommeted holes, clamps, threaded studs, and safety wire lacing studs.

TURBOPUMP

The turbopump is a direct-drive unit consisting of an oxidizer pump, a fuel pump, and a turbine mounted on a common shaft. The turbopump delivers fuel and oxidizer to the gas generator and the thrust chamber. LOX enters the turbopump axially through a single inlet in line with the shaft and is discharged tangentially through dual outlets. Fuel enters the turbopump radially through dual inlets and is discharged tangentially through dual outlets. The dual inlet and outlet design provides a balance of radial loads in the pump.

F-1 turbopump R-4

Three bearing sets support the shaft. Matched tandem ball bearings, designated No. 1 and No. 2, provide shaft support between the oxidizer and fuel pumps. A roller bearing No. 3 provides shaft support between the turbine wheel and the fuel pump. The bearings are cooled with fuel during pump operation. A heater block provides the outer support for No. 1 and No. 2 bearings, and is used during LOX chilldown of the oxidizer pump to prevent freezing of the bearings. A gear ring installed on the shaft is used in conjunction with the torque gear housing for rotating the pump shaft by hand, and also is used in conjunction with a magnetic transducer for monitoring shaft speed. There are nine carbon seals in the turbopump: primary oxidizer seal, oxidizer intermediate seal, lube seal No. 1 bearing, lube seal No. 2 bearing, primary fuel seal, fuel inlet seal, fuel inlet oil seal, hot-gas secondary, and hot-gas primary seal. The main shaft and the parts attaching directly to it are dynamically balanced prior to final assembly on the turbopump.

Oxidizer Pump

The oxidizer pump supplies oxidizer to the thrust chamber and gas generator at a flowrate of 24,811 gpm. The pump consists of an inlet, an inducer, an impeller, a volute, bearings, seals and spacers. Oxidizer is introduced into the pump through the inlet which is connected by duct to the oxidizer tank. The inducer in the inlet increases the pressure of the oxidizer as it passes into the impeller to prevent cavitation. The impeller accelerates the oxidizer to the desired pressure and discharges it through diametrically opposed outlets into the high-pressure oxidizer lines leading to the thrust chamber and gas generator. The oxidizer inlet, which attaches to a duct leading to the vehicle oxidizer tank is bolted to the oxidizer volute. Two piston rings seated between the inlet and the volute expand and contract with temperature changes to maintain an effective seal between the high and low pressure sides of the inlet. Holes in the low-pressure side of the inlet allow leakage past the ring seals to flow into the suction side of the inducer, thus maintaining a low pressure. The oxidizer volute is secured to the fuel volute with pins and bolts which prevent rotational and axial movement. The primary oxidizer seal and spacer located in the oxidizer volute prevent fuel from leaking into the primary oxidizer seal drain cavity. The oxidizer intermediate seal directs a purge flow into the primary seal and No. 3 drain cavities where the purge acts as a barrier to permit positive separation of the oxidizer and bearing lubricants.

Fuel Pump

The fuel pump supplies fuel to the thrust chamber and gas generator at a flowrate of 15,471 gpm. The pump consists of an inlet, an inducer, an impeller, a volute, bearings, seals, and spacers. Fuel is introduced into the pump from the vehicle fuel tank through the inlet. The inducer in the inlet increases the pressure of the fuel as it passes into the impeller to prevent cavitation. The impeller accelerates the fuel to the desired pressure and discharges it through two diametrically opposed outlets into the high-pressure fuel lines leading to the thrust chamber and gas generator. The fuel volute is bolted to the inlet and to a ring, which is pinned to the oxidizer volute. A wear-ring installed on the volute mates against the impeller. The cavity formed between the volute and the impeller is called the balance cavity. Pressure in the balance cavity exerts a downward force against the fuel impeller and counterbalances the upward force of the oxidizer impeller to control the amount of shaft axial force applied to the No. 1 and No. 2 bearings. Leakage between the impeller inlet and the discharge is controlled by a wear-ring, which mates with the impeller and acts as an orifice. The fuel volute provides support for the bearing retainer, which supports the No. 1 and No. 2 bearings and houses the bearing heater. The No. 3 seal, which is installed between the oxidizer intermediate seal and the No. 1 bearing, prevents lubricating fuel for the bearings from contacting the oxidizer. If fuel should pass the seal, purge flow from the oxidizer intermediate seal will expel the fuel overboard. On the fuel side of the No. 2 bearing, the No. 4 lube seal contains the lubricant within the bearing cavity. The remaining seal in the fuel volute is the primary seal and contains fuel under pressure in the balance cavity, maintains the desired balance cavity pressure, and keeps high-pressure fuel out of the low-pressure side.

Turbine

The turbine, producing 55,000 brake horsepower, drives the fuel and oxidizer pumps. It is a two-stage, velocity-compounded turbine consisting of two rotating impulse wheels separated by a set of stators. The turbine mounts on the fuel pump end of the turbopump so that the two elements of the turbo-pump having the greatest operating temperature extremes (1500 Fahrenheit for the turbine and -300 Fahrenheit for the oxidizer pump) are separated. Hot gas from the gas generator enters the turbine at a flowrate of 170 pounds per second through the inlet manifold and is directed through the first-stage nozzle onto the 119-blade first-stage wheel. The hot gas then passes through the second-stage stators onto the 107-blade second-stage wheel, and then into the heat exchanger. This flow of hot gas rotates the turbine, which in turn rotates the propellant pumps. Turbine speed during mainstage operation is 5,550 rpm.

Bearing Coolant Control Valve

This valve, which incorporates three 40-micron filters, three spring-loaded poppets, and a restrictor, performs two functions. Its primary function is to control the supply of coolant fuel to the turbopump bearings. Its secondary function is to provide a means of preserving the turbopump bearings between static firings or during engine storage. During engine firing, the coolant poppet opens and delivers filtered fuel to the turbopump bearing coolant jets, and the restrictor provides the proper turbopump bearing jet pressure.

GAS GENERATOR SYSTEM

The gas generator system provides the hot gases for driving the velocity-compounded turbine, which drives the fuel and oxidizer pumps. The system consists of a gas generator valve, an injector, a combustion chamber, and propellant feed lines connecting the No. 2 turbopump fuel and oxidizer outlet lines to the gas generator. The propellants are supplied to the gas generator from the No. 2 turbo-pump fuel and oxidizer outlet lines. The gas generator mixture ratio, relative to the engine mixture ratio, is fuel-rich. This provides a lower combustion temperature in the uncooled gas generator and in the turbine. Propellants enter the gas generator through the valve and injector and are ignited in the combustion chamber by dual pyrotechnic igniters. The gas generator valve is hydraulically operated by fuel pressure from the hydraulic control system.

Gas Generator Valve

The gas generator valve is a hydraulically operated valve which controls and sequences entry of propellants into the gas generator. Hydraulic fuel is recirculated through a passage in the valve housing to maintain seal integrity and to prevent the fuel in the fuel ball housing from freezing. Fuel is also recirculated through a passage in the piston between the opening port and the closing port to prevent the piston O-ring from freezing.

Gas Generator Combustion Chamber

The gas generator combustion chamber provides a space for burning propellants and exhausts the gases from the burning propellants into the turbopump turbine manifold. It is a single-wall chamber located between the gas generator injector and the turbo-pump inlet.

PROPELLANT FEED CONTROL SYSTEM

The propellant feed system transfers LOX and fuel from the propellant tanks into the pumps which discharge into the high-pressure ducts leading to the gas generator and the thrust chamber. The system consists of two oxidizer valves, two fuel valves, a bearing coolant control valve, two oxidizer dome purge check valves, a gas generator and pump seal purge check valve, turbopump outlet lines, orifices, and lines connecting the components. High-pressure fuel is supplied from the propellant feed system of the engine to the vehicle-contractor-supplied thrust vector control system.

Gas Generator Assembly including Control Valves R -5

Gas Generator Injector

The gas generator injector directs fuel and oxidizer into the gas generator combustion chamber. It is a flat-faced, multi-orificed injector incorporating a dome, a plate, a ring manifold, five oxidizer rings, five fuel rings, and a fuel disc. The gas generator valve and the gas generator injector fuel inlet housing tee are mounted on the injector. Fuel enters the injector through the gas generator fuel inlet housing tee from the gas generator valve. The fuel is directed through internal passages in the plate and injected into the combustion chamber through orifices in the fuel rings and the disc. Some of the orifices in the outer fuel ring also provide a cooling film of fuel for the combustion chamber wall. Oxidizer enters the injector through the oxidizer inlet manifold from the gas generator valve. The oxidizer is directed from the oxidizer manifold through internal passages in the plate and is injected into the combustion chamber through the orifices in the oxidizer rings.

Propellant Feed—The main LOX valve and high-pressure line are shown at left. At right are the main fuel valve and high-pressure line. R -3

Oxidizer Valves

Two identical oxidizer valves, designated No. 1 and No. 2, control LOX flow from the turbopump to the thrust chamber oxidizer dome and sequence the hydraulic fuel to the opening port of the gas generator valve. When the valves are in the open position at rated engine pressures and flowrates, neither will close if the hydraulic fuel opening pressure is lost. Each of the oxidizer valves is a hydraulically actuated, pressure-balanced, poppet type, and contains a mechanically actuated sequence valve. A spring-loaded gate valve permits reverse flow for recirculation of the hydraulic fluid with the propellant valves in the closed position, but prevents fuel from passing through until the oxidizer valve is open 16.4 per cent. As the oxidizer valve reaches this position, the piston shaft opens the gate, allowing fuel to flow through the sequence valve, which in turn opens the gas generator valve.

LOX Distribution - Oxidizer is distributed by the LOX dome (lower center). Main LOX valves are shown at left and right with the engine interface panel above. R-6

A position indicator provides relay logic in the engine electrical control circuit and provides instrumentation for recording movement of the oxidizer valve poppet. The two oxidizer dome purge check valves, mounted on each of the oxidizer valves, allow purge gas to enter the oxidizer valves, but prevent oxidizer from entering the purge system.

FUEL VALVES

Two identical fuel valves, designated No. 1 and No. 2, are mounted 180 degrees apart on the thrust chamber fuel inlet manifold and control the flow of fuel from the turbopump to the thrust chamber. When the valves are in the open position at rated engine pressures and flowrates, they will not close if hydraulic fuel pressure is lost. Position indicators in the fuel valves provide relay logic in the engine electrical control circuit and instrumentation for recording movement of the valve poppets.

Thrust-OK Pressure Switches

Three pressure switches, mounted on a single manifold located on the thrust chamber fuel manifold, sense fuel injection pressure. These thrust-OK pressure switches are used in the vehicle to indicate that all five engines are operating satisfactorily. If pressure in the fuel injection cavity decreases, the switches deactuate, breaking the contact and interrupting the thrust-OK output signal.

PRESSURIZATION SYSTEM

The pressurization system heats GOX and helium for vehicle tank pressurization. The pressurization system consists of a heat exchanger, a heat exchanger check valve, a LOX flowmeter, and various heat exchanger lines. The LOX source for the heat exchanger is tapped from the thrust chamber oxidizer dome, and the helium is supplied from the vehicle. LOX flows from the thrust chamber oxidizer dome through the heat exchanger check valve, LOX flowmeter, and the LOX line to the heat exchanger.

Heat Exchanger

The heat exchanger heats GOX and helium with hot turbine exhaust gases, which pass through the heat exchanger over the coils. The heat exchanger consists of four oxidizer coils and two helium coils installed within the turbine exhaust duct. The heat exchanger is installed between the turbopump manifold outlet and the thrust chamber exhaust manifold inlet. The shell of the heat exchanger contains a bellows assembly to compensate for thermal expansion during engine operation.

Heat Exchanger Check Valve

The heat exchanger check valve prevents GOX or vehicle prepressurizing gases from flowing into the oxidizer dome. It consists of a line assembly and a swing check valve assembly. It is installed between the thrust chamber oxidizer dome and the heat exchanger LOX inlet line.

LOX Flowmeter

The LOX flowmeter is a turbine-type, volumetric, liquid-flow transducer incorporating two pickup coils. Rotation of the LOX flowmeter turbine generates an alternating voltage at the output terminals of the pickup coils.

Heat Exchanger lines

LOX and helium are routed to and from the heat exchanger through flexible lines. The LOX and helium lines terminate at the vehicle connect interface. The LOX line connects the heat exchanger to the heat exchanger check valve.

ENGINE INTERFACE PANEL

The engine interface panel, mounted above the turbopump LOX and fuel inlets, provides the vehicle connect location for electrical connectors between the engine and the vehicle. It also provides the attachment point for the vehicle flexible heat-resistant curtain. The panel is fabricated from heat-resistant stainless-steel casting made in three sections and assembled by rivets and bolts.

ELECTRICAL SYSTEM

The electrical system consists of flexible armored wiring harnesses for actuation of engine controls and the flight instrumentation harnesses.

HYDRAULIC CONTROL SYSTEM

The hydraulic control system operates the engine propellant valves during the start and cutoff sequences. It consists of a hypergol manifold, a check- out valve, an engine control valve, and the related tubing and fittings.

Hypergol Manifold

The hypergol manifold directs hypergolic fluid to the separate igniter fuel system in the thrust chamber injector. It consists of a hypergol container, an ignition monitor valve, a position switch, and an igniter fuel valve. The hypergol container position switch, and igniter fuel valve are internal parts of the hypergol manifold. A spring-loaded, cam-lock mechanism incorporated in the hypergol manifold prevents actuation of the ignition monitor valve until after the upstream hypergol cartridge diaphragm bursts. The same mechanism actuates a position switch that indicates when the hypergol cartridge is installed. The igniter fuel valve is a spring-loaded, cracking check valve that opens and allows fuel to flow into the hypergol container. The hypergol cartridge diaphragms are ruptured by the resultant pressure surge when the igniter fuel valve opens.

Ignition Monitor Valve

The ignition monitor valve is a pressure-actuated, three-way valve mounted on the hypergol manifold. It controls the opening of the fuel valves and permits them to open only after satisfactory combustion has been achieved in the thrust chamber. When the hypergol cartridge is installed in the hypergol manifold, a cam-lock mechanism prevents the ignition monitor valve poppet from moving from the closed position. The ignition monitor valve has six ports: a control port, an inlet port, two outlet ports, a return port, and an atmospheric reference port. The control port receives pressure from the thrust chamber fuel manifold. The inlet port receives hydraulic fuel pressure for opening the fuel valves. When the ignition monitor valve poppet is in the deactuated position, hydraulic fuel from the inlet port is stopped at the poppet seat. When the hypergol cartridge diaphragm bursts, the spring- loaded cam-lock retracts to permit the ignition monitor valve poppet unrestricted motion. When thrust chamber pressure (directed to the control port from the thrust chamber fuel manifold) increases, the ignition monitor valve poppet moves to the open (actuated) position and hydraulic fuel is directed through the outlet ports to the fuel valves.

Checkout Valve

The checkout valve consists of a ball, a poppet, and an actuator. The checkout valve provides for ground checkout of the ignition monitor valve and fuel valves and prevents the ground hydraulic return fuel, used during checkout, from entering the engine system and consequently the vehicle fuel tank. When performing the engine checkout or servicing, the checkout valve ball is positioned so fuel entering the engine hydraulic return inlet port will be directed through the ball and out the GSE return port. For engine static firing or flight, the hall is positioned so fuel entering the engine hydraulic return inlet port will be directed through the ball and out the engine return outlet port.

Engine Control Valve

(Hydraulic Filter and Four-Way Solenoid Valve Manifold)

The engine control valve incorporates a filter manifold, a four-way solenoid valve, and two swing check valves. The filter manifold contains three filters. One filter is in the supply system and one each in the opening and closing pressure systems. The filters prevent entry of foreign matter into the four-way solenoid valve or the engine. Two swing check valves are "teed" into the supply system filter. The check valves permit hydraulic system operation from the ground supplied hydraulic fluid for checkout and servicing procedures or engine supplied hydraulic fluid for normal engine operation. The four-way solenoid valve is comprised of a main spool and sleeves to achieve two-directional control of the fluid flow to the main fuel, main oxidizer, and gas generator valve actuators. The spool is pressure-positioned by two three-way slave pilots. Each slave pilot has a solenoid-controlled, normally open, three-way primary pilot. The de-energized position of the engine control valve provides hydraulic closing pressure to all engine propellant valves. Momentary application of 28 VDC to the start solenoid will initiate control valve actuations that culminate in the positioning of the main spool so that hydraulic pressure is applied to the opening port, and the pressure previously applied to the closing port is vented to the return port. An internal passage in the housing maintains common pressure applied between the opening port and start solenoid poppet. This pressure, after start solenoid de-energization, holds the main spool in its actuated position thereby maintaining the pressure directed to the opening port without further application of the start solenoid electrical signal. Momentary application of 28 VDC to the stop solenoid will initiate control valve actuations that culminate in positioning the main spool so that pressure is vented from the opening port and applied to the closing port. The override piston may be actuated at any time by a remote pressure supply, which, in the event of an electrical power loss, would reposition the main spool and apply hydraulic pressure to the closing port. If electrical power and hydraulic power are both removed, the valve will return to the de-energized position by spring force. If hydraulic pressure is then reapplied, pressure will be applied to the closing port. If an electrical signal is simultaneously sent to the start and stop solenoids, the stop solenoid will override the start and return the valve to a deactuated position.

Swing Check Valve

There are two identical swing check valves installed on the engine control valve. They allow the use of ground hydraulic fuel pressure during engine starting transient and engine hydraulic fuel pressure during engine mainstage and shutdown. One check valve is installed in the engine hydraulic fuel supply inlet port, the other in the ground hydraulic fuel supply inlet port.

FLIGHT INSTRUMENTATION SYSTEM

The flight instrumentation system consists of pressure transducers, temperature transducers, position indicators, a flow measuring device, power distribution junction boxes, and associated electrical harnesses, and permits monitoring of engine performance. The basic flight instrumentation system is composed of a primary and an auxiliary system. The primary instrumentation system is critical to all engine static firings and subsequent vehicle launches; the auxiliary system is used during research, development, and acceptance portions of the engine static test program and initial vehicle flights. The flight instrumentation system components, including both the primary and auxiliary systems, are listed below:

Primary Instrumentation

Fuel turbopump inlet No. 1 pressure
Fuel turbopump inlet No. 2 pressure
Common hydraulic return pressure
Oxidizer turbopump bearing jet pressure
Combustion chamber pressure
Gas generator chamber pressure
Oxidizer turbopump discharge No. 2 pressure
Fuel turbopump discharge No. 2 pressure
Oxidizer pump bearing No. 1 temperature
Oxidizer pump bearing No. 2 temperature
Turbopump bearing temperature
Turbopump inlet temperature
Turbopump speed

Auxiliary Instrumentation

Oxidizer turbopump seal cavity pressure
Turbine outlet pressure
Heat exchanger helium inlet pressure
Heat exchanger outlet pressure
Oxidizer turbopump discharge No. 1 pressure
Heat exchanger LOX inlet pressure

Heat exchanger GOX outlet pressure
Fuel turbopump discharge No. 1 pressure
Engine control opening pressure
Engine control closing pressure
Heat exchanger LOX inlet temperature
Heat exchanger GOX outlet temperature
Heat exchanger helium outlet temperature
Fuel pump inlet No. 2 temperature
Heat exchanger LOX inlet flowrate

Primary and Auxiliary Junction Box

There are two electrical junction boxes in the flight instrumentation system. The primary junction box has provisions for eight electrical connectors, and the auxiliary junction box for five. Both junction boxes are welded closed and pressurized with an inert gas to prevent possible entry of contaminants and moisture.

ENGINE OPERATION

The engine requires a source of pneumatic pressure electrical power, and propellants for sustained engine operation. A ground hydraulic pressure source, thrust chamber prefill, gas generator and turbine exhaust igniters, and hypergolic fluid are required to start the engine. When the start button is actuated, the checkout valve moves to transfer the hydraulic fuel return from the ground line to the turbopump low pressure fuel inlet. The high level oxidizer purge is initiated to the gas generator and thrust chamber LOX dome. The gas generator and turbine exhaust gas igniters fire, and the engine control valve start solenoid is energized. Hydraulic pressure is directed to the opening port of the oxidizer valves. The oxidizer valves are part way open, and the hydraulic pressure is directed to the gas generator valve opening port. The gas generator valve opens, propellants under tank pressure enter the gas generator combustion chamber, and the propellant mixture is ignited by the gas generator igniters. The exhaust gas is ducted through the turbopump turbine, the heat exchanger and the thrust chamber exhaust manifold into the nozzle extension walls where the fuel-rich mixture is ignited by the turbine exhaust gas igniters. As the turbine accelerates the fuel and the oxidizer pumps, the pump discharge pressures increase and propellants at increasing flowrates are supplied to the gas generator. Turbopump acceleration continues and, as the fuel pressure increases, the igniter fuel valve opens and allows fuel pressure to build up against the hypergol cartridge burst diaphragm.

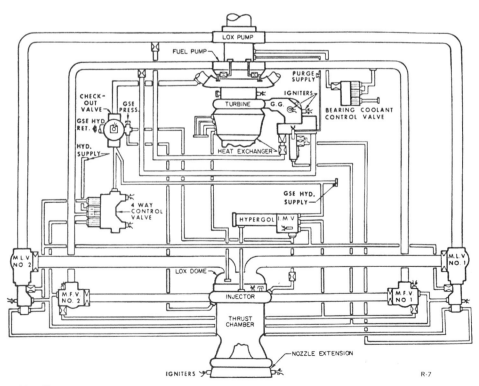

F-1 Propulsion System

The hypergol diaphragms burst under the increasing fuel pressure. Hypergolic fluid, followed by the ignition fuel, enters the thrust chamber. When hypergolic fluid enters the thrust chamber and contacts the oxidizer, spontaneous combustion occurs, establishing thrust chamber ignition. Thrust chamber pressure is transmitted through the sense line to the diaphragm of the ignition monitor valve. When the thrust chamber pressure increases, the ignition monitor valve actuates and allows hydraulic fluid flow to the opening port of the fuel valves. The fuel valves open and fuel is admitted to the thrust chamber.

Fuel enters the thrust chamber fuel inlet manifold and passes through the thrust chamber tubes for cooling purposes and then through the injector into the thrust chamber combustion zone. As the thrust chamber pressure increases, the thrust-OK pressure switches are actuated indicating the engine is operating satisfactorily. The thrust chamber pressure continues to increase until the gas generator reaches rated power, controlled by orifices in the propellant lines feeding the gas generator. When engine fuel pressure increases above the ground- supplied hydraulic pressure, the hydraulic pressure supply source is transferred to the engine.

Hydraulic fuel is circulated through the engine components and then returned through the engine control valve and checkout valve into the turbopump fuel inlet. The ground hydraulic source facility shutoff valve is actuated to the closed position when the fuel valves open. This allows the engine hydraulic system to supply the hydraulic pressure during the cutoff sequence.

ENGINE CUTOFF

When the cutoff signal is initiated, the LOX dome operational oxidizer purge comes on, and the engine control valve stop solenoid is energized. Hydraulic pressure holding open the gas generator valves, the oxidizer valves, and the fuel valves is routed to return. Simultaneously, hydraulic pressure is directed to the closing ports of the gas generator valve, the oxidizer valves, and the fuel valves. The checkout valve is actuated and, as propellant pressures decay, the high level oxidizer purge begins to flow; then the igniter fuel valve and the ignition monitor valve close. Thrust chamber pressure will reach the zero level at about the same time the oxidizer valves reach full-closed.

SECOND STAGE FACT SHEET

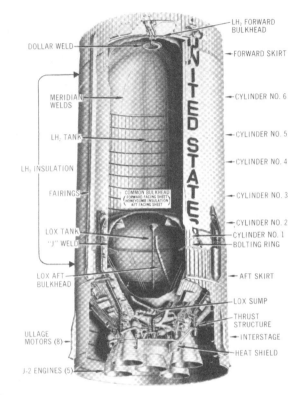

WEIGHT: 95,000 lb. (dry)
DIAMETER: 33 ft.
HEIGHT: 81 ft. 7 in.
BURN TIME: 6 min. approx.
(actually 395 sec.)
VELOCITY: 15,300 miles per hour
at burnout (approx.)
ALTITUDE AT BURNOUT: 114.5 miles

MAJOR STRUCTURAL COMPONENTS
1,037,000 lb. (loaded)
AFT INTERSTAGE THRUST STRUCTURE COMMON BULKHEAD LH_2 FORWARD BULKHEAD
AFT SKIRT AFT LOX BULKHEAD LH_2 CYLINDER WALLS FORWARD SKIRT

MAJOR SYSTEMS
PROPULSION: Five J-2 engines
Thrust: More than 1,000,000 Lb. (225,000 maximum each engine)
Propellant: LH_2 - 260,000 gal. (153,000 lb.)
LOX - 83.000 gal. (789,000 lb.)
ELECTRICAL: 6 electrical bus systems, four 28-volt DC flight batteries, and motor-operated power transfer switches
ORDNANCE: Provides, in operational sequence, ignition of eight ullage motors before ignition of five main engines, explosive separation of second stage interstage skirt, explosive separation of second stage from third stage, and ignition of four retrorockets to decelerate second stage for complete separation
MEASUREMENT: Instrumentation, telemetry, and radio frequency subsystems
THERMAL CONTROL: A ground-operated system that provides proper temperature control for equipment containers in the forward and aft skirt
FLIGHT CONTROL: Gimbaling of the four outboard J-2 engines as required for thrust vector control, accomplished by hydraulic-powered actuators which are electrically controlled from signals initiated in the flight control computer of the instrument unit (atop the Saturn V third stage)

SECOND STAGE

SECOND STAGE DESCRIPTION

The second stage of the Saturn V is the most powerful hydrogen-fueled launch vehicle under production. Manufactured and assembled by North American Aviation's Space Division, it employs the cryogenic (ultra-low temperature) propellants of liquid hydrogen and liquid oxygen, which must be contained at temperatures of -423 and -297 degrees Fahrenheit, respectively. For the lunar mission, the second stage takes over from the Saturn V's first stage at an altitude of approximately 200,000 feet (38 miles) and boosts its payload of the third stage and Apollo spacecraft to approximately 606,000 feet (114.5 miles). When its five J-2 engines ignite, the stage is pushing more than one million pounds, a load greater than that of any U.S. booster prior to the Saturn program. Speed of the stage ranges from 6,000 miles per hour to 15,300 miles per hour. The beginning of second stage boost is a two-step process. When all the F-1 engines of the first stage have cut off, the first stage separates. Eight ullage rocket motors located around the bottom of the second stage then fire for approximately 4 seconds to give positive acceleration to the stage prior to ignition of the five J-2 engines. About 30 seconds after the first stage separation, the part of the second stage structure on which the ullage rockets are located (the aft interstage) is separated by firing explosive charges. This second separation is a precise maneuver: the 18-foot-high interstage must slip past the engines without touching them. With the stage traveling at great speed, the interstage must clear the engines by only a little more than 3 feet.

The second stage burns for about 6 minutes, pushing its payload into space. At the end of boost, all J-2 engines cut off at once, the stages separate, and the J-2 engine on the third stage begins firing to take it and the Apollo spacecraft into a parking earth orbit. The 81-foot 7-inch second stage is basically a container for its 942,000 pounds of propellant with engines attached at the bottom. Propellants represent more than 90 per cent of the stage's total weight. Despite this great weight of propellant and the stresses the stage must take during launch and boost, the stage is primarily without an internal framework. It is constructed mostly of lightweight aluminum alloys ribbed in such a fashion that it is rigid enough to withstand the pressures to which it is subjected. Special lightweight insulation had to be developed to keep its cryogenic propellants from warming and thus turning to gas and becoming totally useless as propellant. The insulation that helps maintain a difference of about 500 degrees between outside (70 to 80-degree normal Florida temperature) and inside (-423 F of liquid hydrogen) is only about 1½ inches thick around the hydrogen tank. A unique feature of the second stage is its common bulkhead, a single structure which is both the top of the liquid oxygen tank and the bottom of the liquid hydrogen tank. This bulkhead was a critical item in the development of the stage. The relatively thin bulkhead, consisting of two aluminum facing sheets separated by a phenolic honeycomb core insulation, must maintain a temperature difference of 126 degrees between the two sides. The insulation which accomplishes this varies from one-tenth of an inch thickness at the girth to 4-3/4 inches thickness at the apex of the bulkhead. Development of the common bulkhead resulted in a weight saving of approximately 4 tons and more than 10 feet in stage length.

Mating - A completed second stage is mated to a first stage at Kennedy Space Center, Fla. This particular stage was used for facilities checkout. S-2

STRUCTURE

The second stage structure consists of an interstage, which links it with the first stage; a thrust structure and aft skirt assembly, which supports and houses the five J-2 engines; an ellipsoidal liquid oxygen tank; a bolting ring, which attaches the liquid oxygen tank to the second stage structure; six aluminum cylinder walls, which are welded together to form the liquid hydrogen tank; a forward domed bulkhead; and the forward skirt, which connects with the Saturn V third stage. Another important part of the structure is the 60-foot systems tunnel located on the outside of the liquid hydrogen cylinder walls through which all electrical wires between the aft skirt and the forward skirt are routed.

internal circumferential supporting frames and external hat sections positioned vertically to provide structural rigidity. After first stage burnout and initial separation, eight rocket motors attached equidistantly around the interstage are fired for approximately 4 seconds. These motors, called ullage motors (an old brewer's term referring to the gaseous zone in a tank above the liquid), provide positive acceleration and therefore pressure to force the stage's propellants into the feed lines to the J-2 engines. This is called the ullage maneuver. The interstage is separated from the second stage approximately 30 seconds after it separates from the first stage. The two-step separation of the interstage is called dual-plane separation.

Interstage

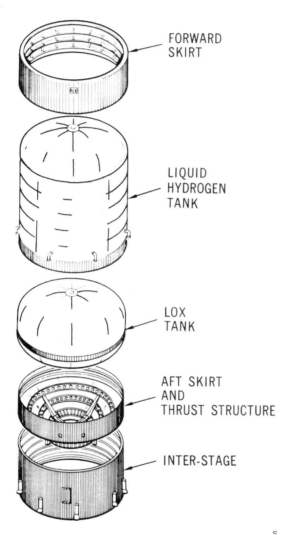

Second Stage Subassemblies

Interstage

The interstage, fabricated at NAA's Tulsa plant, is a semimonocoque structure. Semimonocoque means that the skin has a minimum of internal framework. The interstage is slightly over 18 feet in height and 33 feet in diameter. The structure has

Aft Skirt

Like the interstage, the aft skirt (as well as the thrust structure and forward skirt) is manufactured at NAA's Tulsa facility and delivered to Seal Beach for final assembly. The aft skirt is 7 feet in height and is semimonocoque construction fabricated from aluminum alloy. It is fabricated in four panels with external stringers and subassembled internal frames.

Thrust Structure

The thrust structure consists of four panels of riveted skin-stringer and internal frame construction, which, when assembled, forms an inverted cone decreasing in size from the 33-foot diameter of the upper ring to approximately 18 feet at the lower ring. Four support rings along with an outer skin stiffened with hat sections comprise the basic structure. In addition, eight thrust longerons (two to each panel) extend upward along the conical surface of the thrust structure. The lower circumferential ring rests directly over the line of thrust of each of the four outboard engines while the center engine support beam assembly is directly over the thrust line of the center engine. A rigid heat shield mounted around the five J-2 engines to a frame connecting to the thrust structure protects the base area of the stage against recirculation of hot engine exhaust gases and heat from the exhaust. This heat shield is of lightweight construction protected by low density ablative (heat-resistant) material. Although assembled separately, the aft skirt and thrust structure when joined become a structural entity and together support the five engines and withstand and distribute the thrust and boost structural loads.

Stacking Stage - Aft skirt, thrust structure, and common bulkhead move on transfer table to new station for further buildup of stage. S-5

In addition to engines and engine accessories, the interstage, aft skirt, and thrust structure house electrical and mechanical equipment such as signal conditioners and controllers, telemetry electronics, flight control electronics, service and connecting umbilicals, electrical power control units, power distribution panels and batteries, inverters, propellant management electronics, propellant plumbing, ordnance installations, and hydraulic pumps and accumulators. Equipment that is not required after second-plane separation is in the interstage which is separated 30 seconds after ignition. Equipment necessary for flight operations is located on the aft skirt, thrust structure, and forward skirt.

Liquid Oxygen Tank

The liquid oxygen (LOX) tank is an ellipsoidal container 22 feet high and fabricated from ellipsoidal-shaped top and aft halves. The top half of the LOX tank is known as the common bulkhead and is actually two bulkheads separated by phenolic honeycomb insulation and bonded together to form both the upper portion of the liquid oxygen tank and the lower portion of the liquid hydrogen tank.

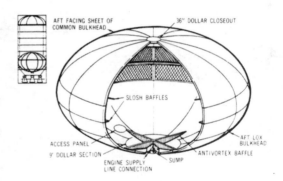

Second Stage LOX Tank

All of the LOX tank bulkheads are formed by welding together 12 high-energy-formed curved sections (gores), each approximately 20 feet long and 8 feet wide. When the gores are welded together, an opening is formed at the apex of the bulkhead. The apex is closed by welding the 12 gores to a circular section called a dollar section.

AFT LOX BULKHEAD

The aft LOX bulkhead, like the aft facing sheet of the common bulkhead, is composed of 12 thin aluminum gores welded to mechanically milled waffle panels. The waffle panels are sheets into which diagonal ribs are machined to form a series of diamonds. The waffle panels are used around the middle (widest part of the LOX tank) to

Tank Fabrication - Workmen close out dollar section of propellant tank. S-11

provide structural strength. Baffles adjacent to the aft facing sheet of the common bulkhead prevent wave action (sloshing) during flight. At the lower apex of the LOX tank, anti-vortex baffles, consisting of a 14-foot cruciform (four fins arranged in a cross) and 12 smaller baffles, are installed over the sump and engine supply line connections. The smaller baffles are essentially thin metal plates extending from the center of the cruciform, three between each pair of fins.

COMMON BULKHEAD

The common bulkhead may be likened to two giant domes, one placed inside the other, open end down, with a layer of insulation sandwiched between. The top dome is called the forward facing sheet and the bottom, the aft facing sheet. The forward facing sheet has a J-shaped periphery, which is welded to the No. 1 liquid hydrogen tank cylinder. In final assembly, a 15-inch, 12-section bolting ring is bolted to the aft skirt and the No. 1 liquid hydrogen tank cylinder. A total of 636 bolts attach the bolting ring to the liquid oxygen tank. Insulating and joining the forward and aft facing sheets into a common bulkhead is a process of several operations. First the aft facing sheet is placed on a bonding fixture and numerous sections of honeycomb phenolic insulation are fitted and tapered to exact but varying thicknesses. Then the insulation is cemented to the aft facing sheet in a multi-stage bonding operation which includes chemical processing of the aft facing sheet, application of adhesive, and pressurizing and curing in the autoclave. After mating the forward facing sheet over the insulated aft facing sheet, impression checks are made to assure a perfect fit. The forward facing sheet is then chemically processed, the insulation placed on the exposed top of the aft facing sheet is prepared with adhesive, and the entire bulkhead assembly is joined and placed in the autoclave for pressurizing and curing. In both bonding operations, checks are performed with ultrasonic equipment to ensure that the adhesive has completely covered the surface.

Bulkhead - Common bulkhead shows aft facing sheet (in sling preparatory to mating). S-8

Gore section of aft facing sheet of common bulkhead before assembly S-13

Liquid Hydrogen Tank

The liquid hydrogen cylinder walls comprise the main bulk of the second stage. Five of the cylinder walls measure slightly over 8 feet in height each, while the sixth, the No. 1 cylinder, is 27 inches high. Each of the six cylindrical sections is comprised of four curved, machined aluminum skins. Numerically machine-milled into the inside of the curved skins are stringers and ring frames. Riveted to the circumferential ring frames are flanged aluminum frames which extend inward for approximately 7 inches. In addition to structural rigidity, the frames act as slosh baffles for the liquid hydrogen. Special lightweight insulation and insulating techniques had to be developed to contain the cryogenic propellants of the second stage. The stage insulation helps maintain the liquid hydrogen at -423 degrees Fahrenheit and the liquid oxygen at -297 degrees Fahrenheit.

Honeycomb Insulation S-7

Systems Tunnel

The systems tunnel is semicircular, approximately 22 inches wide, and almost 60 feet long. It is attached vertically to the outside wall cylinders, protecting and supporting instrumentation, wiring, and tubing, which connect system components located at both ends of the stage.

Insulating - Workers apply insulation to LH2 cylinder panel. S-10

Insulation

The insulation varies in thickness in different parts of the stage. On the cylinder walls of the liquid hydrogen tank, it is only about 1½ inches thick. On the common bulkhead, which separates the liquid hydrogen and liquid oxygen tanks, it varies from a tenth of an inch thick at the edges to about 4-3/4 inches thick at the apex of the bulkhead.

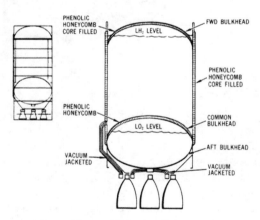

Second Stage Insulation

The LH2 tank wall insulation is formed of a phenolic honeycomb filled with a heat-resistant foam of isocyanate. The honeycomb is sealed top and bottom with a phenolic laminate and a layer of Tedlar plastic film. The helium is forced through passages (grooves) cut into the foam and honeycomb. The purge is continuous from the start of hydrogen loading until just before launch.

Forward Skirt Assembly

Of semimonocoque construction, the forward skirt is assembled from four curved sections with a height of 11½ feet and has four internal support rings. Hat sections attached vertically to the outer skin stiffen the completed assembly and provide structural support for the third stage and Apollo payload. The forward ring has provisions for attachment to the mating ring of the third stage while skin and vertical members of the skirt attach to the forward end of the liquid hydrogen tank structure.

Final Assembly

The second stage of the Saturn V launch vehicle assumes its shape in the vertical assembly building of NAA's Seal Beach facility. Assembly in the vertical position is based on a building-block concept. In this position assembly loading, circumferential exactness, and station locating is benefited by the even gravitational force exerted during each assembly operation. Constant checks and verification of station planes and stage alignment are maintained during each joining procedure by special scopes, levels, and traditional plumb bobs. Another reason for vertical assembly involves the welding of cylinders and bulkhead. If the stage were welded while in a horizontal position, temperature diversion over the circumference could result in harmful expansion near the top of the stage. To facilitate movement of the huge components and of the stage itself, a motorized transfer table rolls from outside to inside the building. Essentially, the assembly sequence begins with the welding of the lower two cylinders. Then the common bulkhead is welded to that assembly. Next the uppermost cylinder is welded to the LH2 forward bulkhead. The aft LOX bulkhead and the aft facing sheet of the common bulkhead are welded together to form the liquid oxygen tank, and the thrust structure and aft skirt are then assembled to it. The remaining cylinders are then welded to the stage, and the forward skirt is then mated to the stage stack. The interstage is fit-checked to the thrust structure before interstage systems are installed. Throughout the assembly and welding operations, hydrostatic, X-ray, dye penetrant, and other tests and quality control devices are performed to ensure that. specifications are met. The liquid hydrogen portion of the second stage as well as the liquid oxygen tank are given a thorough cleaning after assembly. After each bulkhead is welded to its components, it is hydrostatically tested. After completion of stack weld operations, the entire stage is pneumostatically tested. After completion of these tests, the liquid hydrogen and liquid oxygen tanks are thoroughly cleaned. After assembly, the stage is moved to a vertical checkout building for final checks on all stage systems.

Vertical Assembly of Stage S-14

Second Stage Forward Skirt S-12

Move is made - Flight stage is moved onto transporter to new station in vertical assembly building at Seal Beach. S-15

Repositioning - Second stage is turned horizontally for checkout operation. S-16

Engine Installation - J-2 engines are mounted in stage. S-17

PROPELLANT SYSTEM

The propellant system is composed of seven subsystems: purge, fill and replenish, venting, pressurization, propellant feed, recirculation, and propellant management.

Purge Subsystem

The purge subsystem uses helium gas to clear the propellant tanks of contaminants before they are loaded. The important contaminants are oxygen in the liquid hydrogen tank (liquid hydrogen will freeze oxygen which is impact-sensitive) and moisture in the liquid oxygen tank. The tanks are purged with helium gas from ground storage tanks. The tanks are alternately pressurized and vented to dilute the concentration of contaminants. The operation is repeated until samples of the helium gas emptied from the tanks show that contaminants have been removed or reduced to a safe level.

Fill and Replenish Subsystem

Filling of the propellant tanks on the second stage is a complex and precise task because of the nature of the cryogenic liquids and the construction of the stage. Because the metal of the stage is at normal outside temperature, it must be chilled gradually before pumping the ultra-cold propellants into the tanks. The filling operation thus starts with the introduction of cold gas into the tanks, lines, valves, and on transfer table from other components that will come into contact with the cryogenic fluids. The cold gas is circulated until the metal has become chilled enough to begin pumping in the propellants. The filling and replenishing subsystem operation consists of five phases:

Stage Complete - Flight stage assembly building to checkout building. S-18

Channel Installed - Feed line from LH$_2$ tank to one of the five engines is installed. S-19

Chilldown—Propellants are first pumped into the tank at the rate of 500 gallons per minute for LOX and 1,000 gallons per minute for LH$_2$. Despite the preliminary chilling by cold gas, the tanks are still so much warmer than the propellants that much of the latter boils off (converts to gaseous form) when it first goes into the tank. Filling continues at this rate until enough of the propellants remain liquid so that the tanks are full to the five per cent level.

Fast Fill—As soon as tank sensors report that the liquid has reached the five per cent level, the filling rate is increased to 5,000 gallons per minute for LOX and 10,000 gallons per minute for LH2. This rate continues until the liquid level in the tank reaches the 98 per cent level.

Slow-Fill—Propellant tanks are filled at the rate of 1,000 gallons per minute for both LOX and LH$_2$ until the 100 per cent level is reached.

Replenishment—Because filling begins many hours before a scheduled liftoff and the cryogenic liquids are constantly boiling off, filling continues almost up to liftoff (160 seconds before liftoff for LOX and 70 seconds before liftoff for LH$_2$. Tanks are filled at the rate of up to 200 gallons per minute for LOX and up to 500 gallons per minute for LH$_2$, depending on signals from sensors in the tanks on the liquid level.

101 Per Cent Shutdown—A sensor in each tank will send a signal to indicate that the 101 per cent level (over the proper fill level) has been reached; this signal causes immediate shutdown of filling operations.

Filling is accomplished through separate connections, lines, and valves. The ground part of the connections is covered by special shrouds in which helium is circulated during filling operations. This provides an inert atmosphere around the coupling between the ground line and the tanks. The coupling of the fill line and the tanks is engaged manually at the start of filling operations: it is normally disengaged remotely by applying pneumatic pressure to the coupling lock and actuating a push- off mechanism. A backup method involves a remotely attached lanyard in which the vertical rise of the vehicle will unlock the coupling. The fill valves are designed so that loss of helium pressure or electrical power will automatically close them. Liquid oxygen is the first propellant to be loaded onto the stage. It is pumped from ground storage tanks. Liquid hydrogen is transferred to the stage by pressurizing the ground storage tanks with gaseous hydrogen. The liquid hydrogen tank is chilled before the liquid oxygen is loaded to avoid structural stresses. After filling is completed, the fill valves and the liquid oxygen debris valves in the coupling are closed, but the liquid hydrogen debris valve is left open. The liquid oxygen fill line is then drained and purged with helium. The liquid hydrogen line is purged up to the coupling. When a certain signal is received (first stage thrust-commit), the liquid hydrogen debris valve is closed and the coupling is separated from the stage. The tanks can be drained by pressurizing them, opening the valves, and reversing the filling operation.

Venting Subsystem

The venting subsystem is used during loading and flight operations. While the propellant tanks are being loaded, the vent valves (two for each tank) are opened by electrical signals from ground equipment to allow the gas created by propellant boil-off to leave the tanks. The valves are spring-loaded to be normally closed, hut a relief valve will open them if pressure in the tanks reaches an excessive level. Each valve is capable of venting enough gas to relieve the pressure in its tank: two are provided in each tank as backup.

Pressurization

Pressurization of the propellant tanks is a three-stage process. Before launch, pressurization is obtained with gaseous helium from ground support equipment. After J-2 engine start, the pressurization is maintained with gaseous oxygen and gaseous hydrogen converted from the liquid oxygen and hydrogen.

through a heat exchanger before reaching the combustion chamber, converted to a gaseous state, and diverted to a pressurization line and a regulator where it flows back to pressurize the liquid oxygen tank. The flow of pressurant gas into the LH_2 tank is automatically stepped up after 250 seconds of a J-2 engine firing, and the greater flow of gas and the increase of pressure continues for the rest of the firing period.

Propellant Feed Subsystem

The purpose of the propellant feed subsystem is simple: transfer the liquid propellants from their tanks to the five J 2 engines. Each tank has five prevalves which control or stop the flow through separate feed lines to the engines. Solenoids control helium pressure to open and close the valves; if pressure or electrical power is lost, the valves will automatically stay open. The feed lines (except the center engine liquid oxygen line) are 8 inches in diameter and are vacuum-jacketed and insulated. The center engine liquid oxygen line is also 8 inches in diameter but is not insulated. Thermocouples (temperature measuring devices) in the vacuum jackets permit periodic vacuum checks; rupture discs in the jackets relieve excessive pressure. The feed lines also have bellows to allow for thermal expansion and freedom of movement.

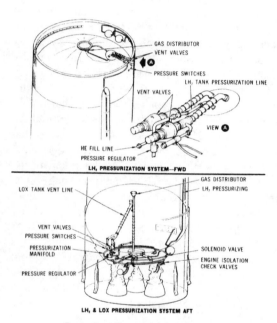

Propellant Pressurization System

Before launch, both the liquid hydrogen and liquid oxygen tanks are pressurized with gaseous helium which flows directly into the stage pressurization lines from ground storage tanks. Pressure switches, which sense ullage pressure, maintain the required pressurization level (37 to 39 psia for liquid oxygen and 31 to 33 psia for liquid hydrogen). This pressurization is maintained until liftoff: boil-off of the liquid propellants maintains adequate pressure until the stage's engines are ignited. A fitting on the upper manifold of the engines permits gaseous hydrogen (converted from its liquid state) to flow through a common manifold, a pressurization line, and regulator hack to pressurize the liquid hydrogen tank to the desired levels. Some of the liquid oxygen is bled

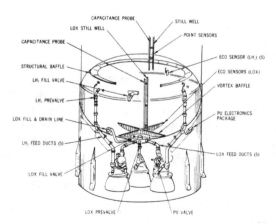

Feed System Components

Recirculation Subsystem

The main purpose of the recirculation subsystem is to keep the liquid propellants in the engine pumps. The subsystem, by keeping the propellants moving through lines, valves, and pumps, also keeps these parts chilled.

The LH$_2$ recirculation subsystem pumps the propellants through the feed lines and valves and back to the LH$_2$ tank through a single return line. The pumps are powered from a 56-volt DC battery system located in the interstage; the batteries are ejected with the interstage approximately 30 seconds after first plane separation. Before liftoff, power for the LH$_2$ recirculation subsystem is supplied by ground equipment. The LOX recirculation system works on the basis of a thermal siphon; heat entering the system is used to provide pumping action by means of fluid density differences across the system. Helium gas is used to supplement the density differences and thereby improve the pumping action. Recirculation of oxygen begins at the start of tanking; LH$_2$ recirculation begins just before launch. The propellants continue to circulate through first stage firing and up until just before the first stage and second stage separate. While the subsystems are operating, the LH$_2$ prevalves which lead to the combustion chambers are closed; as soon as the recirculation subsystem stops, the LH$_2$ prevalves open and the engines ignite.

Propellant Management System

The propellant management system controls loading, flow rates, and measurement of the propellants. It includes propellant utilization, propellant loading, propellant mass indication, engine cutoff, and propellant level monitoring subsystems.

PROPELLANT UTILIZATION SUBSYSTEM

The propellant utilization subsystem controls the flow rates of liquid hydrogen and liquid oxygen in such a manner that both will be depleted simultaneously. It controls the mixture ratio so as to minimize propellant residuals (propellant left in the tanks) at engine cutoff. Propellant utilization bypass valves at the liquid oxygen turbopump outlets control flow of liquid oxygen in relation to the liquid hydrogen remaining. Control of the engine mixture ratio increases the stage's payload capability. The propellant utilization subsystem is interrelated with the propellant loading subsystem and uses some of the same tank sensors and ground checkout equipment.

PROPELLANT LOADING SUBSYSTEM

The loading subsystem is used to control propellant loading and maintain the quantity of propellants in the tanks. Capacitance probes (sensors) running the full length of the propellant tanks sense liquid mass in the tanks and send signals to an airborne computer, which relays them to a ground computer to control loading. They also send signals to an airborne computer for the propellant utilization subsystem's control of flow rates.

PROPELLANT MASS INDICATION SUBSYSTEM

The propellant mass indication subsystem is integrated with the propellant loading subsystem and is used to send signals to the flight telemetry system for transmission to the ground. It utilizes propellant loading sensors to determine propellant levels.

ENGINE CUTOFF SUBSYSTEM

The main function of the engine cutoff subsystem is to signal the depletion point of either propellant. It is an independent subsystem and consists of five sensors in each propellant tank and associated electronics. The sensors will initiate a signal to shut down the engines when two out of five sensors in the same tank signal that propellant is depleted.

PROPELLANT LEVEL MONITORING SUBSYSTEM

The propellant level monitoring subsystem checks the level of propellants in both tanks to provide checkpoints for the sensors used in the propellant utilization and loading subsystems and to monitor propellant levels during firing. These functions are performed by sensors mounted on continuous stillwells adjacent or parallel to the full-length capacitance probes in each tank. There are 14 sensors on each stillwell to indicate various levels in the tanks.

ULLAGE MOTORS

The solid propellant ullage motors are used to provide artificial gravity by momentarily accelerating the second stage forward after first stage burnout. This moment of forward thrust is required in the weightless environment of outer space to make certain that the liquid propellant is in proper position to be drawn into the pumps prior to starting of the second stage engines.

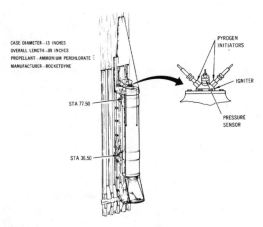

Ullage Motor S-22

Eight ullage motors are utilized on the stage where they are attached around the periphery of the interstage structure between the first and second stages. Each ullage motor measures 12.5 inches in diameter by 89 inches long and each provides 22,500 pounds of thrust for approximately 4 seconds. The motors utilize Flexadyne solid propellant in a formulation developed specifically to provide high performance and superior mechanical properties under operating conditions encountered in space. Ullage motor nozzles are canted 10 degrees to reduce exhaust impingement against the interstage structure.

THERMAL CONTROL SYSTEM

Thermal control is provided by a ground-operated system which maintains proper temperatures for the equipment containers in the forward and aft skirt areas. Tempered air is used to cool the containers before propellant loading. With preparation for loading, the air is changed to nitrogen for container inerting and heating. Separate thermal control systems are provided for the forward and aft skirt areas. Each of the units contains a single manifold connected to each container, individual fixed-flow orifices, and individual relief holes from each container. Container insulation and thermal inertia preclude excessive temperature changes.

FLIGHT CONTROL SYSTEM

Flight control of the second stage is maintained by gimbaling the four rocket thrust engines for thrust vector (direction) control. These are the four outboard engines; the fifth J-2 engine located in the center of the cluster is stationary. Each outboard engine has a separate engine actuation system to provide the force to position the engine. Gimbaling is achieved by hydraulic-powered actuators controlled by electrical signals generated through a flight control computer located in the instrument unit just above the third stage. Hydraulic power for operating each of the gimbaling actuators is supplied by individual engine-driven hydraulic pumps. Each system is self-contained and operates under a pressure of 3,500 psi. The components of each hydraulic system are attached to the thrust structure above each of the outboard engines. The main hydraulic pump is driven by the liquid oxygen turbopump on the respective engine. Two servoactuators that control each engine programmed for gimbaling are located on the engine outboard side. One is on the pitch plane, and the other on the yaw plane. Each actuator will gimbal the engine plus or minus 7 degrees in pitch or yaw and plus or minus 10 degrees in combination to correct for roll errors at a minimum rate of 8 degrees per second. During flight, the guidance system continuously determines an optimum vehicle steering command based on the vehicle's position, velocity, and acceleration. This system, located in the instrument unit, has a guidance signal processor which delivers attitude correction signals to the flight control computer in the instrument unit. These signals are shaped, scaled, and summed electronically. These summed error signals are then directed to the servoactuator amplifiers, which, in turn, drive their respective servoactuators in the second stage. These signals cause the servoactuators to position the engines.

MEASUREMENT SYSTEM

A wide variety of transducers and signal conditioners is used in the instrumentation system, which feeds signals to a high-level telemetering system for transmission to the ground. The various instrumentation sensors monitor pressure, temperature, and propellant flow rates within the tanks. Other sensors record the amount of vibration and noise, and flight position and acceleration. Tied into the measurement system are telemetry and radio frequency subsystems which transmit the performance signals to ground receiving stations for immediate (real-time) and postflight vehicle performance evaluation. Antennas which serve the telemetry and radio frequency subsystems are flush-mounted on the forward skirt and are omni-directional in coverage.

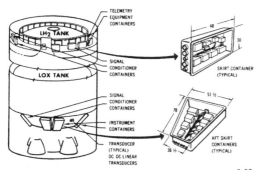

Telemetry System S-23

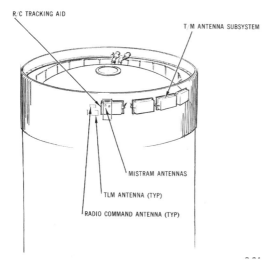

Radio Frequency System S-24

ELECTRICAL SYSTEM

In flight, the second stage electrical system is powered by four 28-volt DC batteries which operate four DC bus systems. The main DC bus powers electrical controls for the pressurization and propellant management systems, J-2 engine control, and the electrical sequence controller. The instrumentation DC bus powers instrumentation and telemetry. The ignition and recirculation bus systems provide the electrical supply for those flight operations. Both the main and instrumentation bus systems are powered by individual 28-volt DC silver-oxide/zinc batteries. The recirculation DC bus system is powered by two series-connected 28-volt DC silver-oxide/zinc batteries. This supplies electrical power to five recirculation pump-motor inverters. The inverters convert the 56 volts to three-phase, 400 cps, quasi-square wave power to the AC induction motors on the liquid hydrogen recirculation pump systems. The ignition system receives its electrical power from a tap on the recirculation system flight batteries through a power transfer switch. Each of the flight electrical power bus systems has a power transfer switch -an electro-mechanical device for transferring the systems from ground service equipment power (prelaunch) to stage battery power for flight. Before flight, the electrical power system and its electrical controls are energized from regulated ground service equipment power.

ORDNANCE SYSTEM

The separation of the first and second stages is a dual-plane separation. With depletion of the first stage propellants, an engine cutoff signal is initiated. A linear-shaped charge utilizing exploding bridge-wire initiators physically severs the two stages. Simultaneously, retrorockets on the first stage are ignited to decelerate the first stage and complete the separation, and in order to assure good propellant flow to the five J-2 engines of the second stage after first stage separation, eight solid-propellant ullage motors located on the second stage's aft interstage are fired to establish positive vehicle acceleration and proper propellant settling. When the outboard engines of the second stage reach 90 per cent of maximum thrust, a signal is transmitted which initiates interstage separation. An explosive charge separates the interstage from the aft skirt of the second stage. Approximately 10 seconds before second stage propellant depletion, a signal activates the separation system which will sever the second stage from the third. An interstage connecting the second and third stage has four retrorockets which are fired to decelerate the second stage.

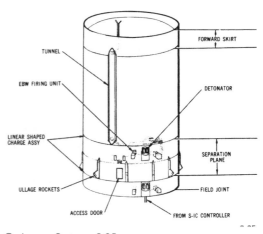

Ordnance System S-25

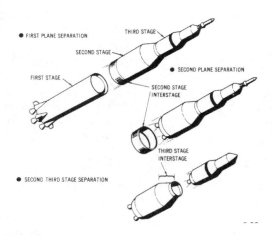

Separation System S-26

with inlet and outlet taps. A gas analyzer determines the concentration of hydrogen in the purge gas (helium) after it has been forced through the insulation, and thus indicates any leakage. From the start of hydrogen loading until launch, the insulation and core of the common bulkhead are continuously purged of hazardous gases. Vacuum equipment is used for evacuation to prevent pressure buildup in the insulation and bulkheads by removing trapped gases. The insulation purge prevents air from entering the insulation in the event of damage during cryogenic operations.

GROUND SUPPORT

Ground support operations play an important part in getting the second stage ready for operation. Among the vital operations in this area are checkout (performed mostly with complex electronic equipment and computerized routines which stimulate stage systems and analyze responses), leak detection and insulation purge, and engine compartment conditioning.

Leak Detection and Insulation Purge

The purpose of this system is to detect hydrogen, oxygen, or air leaks; to dilute and remove leaking gases; and to prevent air from liquefying during tanking operations. Any operation involving liquid hydrogen can be, extremely hazardous; liquid hydrogen in the presence of oxygen can explode or create a fire. The low-temperature atmosphere of liquid hydrogen causes air to liquefy and solidify against the hydrogen tank wall if there is any leak in the tank insulation. The organic portion of the insulation will become impact-sensitive when drenched in liquid air or oxygen; insulation saturated with cryopumped air will add weight to the stage and could cause damage during draining because of a pressure buildup created by the liquefied air returning to a gas. For these reasons, detection, control, and elimination of any hydrogen leaks from the stage and ground equipment are of great importance. The leak detection system checks out the liquid hydrogen tank, tank insulation, and the common bulkhead. The areas to be checked are divided (tank wall, forward bulkhead, and common bulkhead), each

Engine Compartment Conditioning

The purpose of this system is to purge the engine and interstage areas of explosive mixtures and to maintain proper temperature in critical regions of the aft compartment of the second stage. The compartment is purged before tanking and while the propellants are loaded. The system consists of a 13-inch diameter feed line, manifold, ducts, and a series of vents surrounding the engine compartment and skirt area. The system provides temperature control for the hydraulic systems and certain components on the J-2 engines. The purge gas is forced through orifices in the manifold to the following areas requiring warming: the area between the thrust structure and the liquid oxygen tank, the bottom of the thrust structure including the lower surface of the thrust cone, the aft skirt and interstage, and the top surface of the heat shield. The vent holes are located under the supporting hat sections on the outside of the aft skirt; this prevents wind, rain, and dust from entering the engine compartment. The vents are located so that the flow pattern provides good thermal control and expels hazardous gases. The aft skirt and interstage are purged with warm (80 to 250 degrees) nitrogen. The nitrogen is sent through the feed line into the manifold, and then through ducts to the temperature-sensitive areas. By maintaining a 98 per cent nitrogen atmosphere in the engine compartment, desired temperatures are maintained and the danger of fire or explosion resulting from propellant leaks are minimized.

THIRD STAGE FACT SHEET

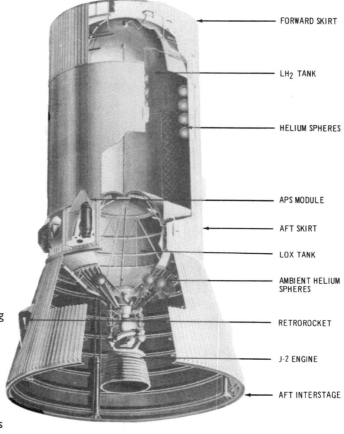

WEIGHT: 34,000 lb. (dry) including 7,700-lb. aft interstage
DIAMETER: 21 ft. 8 in.
HEIGHT: 58 ft. 7 in.
BURN TIME: 1st burn-2.75 min. (approx.)
2nd burn-5.2 min. (approx.)
VELOCITY: 1st burn-17,500 miles per hour at burnout (approx.)
2nd burn-24,500 miles per hour (approx. typical lunar mission escape velocity)
ALTITUDE AT BURNOUT: 115 miles after 1st burn and into a translunar injection on 2nd burn

MAJOR STRUCTURAL COMPONENTS

AFT INTERSTAGE   THRUST STRUCTURE   COMMON BULKHEAD
AFT SKIRT   PROPELLANT TANK   FORWARD SKIRT

MAJOR SYSTEMS

PROPULSION : One bipropellant J-2 engine 262,000 lb. (loaded)
        Total Thrust: 225,000 lb. (maximum)
        Propellants: LH_2 - 63,000 gal. (37,000 lb.)
        LOX- 20,000 gal. (191,000 lb.)
HYDRAULIC: Power for gimbaling J-2 engine
ELECTRICAL: One 56 VDC and three 28 VDC batteries, providing basic power for all electrical functions
TELEMETRY AND INSTRUMENTATION: Five modulation subsystems, providing transmission of flight data to ground stations
ENVIRONMENTAL CONTROL: Provides temperature-controlled environment for components in aft skirt, aft interstage, and forward skirt
ORDNANCE: Provides explosive power for stage separation, retrorocket ignition, ullage rocket ignition and jettison and range safety requirements
FLIGHT CONTROL: Provides stage attitude control and propellant ullage control

THIRD STAGE

STAGE DESCRIPTION

Basically the Saturn V third stage, the S-IVB, is an aluminum air-frame structure powered by a single J-2 engine, which burns liquid oxygen and liquid hydrogen. The engine has a thrust of 225,000 pounds. The structure has a capacity of 228,000 pounds of fuel and oxidizer.

STAGE FABRICATION AND ASSEMBLY

The third stage structure consists of a forward skirt assembly, propellant tank assembly, thrust structure assembly, aft skirt assembly, and aft interstage assembly. The propellant tank assembly consists of a single tank separated by a common bulkhead into a fuel compartment and an oxidizer compartment.

Forward Skirt Assembly

The forward skirt is a cylindrical aluminum skin and stringer structure that provides a hard attach point for the instrument unit. In addition, the forward skirt provides an interior mounting structure for electrical and electronic equipment that requires environmental conditioning, as well as range safety and telemetry antennas mounted around the exterior periphery. Environmental conditioning for electronic equipment is provided by cold plates which utilize a coolant supplied from the IU thermo-conditioning system.

Propellant Tank Assembly

Structural elements of the propellant tank assembly are a cylindrical tank section, common bulkhead, aft dome, and forward dome. Seven segments are machined from aluminum alloy plate to form the tank section. A waffle pattern is then machine-milled into each segment to reduce weight and provide shell stiffness. The formed segments are joined into a complete cylinder by single-pass internal weld on a Pandjiris welding machine. Aft and forward domes are made by forming "orange peel" segments on a stretch press. Orange peel segments are then joined in a dome welder. Each dome assembly rotates in the fixture under a stationary welding head which is automatically positioned by a servo-controlled sensing element. To complete the hemisphere, a 43-inch "dollar" segment is bolted in the top center of the dome. Subsequently, all fittings for various tank connections are installed by machine weld.

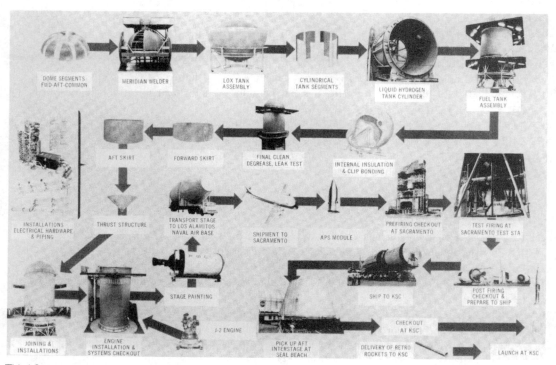

Third Stage production Sequence DAC-16183

COMMON BULKHEAD

The common bulkhead, which forms the physical separation between the LOX and hydrogen tanks, is a 130-inch-radius hemisphere consisting of two aluminum domes separated and insulated by a fiberglass honeycomb core. The honeycomb core is bonded between the two domes under heat and pressure. Edges of two peripheral tees are welded together to provide a seal for the core. Joining of the common bulkhead and the aft dome completes the LOX tank subassembly. A slosh-baffle located within the LOX tank breaks up any wave action of the oxidizer during flight. The baffle is made up of four rings supported by "A" frames.

Thrust Structure Assembly

The thrust structure distributes J-2 engine thrust over the entire tank circumference. In addition, hydraulic system components, propellant feed lines, propellant tank ambient helium pressurization spheres, pneumatic components, and miscellaneous components which support engine operation are mounted on the thrust structure assembly.

Aft Skirt Assembly

The aft skirt is a cylindrical structure fabricated of aluminum, stringer-stiffened skin panels and provides structural interface between the aft interstage and propellant tank assembly. After second stage burnout, the second stage separates from the third stage at a separation plane located on the aft skirt assembly. Two auxiliary propulsion system (APS) engine modules, two ullage rocket modules, stage separation systems, an aft umbilical connector plate, and associated system support equipment are located on the aft skirt assembly.

Aft Interstage Assembly

The aft interstage is a truncated cone-shaped structure fabricated of aluminum skin and stringers. It attaches to the third stage aft skirt and provides the structural interface to the second stage. It also houses the second stage retrorockets.

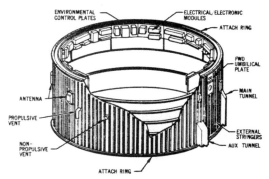

Forward Skirt Assembly D-NRV-28

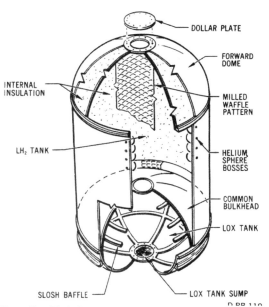

Propellant Tank Assembly D-PB-110

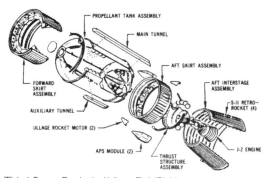

Third Stage Exploded View D-NRV-1

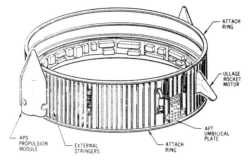

Aft Skirt Assembly D-NRV.4

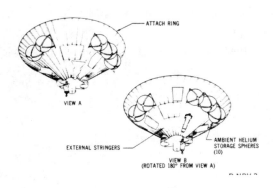

Thrust Structure Assembly D-NRV-3

Propellant Tank Assembly Area at Huntington Beach DAC-17793

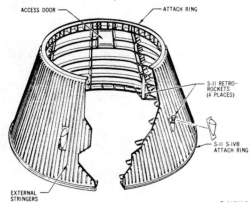

Aft Interstage Assembly D-NRV-2

Final Assembly

Final assembly of the third stage propellant tank structural components is accomplished in the assembly and welding tower. The assembly is then removed from the tower and transported to the insulation chambers building where the LH_2 tank insulating tiles are fitted and installed, a glass cloth liner placed on the insulation, anti sealant is added. Propulsion system components, internally mounted in the LH_2 tank, are installed following the completion of tank inflation. The hydrogen tank contains a slosh baffle and wave deflector system, which contributes to fuel ullage control during flight, and eight cold helium storage spheres for LOX tank pressurization. The structure is then returned to the assembly tower where the forward and aft skirts and thrust structure are installed.

Tank Section in Trim and Welding Jig D-NRV-33

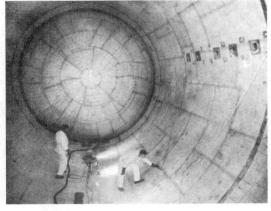

LH_2 Tank - Workmen apply resin to insulation tiles in LH_2 tank. D-NRV-39

Engine Installed-J-2 engine is attached to stage in final assembly tower at Huntington Beach. D-NRV-3 2

Slosh Baffle - Horizontal rings are installed inside LH_2 tank for propellant stabilization during flight. D-NRV-40

Third stage vehicles reach end of assembly sequence with final assembly and checkout in 115-foot vertical towers. D-NRV-42

2 engine. The stage is in a vertical position in the tower where a complete stage checkout of subsystems and systems is conducted except for actual ignition of engine. After satisfactory checkout, the stage is removed from the tower, placed on a dolly, and ground support rings are installed at each end of the stage. It is then painted, weighed, and prepared for shipment to the Douglas Sacramento Test Center for simulated and static firing of APS engines and J-2 engine.

THIRD STAGE SYSTEMS

Major systems required for third stage operation are the propulsion system, flight control system, electrical power and distribution system, instrumentation and telemetry system, environmental control system, and ordnance systems.

Propulsion System

The propulsion system consists of the J-2 engine, propellant system, pneumatic control system, and propellant utilization system. The J-2 engine burns LOX as an oxidizer and LH_2 as fuel at a nominal mixture ratio of 5:1. Both fuel and oxidizer systems utilize tank pressurization systems and have vent and relief capabilities to protect the propellant tanks from over-pressurization. The pneumatic control system regulates and controls both the oxidizer and fuel systems. The propellant utilization (PU) system assures simultaneous and precise fuel and oxidizer depletion by controlling engine mixture ratio.

Final installation of various subsystem components is performed in a checkout tower, along with the installation and alignment of the J-

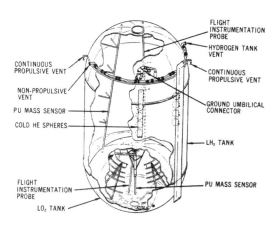

Propulsion System Components D-NRV-5

J-2 ENGINE

The engine system consists of the J-2 engine, propellant feed system, start system, gas generator system, control system, and a flight instrumentation system. The propellant feed system utilizes independently driven, direct-drive fuel and oxidizer turbopumps to supply propellants at the proper mixture ratio to the engine combustion chamber. Additional information on the J-2 engine system may be found in the J-2 Engine section.

PROPELLANT SYSTEM

The propellant system consists of related stage subsystems to support an initial J-2 engine propulsive burn phase and an engine restart capability to provide a second J-2 engine propulsive burn phase for the third stage. It includes the oxidizer system, fuel system, pressurization system, repressurization system, tank venting system, and chill-down recirculation system.

Oxidizer System

LOX is loaded into the LOX tank at a temperature of -297 Fahrenheit through a LOX fill and drain line assembly which disconnects automatically at the time of vehicle liftoff. The LOX tank capacity is 2,828 cubic feet which provides tankage for approximately 191,000 pounds (20,000 gallons) of usable LOX. The tank is pressurized with gaseous helium at 38 to 41 psia, and is maintained at this pressure during liftoff, boost, and stage engine operation.

Fill and Drain-The LOX filling operation consists of purging and chilldown of the tank and filling in four stages: slow fill, fast fill, replenish (topping), and pressurization. The ground-controlled combination vent and relief valve is pneumatically opened at the start of the fill operation. During slow fill, LOX is loaded at a rate of 500 gpm until five per cent of full level is attained; then fast fill at 1,000 gpm is initiated. When 98 per cent of the LOX has been loaded, the fill rate is reduced to a rate of 0 to 300 gpm. When the LOX tank is 100 per cent loaded, the full level is maintained until liftoff by a replenish flowrate of 0 to 30 gpm, as required to compensate for LOX boil-off. If for any reason the LOX tank becomes over-pressurized during fill, such as a vent malfunction, or by an excessive LOX fill flow, a pressure switch signals the LOX ground fill valve to close. The LOX tank is capable of being unloaded by reversing the flow through the fill system under tank pressure and/or from gravity effect. Drain capacity is at 500 gpm at 33 psia.

LOX Tank Pressurization - The LOX tank is pressurized at 38 to 41 psia from a ground supply of cold helium regulated to -360° Fahrenheit. After liftoff and until engine ignition, the LOX tank pressure is maintained from eight helium storage spheres located in the LH2 tank that have been charged to 3,100 ± 100 psi at -360° Fahrenheit. During J-2 engine burn, the engine heat exchanger heats and expands a portion of the helium flow before it is fed into the LOX tank. An ullage tank pressure switch controls in flight pressurization by opening or closing cold helium flow to the heat exchanger as required. In case of pressure switch failure in flight, a pressure switch and plenum chamber act as a backup pressure regulator.

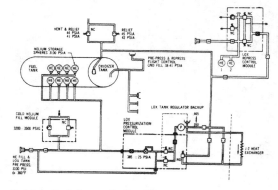

Oxidizer Tank Pressurization System D-NRV-6

LOX Tank Repressurization - During the coast phase, prior to J-2 engine second ignition, the LOX tank pressurization system is inoperative, allowing LOX tank pressure to decay. However, before the second engine ignition the LOX tank is pressurized by the LOX tank repressurization system. The repressurization system utilizes helium from two ambient helium storage spheres located on the thrust structure. The repressurization helium supply is controlled by the LOX repressurization control module. Controlled repressurization is continued until second ignition is accomplished. At second ignition and continuing through the second burn phase, the LOX tank is pressurized with cold helium gas heated by the engine heat exchanger and supplied by spheres contained within the LH$_2$ tank.

LOX Tank Vent-Relief System-The LOX tank vent-relief system consists of a tee assembly with a pneumatically operated vent valve and a backup relief valve. Pneumatic operation is provided by the LOX vent actuation module using helium gas from the pneumatic control system. The vent-relief valve is opened during the ground fill operation and closed prior to pressurization. During fill operations, the vent valve is capable of venting all LOX vapor. The relief valve backup system automatically relieves at 45 psia and reseats at 42 psia. During liftoff and non-powered stage flight, pressure relief or venting is not anticipated. However, the vent system becomes operational in the event of LOX tank over-pressurization.

LOX Feed System-Prior to vehicle liftoff and prior to engine restart for the second burn phase, all LOX feed system components of the J-2 LOX turbopump assembly must be "chilled" to operating temperature for proper operation. Chilldown of the LOX system is accomplished by a closed-loop, forward-flow recirculation system. On command from the IU, a prevalve in the LOX feed duct closes and a bypass shutoff valve opens. An auxiliary, electrically driven centrifugal chilldown pump, mounted in the LOX tank, starts and LOX chilldown circulation begins. LOX is circulated from the LOX tank, through the low pressure feed duct, to the J-2 engine LOX pump and bleed valve, then back to the LOX tank through return lines. The pump is capable of delivering a minimum flowrate of 31 gpm at 25 psia. Recirculation chilldown continues through the boost phase and up to the time for J-2 engine ignition. In the event of an emergency, the chilldown system shutoff valve closes upon command from the IU. A low pressure supply duct supplies LOX from the tank to the engine at a nominal flowrate of 391 pounds per second at -297 Fahrenheit at 25 psia and up. The main LOX feed valve is a 4-inch butterfly-type valve and opens in two distinct steps: the first, a partially opened position; the second, a fully opened position. The LOX feed valve is solenoid-controlled. A signal from the engine sequencer energizes the LOX feed valve, as required, to obtain steady- state operation. During steady-state operation, LOX feed is regulated by a propellant utilization valve which controls the oxidizer flow to the engine. A complete description of engine operation may be found in the J-2 Engine section.

Fuel System
LH_2 is loaded into the insulated fuel tank at a temperature of -423 Fahrenheit through the LH_2 fill and drain valve assembly which is automatically disconnected at time of vehicle liftoff. The tank capacity is 10,446 cubic feet providing tankage for approximately 37,000 pounds (63,000 gallons) of usable fuel. The tank is pressurized from a ground source of helium at 31 to 34 psia. During liftoff, boost, and stage engine operation, pressure is maintained in the fuel tank at 28 to 31 psi.

Fill and Drain-The LH2 loading operation consists of purging, chilldown of the tank, and filling in four stages: slow fill, fast fill, replenish (topping), and pressurization. Immediately prior to LH_2 input into the tank, a combination vent and relief valve is pneumatically opened. Loading is then initiated at 500 gpm until five per cent of full level is reached; then fast fill begins. During fast fill, LH_2 is supplied to the tank at 3,000 gpm. When 98 per cent of the loading is completed, the loading rate is curtailed to between 0 to 500 gpm. When the LH2 tank is 100 per cent loaded, the full level is maintained by a replenish flowrate of 0 to 300 gpm as required to compensate for LH_2 boil-off. During the final topping operation, the fuel tank venting system is closed and the tank is simultaneously pressurized from the ground source of helium. If over-pressurization of the tank should occur during fill or during the boost phase, a relief valve, which is spring-loaded to open at 37 psia and close at 34 psia, is actuated to relieve excess pressure. The LH_2 tank is capable of being unloaded through the fill system. LH_2 unloading is accomplished by reversing the flow through the fill system under tank pressure and/or from gravity effect.

Fuel Tank Pressurization - During initial tank pressurization, an external tank connection is made to a ground supply of helium. The helium is supplied to the fuel tank at -360 Fahrenheit at 600 psig. When the tank ullage pressure reaches a maximum of 31 to 34 psia, a pressure switch sends a signal to deactivate the ground pressurization valve indicating that a satisfactory liftoff pressure has been attained, and pressurization is discontinued.

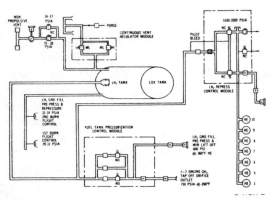

Fuel Tank Pressurization System D-NRV-7

During liftoff and prior to J-2 engine start, additional pressurization is not required, as tank ullage pressure will be maintained from fuel boil-off. At the initiation of J-2 engine start, GH_2 is bled from the J-2 engine at 750 psia, -260 Fahrenheit to provide ullage pressure during fuel depletion. The pressure bled from the engine into the fuel tank is controlled by a fuel tank pressurization control module.

Fuel Tank Repressurization System - During the coast period prior to engine restart, there is no requirement for fuel tank pressurization. Tank pressure will build up within the tank due to LH_2 boil- off which is vented continuously through a propulsive vent system designed to provide a minimum thrust requirement to assure propellant settling. Additional pressure is vented through the fuel tank vent-relief system. Prior to J-2 engine restart, the propulsive vent system and tank vent-relief system is closed in the pressurization control module, The tank is then repressurized between 31 to 34 psia from ambient helium gas stored in seven helium spheres mounted on the thrust structure. Following engine restart, the LH_2 tank is again pressurized throughout the second burn phase with GH_2 bled from the engine.

Fuel Tank Vent-Relief System - Venting of the LH_2 tank is accomplished by a vent and relief system capable of relieving all excess pressure accumulated from over-pressurization or fuel boil-off during fill and flight operation. During fill, vaporization is vented through a self-sealing disconnect located in the forward skirt. During liftoff and flight, the gases are vented overboard through a nonpropulsive exhaust. The venting system consists of an actuation control module, vent valve, and nonpropulsive over- board exhaust. Actuation of the vent valve is commanded from an external ground signal during fill operations and from the flight sequencer during liftoff and flight. The vent valve is designed to open in a maximum of 0.1 second upon command. The relief valve, which provides a backup capability in case of vent valve failure, opens at 38 psia and reseats at 35 psia, and has a flow/relief capability of 2 pounds per second at sea level. A directional control valve directs excessive pressures through the ground disconnect during fill and directs excessive pressures through the nonpropulsive vent during liftoff and flight. The nonpropulsive vent system extends from the directional control valve into two 4-inch vent lines that terminate into two nonpropulsive exhaust ports. The ports are located 180° apart in the forward skirt area. The ports are arranged to direct the exhaust for total thrust cancellation.

LH_2 Continuous Propulsive Vent System—The continuous vent system is used to provide a thrust force required to position propellants at the aft end of each tank during coast. The system consists of a vent line originating at the vent-relief valve, terminating at two low thrust nozzles located 180° apart, and facing aft on the forward skirt. Continuous venting is controlled and regulated by a pneumatically operated continuous propulsive vent module. At the completion of the first burn engine cutoff, APS ullage engines are activated to settle the liquid propellants in the aft end of the tanks during the shutdown phase. LH_2 tank pressure is then vented through the continuous propulsive vent system, providing a continuous propulsive thrust to the stage. This maintains control of the propellants within the tanks. The APS engine are shut off after the transition is complete and the propulsive venting continues throughout the coast phase. The continuous propulsive vent module controls venting from a maximum of 45 pounds to a minimum of approximately 7 pounds.

LH_2 Feed System-Prior to vehicle liftoff and prior to engine restart, all LH_2 feed system components of the J-2 turbopump assembly must be chilled to assure proper operation. Chilldown of the LH_2 system is accomplished by a closed loop, forward-flow recirculation system. On command from the IU, the prevalve in the LH_2 feed duct closes and the chilldown shutoff valve opens. An auxiliary, electrically driven LH_2 chilldown pump, mounted in the LH_2 tank, circulates the LH_2 within the system and is capable of a minimum flowrate of 135 gpm at 6.1 psi. LH_2 is circulated from the LH_2 tank through the low pressure feed duct, through the J-2 engine fuel pump, the fuel bleed valve, and back to the tank through a return line. Recirculation chilldown continues through the boost phase and up to J-2 engine ignition. In the event of an emergency shutdown requirement, the chilldown system shutoff valve is closed upon command from the IU. LH_2 is supplied to the J-2 engine through a vacuum-jacketed, low-pressure duct at a flowrate of 81 pounds per second at −423° Fahrenheit, 28 psia. The duct is located in the fuel tank side wall above the common bulkhead joint and is equipped with bellows to compensate for thermal motion. Signals from the engine sequencer energize the LH_2 feed valve, as required to obtain steady-state operation. A complete description of engine operation may be found in the J-2 Engine section.

PROPELLANT UTILIZATION SYSTEM

The primary function of the PU system is to assure simultaneous depletion of propellants by controlling the LOX flowrate to the J-2 engine. It also provides propellant mass information for controlling the fill and topping valves during propellant loading operations. The system consists of mass sensors, an electronics assembly, and an engine-mounted mixture ratio valve. During loading operations, the mass of propellants loaded is determined within one per cent by the mass sensors. Tank over-fill sensors act as a backup system in the event the loading system fails to terminate fill operations.

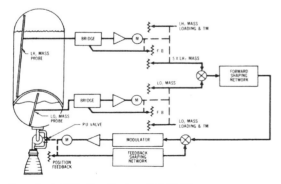

Propellant Utilization System D-NRV-8

Continuous LH_2 and LOX residual readout signals are provided throughout third stage powered flight. Differences between the fuel and oxidizer mass indications, as sensed by the mass sensors, are continually analyzed and are then used to control the oxidizer pump bypass flowrate, which changes the engine mixture ratio correspondingly. The static inverter/converter supplies the analog voltages necessary to operate the PU system. It is commanded "on" and "off" by a switch selector and sequencer combination.

PNEUMATIC CONTROL SYSTEM

The pneumatic control system provides GHe (gaseous helium) pressure to operate all third stage pneumatically operated valves with the exception of those provided as components of the J-2 engine. GHe is supplied from an ambient helium sphere and pressurized from a ground source before propellant fill operations at 3,100 ± 100 psia at 70 Fahrenheit for valve operation. The sphere is located on the thrust structure and is pre-conditioned to above 70 Fahrenheit from the environmental control system before liftoff. The pneumatic control system provides regulated pressure at 475 ± 25 psig for operation of the LH_2 and LOX vent-relief valves during propellant loading, LH_2 directional control valve, LOX and LH_2 fill and drain valves during loading, and the GH_2 engine start system vent-relief valve. It also provides operating pressures for the LH_2 and LOX turbopump turbine purge module, LOX chilldown pump purge module control, LOX and LH_2 prevalves, LOX and LH_2 chilldown shutoff valves, and the LH_2 continuous propulsive vent control module. The pneumatic control subsystem is protected from overpressure by a normally open solenoid valve controlled by a downstream pressure-sensing switch. At pressures greater than 535 + 15, -10 psia, the pressure switch actuates and closes the valve. At pressures below 450 t 15, -10 psia, the pressure switch drops out and the solenoid opens, thus acting as a backup regulator.

Flight Control System

The flight control system provides stage thrust vector steering and attitude control. Steering is achieved by gimbaling the J-2 engine during powered flight. Hydraulic actuator assemblies provide J-2 engine deflection rates proportional to steering signal corrections supplied by the IU. Stage roll attitude during powered flight is controlled by firing the APS attitude control engines.

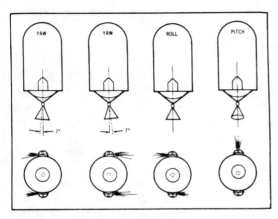

Flight Control System D-NRV-12

HYDRAULIC SYSTEM

The hydraulic system performs engine positioning upon command from the IU. Major components are a J-2 engine-driven hydraulic pump, two hydraulic actuator assemblies, and an accumulator-reservoir assembly. The electrically driven auxiliary hydraulic pump is started before vehicle liftoff to pressurize the hydraulic system. Electric power for the pump is provided by a ground source. At liftoff, the pump is switched to stage battery power. Pressurization of the hydraulic system restrains the J-2 engine in a null position with relation to the third stage center line, preventing pendulum-like shifting from forces encountered during liftoff and boost. During powered flight, the J-2 engine may he gimbaled up to 7° in a square pattern by the hydraulic system upon command from the IU.

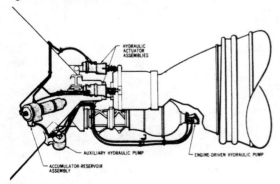

J-2 Engine Hydraulic System Components D-NRV-10

Engine-Driven Hydraulic Pump—The engine-driven hydraulic pump is a variable displacement type pump capable of delivering hydraulic fluid under continuous system pressure and varying volume as required for operation of the hydraulic actuator assemblies. The pump is driven directly from the engine oxidizer turbopump. A thermal isolator in the system controls hydraulic fluid temperature to ensure proper operation.

Auxiliary Hydraulic Pump—The auxiliary hydraulic pump is an electrically driven variable displacement pump which supplies a constant minimum supply of hydraulic fluid to the hydraulic system at all times. The pump is also used to perform preflight engine gimbaling checkouts, hydraulically lock the engine in the null position during boost phase, maintain system hydraulic fluid at operating temperatures during other than the powered phase, and augment the engine-driven hydraulic pump during powered flight. It also pro- vides an emergency backup supply of fluid to the system.

Hydraulic Actuator Assemblies—Two hydraulic actuator assemblies are attached directly to the J-2 engine and the thrust structure and receive IU command signals to gimbal the engine. The actuator assemblies are identical and interchangeable.

Accumulator Reservoir Assembly—The accumulator-reservoir assembly is an integral unit mounted on the thrust structure. The reservoir section is the storage area for hydraulic fluid; the accumulator section supplies peak system fluid requirements and dampens high-pressure surges with- in the system.

AUXILIARY PROPULSION SYSTEM

The APS provides auxiliary propulsive thrust to the stage for three-axis attitude control and for ullage control. Two APS modules are mounted 180° apart on the aft skirt assembly. Two solid propellant rocket motors are mounted 180° apart between the APS modules on the aft skirt assembly and provide additional thrust for ullage control. APS Modules Each APS module contains three 150-pound-thrust attitude control engines and one 70-pound-thrust ullage control engine. The attitude control engines are fired upon command from the IU in short duration bursts for attitude control of the stage during the orbital coast phase of flight. Minimum engine-firing pulse-duration is approximately 70 milliseconds. The attitude control engines are approximately 15 inches long with exit cones approximately 6.5 inches in diameter. Engine cooling is accomplished by an ablative process.

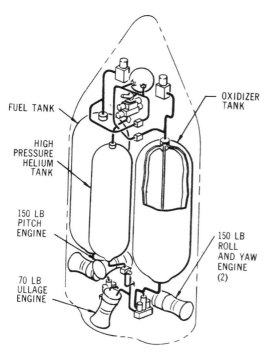

Auxiliary Propulsion System Module D-NRV-13

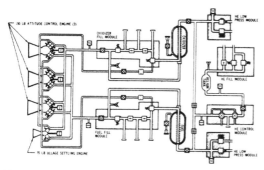

APS Schematic D-NRV-13

The ullage control engines are fired also upon command from the IU during the transition between J-2 engine first burn and the coast phase of flight to prevent undesirable propellant movement within the tanks. Firing continues for approximately 50 seconds until activation of the LH_2 continuous propulsive vent system. The ullage engines are again fired at the end of the third stage coast phase of flight and prior to J-2 engine restart to assure proper propellant positioning at inlets to the propellant feed lines during propellant tank repressurization. The ullage control engines are similar to the attitude control engines and are approximately 15 inches long with an exit cone approximately 5.75 inches in diameter. Engine cooling is accomplished by an ablative process. Each APS module contains an oxidizer system, fuel system, and pressurization system. The modules are self-contained and easily detached for separate checkout and environmental testing. An ignition system is unnecessary because fuel and oxidizer are hypergolic (self-igniting). Nitrogen tetroxide (N_2O_4), the oxidizer, is stable at room temperature. Separate fuel and oxidizer tanks of the expulsion bellows type are mounted within the APS module along with a high-pressure helium bottle, which provides pressurization for both the propellant tanks and the associated plumbing and control systems. The fuel, monomethyl hydrazine ($CH_3N_2H_3$), is stable to shock and extreme heat or cold. The APS module carries approximately 115 pounds of usable fuel and about 150 pounds of usable oxidizer.

Ullage Control

Two solid propellant Thiokol TX-280 rocket motors, each rated at 3,390 pounds of thrust, are ignited during separation of the second and third stages for ullage control approximately 4 seconds before J-2 ignition. This thrust produces additional positive stage acceleration during separation and positions LOX and LH_2 propellants toward the aft end of the tanks. In addition, propellant boil-off vapors are forced to the forward end where they are safely vented overboard. Tank outlets are covered to ensure a net positive suction head (NPSH) to the propellant pumps, thus preventing possible pump cavitation during J-2 engine start. Ullage rockets ignite upon command from the stage sequencer and fire for approximately 4 seconds. At about 12 seconds from ignition, the complete rocket motor assemblies, including bracketry, are jettisoned from the stage, upon command from the stage sequencer.

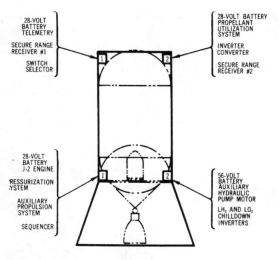

Third Stage Basic Electrical Power and Distribution System D-NRV-15

transmission of stage instrumentation information to ground receiving stations. Five transmitters, using two separate antenna systems, are capable of returning information on 45 continuous output data channels during third stage flight. The telemetry transmission links consist of five systems using three basic modulation schemes: Pulse Amplitude Modulated/FM/FM (PAM/FM/FM) ; Single Side- band/FM (SS/FM); and Pulse Code Modulated/FM (PCM/FM). There are three separate systems using PAM/FM/FM modulation. A Digital Data Acquisition System (DDAS) airborne tape recorder stores sampled data normally lost during staging and over-the-horizon periods of orbital missions, and plays back information when in range of ground stations.

Electrical Power and Distribution System

Four battery-powered systems provide electrical requirements for third stage operation. Forward Power System No. 1 includes a 28 VDC battery and power distribution equipment for telemetry, secure range receiver No. 1, forward battery heaters, and a power switch selector located in the forward skirt area. Forward Power System No. 2 includes a 28 VDC battery and power distribution equipment for the PU assembly, inverter-converter, and secure range receiver No. 2. Aft Power System No. 1 includes a 28 VDC battery and power distribution equipment for the J-2 engine, pressurization systems, APS modules, TM signal power, aft battery heaters, hydraulic system valves, and stage sequencer. Aft Power System No. 2 includes a 56 VDC battery and power distribution equipment for the auxiliary hydraulic pump, oxidizer chilldown inverter, and fuel chilldown inverter. Silver-oxide, zinc batteries used for electrical power and distribution systems are manually activated. The batteries are "one-shot" units, and not inter- changeable due to different load requirements. Electrical power and distribution systems are switched from ground power to the batteries by command through the aft umbilical prior to liftoff.

Telemetry and Instrumentation System

Radio frequency telemetry systems are used for

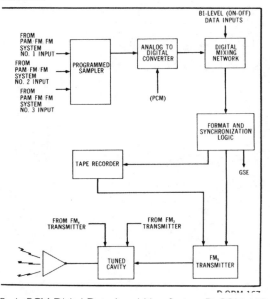

Basic PCM Digital Data Acquisition System D-ORM-167

PAM/FM/FM SYSTEMS Transducer input signals constitute the PAM input. The PAM systems use an electronically switched network that samples up to 30 channels of transducer inputs at 120 times a second. Deviations in transducer input voltages are represented as output pulses of varying amplitude for subsequent evaluation.

SS/FM SYSTEM
The SS/FM system is reserved for pertinent research requirements. Vibration and acoustical data needed for manned flight development will be transmitted by this system.

PCM/FM SYSTEM

The PCM/FM system (DDAS) is used during automatic checkout to provide data for the ground check-out computer. The system is also used to provide precise information concerning stage environment and performance of systems during flight.

ENVIRONMENTAL CONTROL SYSTEM

AFT SKIRT AND INTERSTAGE THERMO CONDITIONING AND PURGE

The thermo conditioning and purge system purges the aft skirt and aft interstage of combustible gases and distributes temperature controlled air or gaseous nitrogen around electrical equipment in the aft skirt during the vehicle countdown. The purging gas, supplied from a ground source through the umbilical, passes over the electrical equipment and flows into the aft interstage area. Some of the gas is directed through each of the auxiliary propulsion modules and exhausts into the interstage. A duct from the skirt manifold directs air or GN_2 to a thrust structure manifold. From the thrust structure manifold supply duct, a portion of air or GN_2 is directed to a shroud covering the hydraulic accumulator reservoir. Temperature control is accomplished by two dual-element thermistor assemblies located in the gaseous exhaust stream of each of the auxiliary propulsion modules. Elements are wired in series to sense average temperature. Two series circuits are formed, each circuit utilizing one element from each thermistor assembly. One series is used for temperature control, the other for temperature recording.

FORWARD SKIRT THERMO CONDITIONING

Electrical equipment in the third stage forward skirt area is thermally conditioned by a heat transfer system, using "cold plates" on which electronic components are mounted, and through which coolant fluid circulates. Coolant is pumped through the system from the IU and returned. Heat from electrical equipment attached to the cold plates is dissipated by conduction through the mounting feet and the cold plates to the fluid. Refer to the Instrument Unit section for a complete description of the IU environmental conditioning system.

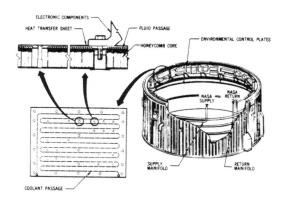

Forward Skirt Environmental Control System D-NRV-19

FORWARD SKIRT AREA PURGE

The forward skirt area is purged with gaseous nitrogen to minimize fire and explosion hazards while propellants are loaded or stored in the stage. Gaseous nitrogen is supplied and remotely controlled from a ground source.

Ordnance Systems

The ordnance systems perform stage separation, retrorocket ignition, ullage control rocket ignition and jettison, and range safety functions.

STAGE SEPARATION SYSTEM

The stage separation system consists of a severable tension strap, mild detonating fuse (MDF), exploding bridgewire, (EBW), detonators and EBW firing units.

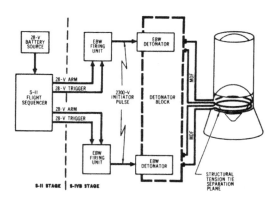

Separation System D-NRV-2

The severable tension strap houses two redundant MDF cords in a "V" groove circumventing the stage between the aft skirt and aft interstage at the separation plane. Ignition of the MDF cords is triggered by a signal from the second stage sequencer through the EBW and EBW firing units about 3 seconds after second stage engine cutoff. The MDF consists of a flexible metal sheath surrounding a continuous core of high explosive material. Once detonated, the explosive force of the MDF occurs at a rate of 23,000 feet per second. The EBW detonator is fired to initiate the MDF explosive train. A 2,300 VDC pulse is applied to a small resistance wire and a spark gap. The high voltage electrical arc across the spark gap ignites a charge of high explosive material which in turn detonates the MDF. The high voltage pulse requirement for ignition renders this system safe from random ground or vehicle electrical power. Upon command, each EBW firing unit supplies high voltage and current required to fire a specific EBW detonator.

RETROROCKET IGNITION SYSTEM

Four solid propellant retrorockets are mounted equidistant around the aft interstage assembly, and when ignited, assure clean separation of the third stage from the second stage by decelerating or braking the spent booster. Each retrorocket is rated for a nominal thrust of 35,000 pounds, weight of 384 pounds, and burn time of about 1.5 seconds.

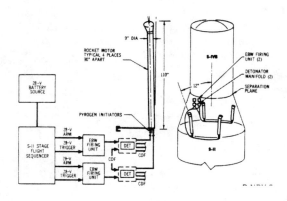

Retrorocket System D-NRV-2

A signal from the second stage initiates two EBW firing units located on the aft interstage. The EBW firing units ignite two detonator manifolds, which in turn ignite the retrorockets through redundant pairs of confined detonating fuse (CDF) and pyrogen initiators.

ULLAGE CONTROL ROCKET IGNITION AND JETTISON SYSTEM

Two solid propellant ullage rockets, located on the third stage aft skirt just forward of the stage separation plane, are ignited on signal from the stage sequencer by EBW initiators. After firing, the burned-out ullage rocket casings and fairings are jettisoned to reduce stage weight. Upon command from the stage sequencer, two forward and aft frangible nuts, which secure each rocket motor and its fairing to the stage, are detonated by confined detonating fuse (CDF) to free the entire assembly from the vehicle.

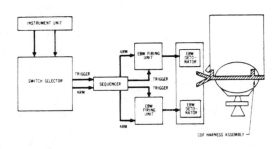

Ullage Rocket System D-NRV-22

RANGE SAFETY SYSTEM

The range safety system terminates vehicle flight upon command of the range safety officer. Redundant systems are used throughout to provide maximum reliability. Four antennas, mounted around the periphery of the third stage forward skirt assembly, feed two redundant secure range receivers located in the for- ward skirt assembly. Both receivers have separate power supplies and circuits. A unique combination of coded signals must he transmitted, received, and decoded to energize this destruct system. A safety and arming device prevents inadvertent initiation of the explosive train by providing a positive isolation of the EBW detonator and explosive train until arming is commanded. Visual and remote indications of SAFE and ARMED conditions are displayed at all times at the firing center. Upon proper command, EBW firing units activate EBW detonators. A CDF, detonated by the safety and arming device, explodes a flexible linear-shaped charge which cuts through the tank skin to disperse both fuel and oxidizer.

J-2 ENGINE FACT SHEET

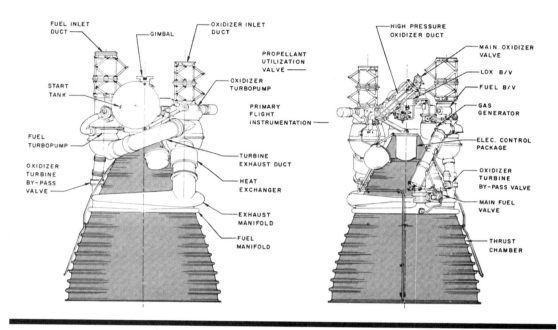

LENGTH	11 ft. 1 in.
WIDTH	6 ft. 8 in.
NOZZLE EXIT DIAMETER	6 ft. 5 in.
THRUST (altitude)	225,000 Lb.
SPECIFIC IMPULSE	424 sec. (427 at 5:1 mixture ratio)
RATED RUN DURATION	500 sec.
FLOWRATE: Oxidizer	449 Lb./sec (2,847 gpm)
Fuel	81.7 Lb./sec (8,365 gpm)
MIXTURE RATIO	5.5:1 oxidizer to fuel
CHAMBER PRESSURE (Pc)	763 psia
WEIGHT, DRY, FLIGHT CONFIGURATION	3,480 Lb.
EXPANSION AREA RATIO	27.5 :1
COMBUSTION TEMPERATURE	5,750°F

Note: J-2 engines will be uprated to a maximum of 230,000 pounds of thrust for later vehicles.

J-2 ENGINE

Engine Description

The Rocketdyne J-2 engine is a high performance, upper stage, propulsion system utilizing liquid hydrogen and liquid oxygen propellants and developed a maximum vacuum thrust of 225,000 pounds. All J-2 engines are identical when delivered and may he allocated to either the second or third stage. Each engine is equipped to be restarted in flight. However, the restart capability will be utilized only in the third stage. The single J-2 engine used in the third stage is gimbal-mounted so that it can be moved in flight and used to steer the stage. Five J-2 engines are arranged in a cluster in the second stage. The four outboard engines of the five-engine cluster are gimbal-mounted to provide the vehicle with pitch, yaw, and roll control. The center engine is mounted in a fixed position. Major systems of the J-2 engine include a thrust chamber and gimbal assembly system, propellant feed system, gas generator and exhaust system, electrical and pneumatic control system, start tank assembly system, and flight instrumentation system.

Thrust Chamber and Gimbal System

The J-2 engine thrust chamber serves as a mount for all engine components. It is composed of the following subassemblies: thrust chamber body, injector and dome assembly, gimbal bearing assembly, and augmented spark igniter. Thrust is transmitted through the gimbal mounted on the thrust chamber assembly dome to the vehicle thrust frame structure. The thrust chamber injector receives the propellants from a dual turbopump system (oxidizer and fuel) under pressure, mixes the propellants, and burns them to impart a high velocity to the expelled combustion gases to produce thrust.

THRUST CHAMBER

The thrust chamber is constructed of stainless steel tubes of 0.0l2-inch wall thickness. Tubes with thin walls are required for heat transfer purposes. The thrust chamber tubes are stacked longitudinally and furnace-brazed to form a single unit. The chamber is hell-shaped with a 27.5 to 1 expansion area ratio for efficient operation at altitude, and is regeneratively cooled by the fuel.

Fuel enters from a manifold located midway between the thrust chamber throat and the exit at a pressure of more than 1,000 psi. In cooling the chamber the fuel makes a one-half pass downward through 180 tubes and is returned in a full pass up to the thrust chamber injector through 360 tubes. (See schematic drawing.)

DOME

The injector and oxidizer dome assembly is located a t the top of the thrust chamber. The dome provides a manifold for the distribution of the liquid oxygen to the injector and serves as a mount for the gimbal bearing and the augmented spark igniter.

THRUST CHAMBER INJECTOR

The thrust chamber injector atomizes and mixes the propellants in a manner to produce the most efficient combustion. Six hundred and fourteen hollow oxidizer posts are machined to form an integral part of the injector. Fuel nozzles are threaded and installed over the oxidizer posts forming concentric orifices. The injector face is porous and is formed from layers of stainless steel wire mesh and is welded at its periphery to the injector body. Each fuel nozzle is swaged to the face of the injector. The injector receives liquid oxygen through the dome manifold and injects it through the oxidizer posts into the combustion area of the thrust chamber. The fuel is received from the upper fuel manifold in the thrust chamber and injected through the fuel orifices which are concentric with the oxidizer orifices. The propellants are injected uniformly to ensure satisfactory combustion.

GIMBAL

The gimbal is a compact, highly loaded (20,000 psi) universal joint consisting of a spherical, socket-type bearing with a Teflon/fiberglass composition coating that provides a dry, low-friction bearing surface. It also includes a lateral adjustment device for aligning the chamber with the vehicle. The gimbal transmits the thrust from the injector assembly to the vehicle thrust structure and pro- vides a pivot bearing for deflection of the thrust vector, thus providing flight attitude control of the vehicle. The gimbal is mounted on the top of the injector and oxidizer dome assembly.

AUGMENTED SPARK IGNITER

The augmented spark igniter (ASI) is mounted to the injector face. It provides the flame to ignite the propellants in the thrust chamber. When engine start is initiated, the spark exciters energize two spark plugs mounted in the side of the igniter chamber. Simultaneously, the control system starts the initial flow of oxidizer and fuel to the spark igniter. As the oxidizer and fuel enter the combustion chamber of the ASI, they mix and are ignited. Mounted in the ASI is an ignition monitor which indicates that proper ignition has taken place. The ASI operates continuously during entire engine firing, is uncooled, and is capable of multiple reignitions under all environmental conditions.

Propellant Feed System

The propellant feed system consists of separate fuel and oxidizer turbopumps, main fuel valve, main oxidizer valve, propellant utilization valve, fuel and oxidizer flowmeters, fuel and oxidizer bleed valves, and interconnecting lines.

J-2 Assembly - Hydrogen fueled J-2 rocket engines for upper stages of Saturn V vehicles are completed on this assembly line. J-2 develops a maximum thrust of 225,000 pounds. R-10

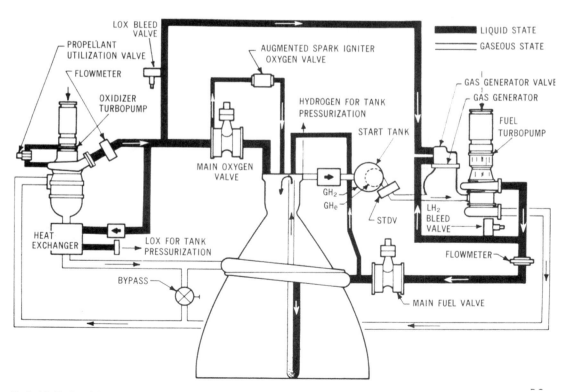

Basic J-2 Engine Schematic R-8

FUEL TURBOPUMP

The fuel turbopump, mounted on the thrust chamber, is a turbine-driven, axial flow pumping unit consisting of an inducer, a seven-stage rotor, and a stator assembly. It is a high-speed pump operating at 27,000 rpm, and is designed to increase hydrogen pressure from 30 psia to 1,225 psia through high- pressure ducting at a flowrate which develops 7,800 brake horsepower. Power for operating the turbopump is provided by a high-speed, two-stage turbine. Hot gas from the gas generator is routed to the turbine inlet manifold which distributes the gas to the inlet nozzles where it is expanded and directed at a high velocity into the first stage turbine wheel. After passing through the first stage turbine wheel, the gas is redirected through a ring of stator blades and enters the second stage turbine wheel. The gas leaves the turbine through the exhaust ducting. Three dynamic seals in series prevent the pump fluid and turbine gas from mixing. Power from the turbine is transmitted to the pump by means of a one-piece shaft.

OXIDIZER TURBOPUMP

The oxidizer turbopump is mounted on the thrust chamber diametrically opposite the fuel turbopump. It is a single-stage centrifugal pump with direct turbine drive. The oxidizer turbopump increases the pressure of the liquid oxygen and pumps it through high-pressure ducts to the thrust chamber. The pump operates at 8,600 rpm at a discharge pressure of 1,080 psia and develops 2,200 brake horse-power. The pump and its two turbine wheels are mounted on a common shaft. Power for operating the oxidizer turbopump is pro- vided by a high-speed, two-stage turbine which is driven by the exhaust gases from the gas generator. The turbines of the oxidizer and fuel turbopumps are connected in a series by exhaust ducting that directs the discharged exhaust gas from the fuel turbopump turbine to the inlet of the oxidizer turbo-pump turbine manifold. One static and two dynamic seals in series prevent the turbopump oxidizer fluid and turbine gas from mixing. Beginning the turbopump operation, hot gas enters the nozzles and, in turn, the first stage turbine wheel. After passing through the first stage turbine wheel, the gas is redirected by the stator blades and enters the second stage turbine wheel. The gas then leaves the turbine through exhaust ducting, passes through the heat exchanger, and exhausts into the thrust chamber through a manifold directly above the fuel inlet manifold. Power from the turbine is transmitted by means of a one-piece shaft to the pump. The velocity of the

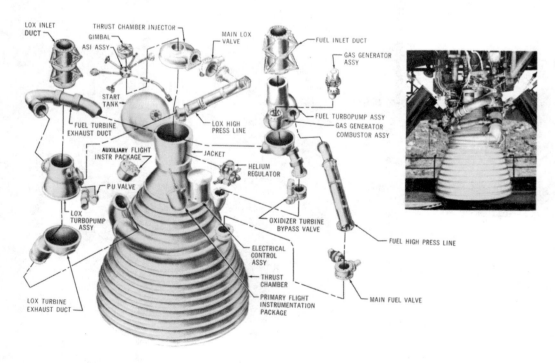

J-2 Major Component Breakdown R-8

liquid oxygen is increased through the inducer and impeller. As the liquid oxygen enters the outlet volute, velocity is converted to pressure and the liquid oxygen is discharged into the outlet duct at high pressure. Bearings in the liquid hydrogen and liquid oxygen turbopumps are lubricated by the fluid being pumped because the extremely low operating temperature of the engine precludes use of lubricants or other fluids.

MAIN FUEL VALVE

The main fuel valve is a butterfly-type valve, spring-loaded to the closed position, pneumatically operated to the open position, and pneumatically assisted to the closed position. It is mounted between the fuel high-pressure duct from the fuel turbopump and the fuel inlet manifold of the thrust chamber assembly. The main fuel valve controls the flow of fuel to the thrust chamber. Pressure from the ignition stage control valve on the pneumatic control package opens the valve during engine start. As the gate starts to open, it allows fuel to flow to the fuel inlet manifold.

MAIN OXIDIZER VALVE

The main oxidizer valve (MOV) is a butterfly-type valve, spring-loaded to the closed position, pneumatically operated to the open position, and pneumatically assisted to the closed position. It is mounted between the oxidizer high-pressure duct from the oxidizer turbopump and the oxidizer inlet on the thrust chamber assembly. Pneumatic pressure from the normally closed port of the mainstage control solenoid valve is routed to both the first and second stage opening actuators of the main oxidizer valve. Application of opening pressure in this manner, together with controlled venting of the main oxidizer valve closing pressure through a thermal-compensating orifice, provides a controlled ramp opening of the main oxidizer valve through all temperature ranges. A sequence valve, located within the MOV assembly, supplies pneumatic pressure to the opening control part of the gas generator control valve and through an orifice to the closing part of the oxidizer turbine bypass valve.

PROPELLANT UTILIZATION VALVE

The propellant utilization (PU) valve is an electrically operated, two-phase, motor-driven, oxidizer transfer valve and is located at the oxidizer turbo-pump outlet volute. The propellant utilization valve ensures the simultaneous exhaustion of the contents of the propellant tanks. During engine operation, propellant level sensing devices in the vehicle propellant tanks control the valve gate position for adjusting the oxidizer flow to ensure simultaneous exhaustion of fuel and oxidizer. An additional function of the PU valve is to provide thrust variations in order to maximize payload. The second stage, for example, operates with the PU valve in the closed position for more than 70 per cent of the firing duration. This valve position provides 225,000 pounds of thrust at a 5.5:1 propellant (oxidizer to fuel by weight) mixture ratio. During the latter portion of the flight, the PU valve position is varied to provide simultaneous emptying of the propellant tanks. The third stage also operates at the high-thrust level for the majority of the burning time in order to realize the high thrust benefits. The exact period of time at which the engine will operate with the PU valve closed will vary with individual mission requirements and propellant tanking levels. When the PU valve is fully open, the mixture ratio is 4.5:1 and the thrust level is 175,000 pounds. The propellant utilization valve and its servomotor are supplied with the engine. A position feedback potentiometer is also supplied as a part of the PU valve assembly. The PU valve assembly and a stage or a facility-mounted control system make up the propellant utilization system.

FUEL AND OXIDIZER FLOWMETERS

The fuel and oxidizer flowmeters are helical-vaned, rotor-type flowmeters. They are located in the fuel and oxidizer high-pressure ducts. The flowmeters measure propellant flowrates in the high-pressure propellant ducts. The four-vane rotor in the hydrogen system produces four electrical impulses per revolution and turns approximately 3,700 revolutions per minute at nominal flow. The six-vane rotor in the liquid oxygen system produces six electrical impulses per revolution and turns at approximately 2,600 revolutions per minute at nominal flow.

PROPELLANT BLEED VALVES

The propellant bleed valves used in both the fuel and oxidizer systems are poppet-type which are spring-loaded to the normally open position and pressure-actuated to the closed position. Both propellant bleed valves are mounted to the bootstrap lines adjacent to their respective turbopump discharge flanges. The valves allow propellant to circulate in the propellant feed system lines to achieve proper operating temperature prior to engine start. The bleed valves are engine controlled. At engine start, a helium control solenoid valve in the pneumatic control package is energized allowing pneumatic pressure to close the bleed valves, which remain closed during engine operation.

Gas Generator and Exhaust System

This system consists of the gas generator, gas generator control valve, turbine exhaust system and exhaust manifold, heat exchanger, and oxidizer turbine bypass valve.

GAS GENERATOR

The gas generator is welded to the fuel pump turbine manifold, making it an integral part of the fuel turbopump assembly. It produces hot gases to drive the fuel and oxidizer turbines and consists of a combustor containing two spark plugs, a control valve containing fuel and oxidizer ports, and an injector assembly. When engine start is initiated, the spark exciters in the electrical control package are energized, providing energy to the spark plugs in the gas generator combustor. Propellants flow through the control valve to the injector assembly and into the combustor outlet and are directed to the fuel turbine and then to the oxidizer turbine.

GAS GENERATOR CONTROL VALVE

The gas generator control valve is a pneumatically operated poppet-type that is spring-loaded to the closed position. The fuel and oxidizer poppets are mechanically linked by an actuator. The gas generator control valve controls the flow of propellants through the gas generator injector. When the mainstage signal is received, pneumatic pressure is applied against the gas generator control valve actuator assembly which moves the piston and opens the fuel poppet. During the fuel poppet opening, an actuator contacts the piston that opens the oxidizer poppet. As the opening pneumatic pressure decays, spring loads close the poppets.

TURBINE EXHAUST SYSTEM

The turbine exhaust ducting and turbine exhaust hoods are of welded sheet metal construction. Flanges utilizing dual (Naflex) seals are used at component connections. The exhaust ducting conducts turbine exhaust gases to the thrust chamber exhaust manifold which encircles the thrust chamber approximately halfway between the throat and the nozzle exit. Exhaust gases pass through the heat exchanger and exhaust into the main thrust chamber through 180 triangular openings between the tubes of the thrust chamber.

HEAT EXCHANGER

The heat exchanger is a shell assembly, consisting of a duct, bellows, flanges, and coils. It is mounted in the turbine exhaust duct between the oxidizer turbine discharge manifold and the thrust chamber. It heats and expands helium gas for use in the third stage or converts liquid oxygen to gaseous oxygen for the second stage for maintaining vehicle oxidizer tank pressurization. During engine operation, either liquid oxygen is tapped off the oxidizer high-pressure duct or helium is provided from the vehicle stage and routed to the heat exchanger coils.

OXIDIZER TURBINE BYPASS VALVE

The oxidizer turbine bypass valve is a normally open, spring-loaded, gate type. It is mounted in the oxidizer turbine bypass duct. The valve gate is equipped with a nozzle, the size of which is determined during engine calibration. The valve in its open position depresses the speed of the oxygen pump during start, and in its closed position acts as a calibration device for the turbopump performance balance.

Control System

The control system includes a pneumatic system and a solid-state electrical sequence controller pack- aged with spark exciters for the gas generator and the thrust chamber spark plugs, plus interconnecting electrical cabling and pneumatic lines.

PNEUMATIC SYSTEM

The pneumatic system consists of a high-pressure helium gas storage tank, a regulator to reduce the pressure to a usable level, and electrical solenoid control valves to direct the central gas to the various pneumatically controlled valves.

ELECTRICAL SEQUENCE CONTROLLER

The electrical sequence controller is a completely self-contained, solid-state system, requiring only DC power and start and stop command signals.

Pre-start status of all critical engine control functions is monitored in order to provide an "engine ready" signal. Upon obtaining "engine ready" and "start" signals, solenoid control valves are energized in a precisely timed sequence as described in the "Engine Operation" section to bring the engine through ignition, transition, and into main-stage operation. After shutdown, the system automatically resets for a subsequent restart.

Start Tank Assembly System

This system is made up of an integral helium and hydrogen start tank, which contains the hydrogen and helium gases for starting and operating the engine. The gaseous hydrogen imparts initial spin to the turbines and pumps prior to gas generator combustion, and the helium is used in the control system to sequence the engine valves.

HELIUM AND HYDROGEN TANKS

The spherical helium tank is positioned inside the hydrogen tank to minimize engine complexity. It holds 1,000 cubic inches of helium. The larger spherical hydrogen gas tank has a capacity of 7,257.6 cubic inches. Both tanks are filled from a ground source prior to launch and the gaseous hydrogen tank is refilled during engine operation from the thrust chamber fuel inlet manifold for subsequent restart in third stage application.

Flight Instrumentation System

The flight instrumentation system is composed of a primary instrumentation package and an auxiliary package.

PRIMARY PACKAGE

The primary package instrumentation measures those parameters critical to all engine static firings and subsequent vehicle launches. These include some 70 parameters such as pressures, temperatures, flows, speeds, and valve positions for the engine components, with the capability of transmitting signals to a ground recording system or a telemetry system, or both. The instrumentation system is designed for use throughout the life of the engine, from the first static acceptance firing to its ultimate vehicle flight.

AUXILIARY PACKAGE

The auxiliary package is designed for use during early vehicle flights. It may be deleted from the basic engine instrumentation system after the propulsion system has established its reliability during research and development vehicle flights. It contains sufficient flexibility to provide for deletion, substitution, or addition of parameters deemed necessary as a result of additional testing. Eventual deletion of the auxiliary package will not interfere with the measurement capability of the primary package.

Engine Operation

START SEQUENCE

Start sequence is initiated by supplying energy to two spark plugs in the gas generator and two in the augmented spark igniter for ignition of the propellants. Next, two solenoid valves are actuated: one for helium control, and one for ignition phase control. Helium is routed to hold the propellant bleed valves closed and to purge the thrust chamber LOX dome, the LOX pump intermediate seal, and the gas generator oxidizer passage. In addition, the main fuel valve and ASI oxidizer valve are opened, creating an ignition flame in the ASI chamber that passes through the center of the thrust chamber injector. After a delay of 1, 3 or 8 seconds, during which time fuel is circulated through the thrust chamber to condition the engine for start, the start tank discharge valve is opened to initiate turbine spin. The length of the fuel lead is dependent upon the length of the Saturn V first stage boost phase. When the J-2 engine is used in the second stage of the Saturn V vehicle, a one-second fuel lead is necessary. The third stage of the Saturn V vehicle, on the other hand, utilizes a three-second fuel lead for its initial start and an eight-second fuel lead for its restart. After an interval of 0.450 second, the start tank discharge valve is closed and a mainstage control solenoid is actuated to:

1) turn off gas generator and thrust chamber helium purges;
2) open the gas generator control valve (hot gases from the gas generator now drive the pump turbines);
3) open the main oxidizer valve to the first position (14 degrees) allowing LOX to flow to the LOX dome to burn with the fuel that has been circulating through the injector;
4) close the oxidizer turbine bypass valve (a portion of the gases for driving the oxidizer turbopump were bypassed during the ignition phase);

5) gradually bleed the pressure from the closing side of the oxidizer valve pneumatic actuator controlling the slow opening of this valve for smooth transition into mainstage. Energy in the spark plugs is cut off and the engine is operating at rated thrust. During the initial phase of engine operation, the gaseous hydrogen start tank will be recharged in those engines having a restart requirement. The hydrogen tank is repressurized by tapping off a controlled mixture of liquid hydrogen from the thrust chamber fuel inlet manifold and warmer hydrogen from the thrust chamber fuel injection manifold just before entering the injector.

FLIGHT MAINSTAGE OPERATION

During mainstage operation, engine thrust may he varied between 175,000 and 225,000 pounds by actuating the propellant utilization valve to increase or decrease oxidizer flow as described in the section "PU Valve". This is beneficial to flight trajectories and for overall mission performance to make greater payloads possible.

CUTOFF SEQUENCE

When the engine cutoff signal is received by the electrical control package, it de-energizes the main-stage and ignition phase solenoid valves and energizes the helium control solenoid de-energizer timer. This, in turn, permits closing pressure to the main fuel valve, main oxidizer valve, gas generator control valve, and augmented spark igniter valve. The oxidizer turbine bypass valve and propellant bleed valves open and the gas generator and LOX dome purges are initiated.

ENGINE RESTART

To provide third stage restart capability for the Saturn V, the J-2 gaseous hydrogen start tank is refilled in 60 seconds during the previous firing after the engine has reached steady-state operation. (Refill of the gaseous helium tank is not required because the original ground fill supply is sufficient for three starts.) Prior to engine restart, the stage ullage rockets are fired to settle the propellants in the stage propellant tanks, ensuring a liquid head to the turbopump inlets. Also, the engine propellant bleed valves are opened, the stage recirculation valve is opened, the stage prevalve is closed, and a LOX and LH_2 circulation is effected through the engine bleed system for five minutes to condition the engine to the proper temperature to ensure proper engine operation. Engine restart is initiated after the "engine ready" signal is received from the stage. This is similar to the initial "engine ready". The hold time between cutoff and restart is from a minimum of 1½ hours to a maximum of 6 hours, depending upon the number of earth orbits required to attain the lunar window for translunar trajectory.

INSTRUMENT UNIT FACT SHEET

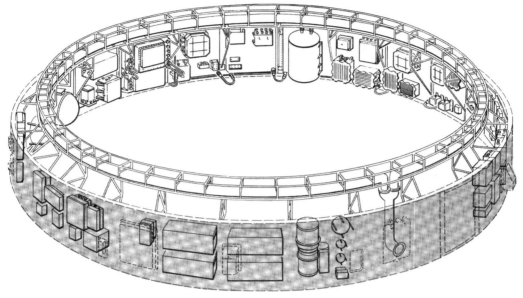

IBM-DR-27

DIAMETER: 260 in.
HEIGHT: 36 in.
WEIGHT: 4,500 Lb. (approx.)

MAJOR SYSTEMS

ENVIRONMENTAL CONTROL SYSTEM : Provides cooling for electronic modules and components within the IU and forward compartments of third stage

GUIDANCE AND CONTROL SYSTEM: Determines course of Saturn V through space and adapts that course to fulfill mission requirements

INSTRUMENTATION SYSTEM: Measures vehicle conditions and reactions during mission and transmits this information to ground for subsequent analysis, as well as providing for ground station-to-vehicle communication

ELECTRICAL SYSTEM: Provides basic operating power for all electronic and electrical equipment in the IU; also monitors vehicle performance and may initiate automatic mission abort if an emergency arises

STRUCTURAL SYSTEM: Serves as a load bearing part of the launch vehicle, supporting both the components within the IU and the spacecraft; composed of three 120-degree segments of thin-wall aluminum alloy face sheets bonded over a core of aluminum honeycomb about an inch thick

INSTRUMENT UNIT

INSTRUMENT UNIT DESCRIPTION

The instrument unit (IU) for Saturn V was designed by NASA at MSFC and was developed from the Saturn I IU. Overall responsibility for the IU has been assigned to IBM's Federal Systems Division for fabrication and assembly of the unit, system testing, and integration and checkout of the unit with the launch vehicle. IBM also assembles and delivers computer programming necessary to sup- port the IU, These programs are used:

1. In IBM's automated systems checkout computer complex in Huntsville. This system verifies IU system integrity prior to release of the IU to NASA.
2. In IBM's simulation laboratory in Huntsville to verify the flight readiness of the IU's launch vehicle digital computer program, as well as the passive filters contained in the IU's analog flight control computer.
3. To operate the automated launch computer complex at John F. Kennedy Space Center. This computer complex is used to automatically checkout the flight readiness of the vehicle prior to liftoff.
4. To operate the IU's launch vehicle digital computer in flight, as well as programming that will be used by NASA, for postflight analysis of vehicle environment and performance data.

The IU is Saturn V's nerve center. It contains the electronic and electrical equipment needed for guidance, tracking, and origination and communication of vehicle environmental and performance data. The IU also contains environmental control equipment for temperature control, batteries, and power supplies to furnish operating power for electronic equipment. The stage structure is 260 inches in diameter and 36 inches high and becomes a load-bearing part of the vehicle. It supports the components within the IU and the weight of the spacecraft.

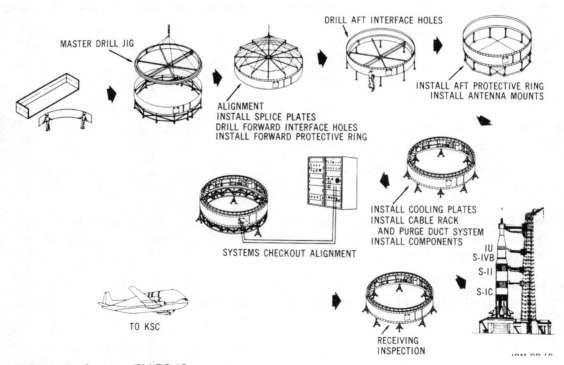

IU Production Sequence IBM-DR-10

INSTRUMENT UNIT FABRICATION

The structure is manufactured in three, 120-degree segments, each consisting of thin-wall aluminum honeycomb. An aluminum alloy channel ring, bonded to the top and bottom edge of each segment, provides mating surfaces between the IU, the third stage, and the payload adapter. Mounted, inner skin brackets provide attachment points for the environmental control system's cold plates or for cold plate installation.

Segments are aligned and joined by splice plates bolted both inside and outside the joints. A spring-loaded umbilical door provides access to electrical connections between IU equipment and ground test areas. A larger access door, bolted in place, permits personnel to enter the IU after vehicle mating.

Assembly of an IU begins when the three curved structural segments, three feet high by 14 feet long, arrive at IBM's Huntsville, Ala., facility. Each segment weighs only 140 pounds.

Structure Segments - Prior to splicing, mounting brackets for thermal conditioning panels can be seen on interior surface of segments. The exterior of the spring-loaded umbilical door and the access door are visible at right center. IBM-DR-22

Extremely accurate theodolites, similar to a surveyor's transit, are used to align the segments in a circle prior to splicing. Metal splicing plates join the three segments, and the holes which permit the IU to be joined to mating surfaces of the launch vehicle are drilled at top and bottom edges of the structure for ease in handling.

Protective rings are bolted to these edges to stiffen the structure. Vehicle antenna holes are cut after splices are bolted. After structure fabrication is completed, module and component assembly operations begin. Temperature transducers are fastened to the inner skin, environmental control system (ECS) cold plates are mounted, and a cable tray is bolted to the top of the structure. Components are mounted on the cold plates and ECS system pumps, storage tanks (called accumulators), heat exchangers, and plumbing are installed. Two nitrogen supply systems are installed: one for gas bearings of the inertial platform and the other for pressurization of the ECS. Finally, ducts, tubing, and electrical cables complete the assembly and the IU now weighing in excess of 4,000 pounds is ready for a long series of tests.

INSTRUMENT UNIT SYSTEMS

Environmental Control System

The ECS cools the electronic equipment in the IU and the forward third stage skirt. Sixteen cold plates are installed in each stage. An antifreeze-like coolant, 60 per cent methanol and 40 per cent water, from a reservoir within the IU is circulated through the cold plates. Heat generated by the mounted components is transferred to the coolant by means of conduction. Prior to liftoff a preflight heat exchanger serviced by ground support equipment transfers heat from the coolant. Approximately 163 seconds after lift-off, ECS's sublimator-heat exchanger takes over the job of temperature control. Some of the more complex components like the guidance computer, flight control computer, and the ST-124-M platform, have coolant fluid circulated through them to provide more efficient heat removal. In the vacuum of space the warmed coolant, after leaving the cold plates, is routed through a device called a sublimator. Water, from an IU reservoir, goes to the sublimator and is exposed through a porous plate to the low temperature and pressure of outer space where it freezes, blocking the pores in the plate. The heat from the coolant, transferred to the plate, is absorbed by the ice converting it directly into water vapor (a process called sublimation).

Splice Joint Operations - Final grinding of a splice joint ensures a smooth surface prior to splice plate assembly. IBM-DR-16

The system is self-regulating. The rate of heat dissipation varies with the amount of heat input, speeding up or slowing down as heat is generated. If the coolant temperature falls below a pre-set level, an electronically controlled valve causes the coolant mixture to bypass the sublimator until the temperature rises sufficiently to require further cooling. Nitrogen gas provides artificial pressure for both coolant solution and sublimator water reservoirs during orbit. A coolant circulating pump along with the necessary valves and piping to control flow complete the environmental control equipment.

GN_2 Storage Sphere - In place next to the ST-124-M inertial platform, the sphere holds 2 cubic feet of gas used for gas bearings of the platform. Also visible are a pressure regulator, heat exchanger for warming gas, and pressure indicators. IBM-DR-21

Environment Control LA mobile clean room protects against contamination during assembly of environmental control system components. Gaseous nitrogen will be circulated from a ground supply through the duct partially assembled in the cable tray to purge the IU following vehicle fueling. IBM-DR-23

Instrument Unit Assembly in IBM Manufacturing Area Splicing operations and assembly of the tubular cable tray are complete, the cold plates have been installed, and installation of components is underway. IBM-DR-19

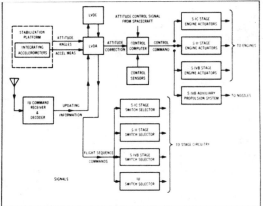

Block Diagram of Guidance and Control System IBM-DR-8

Guidance and Control

The IU's guidance and flight control systems navigate (determine vehicle position and velocity), guide (determine attitude correction signals), and control (determine and issue control commands to the engine actuators) the Saturn V vehicle. Completely self-contained, these systems measure acceleration and vehicle attitude, determine velocity and position and their effect on the mission, calculate attitude correction signals, and determine and issue control commands to the engine actuators. All this is done to place the vehicle in a desired attitude to reach the required velocity and altitude for mission completion. Major components are an inertial platform, the launch vehicle digital computer (LVDC), the launch vehicle data adapter (LVDA), an analog flight control computer, and control and rate gyros.

powered flight, boost period after initial entry into space, and the coasting period. Atmospheric boost causes the greatest vehicle load because of atmospheric pressure. During this time the guidance system is primarily checking vehicle integrity and is programmed to minimize this pressure.

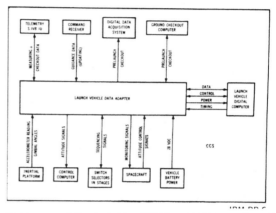

Guidance and Control - The LVDC and LVDA portion of the guidance system is shown in this block diagram. The LVDC receives information from all parts of the vehicle via the LVDA, and in turn issues commands. IBM-DR-6

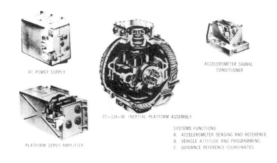

ST-124-M Inertial Platform System IBM-DR-4

Prior to liftoff, launch parameters go to the LVDC. About five seconds before liftoff, the inertial guidance platform and the LVDC are released from ground control. As the vehicle ascends, the guidance platform senses and measures vehicle acceleration and attitude and sends these measurements to the LVDC via the LVDA. The LVDC integrates these measurements with the time since launch to determine vehicle position relative to starting point and destination. It then computes the desired vehicle attitude, using data stored in its memory, and the difference between the desired attitude and the actual becomes the generated attitude correction signal. This signal is sent to the analog flight control computer, where it is combined with information from rate gyros. Using this data, the flight control computer determines and issues the command to gimbal the engines and change the thrust direction. Each mission has at least three phases: atmospheric-

The vehicle maintains liftoff orientation long enough to clear the launch equipment, and then it performs a roll maneuver to get to the flight azimuth direction. The time tilt program is applied after the roll maneuver. The pitch angle is regulated by the tilt program, and is independent of navigation measurements. However, navigation measurements and computations are performed throughout the flight, beginning at the time the platform is released (i.e. five seconds before liftoff). First stage engine cut-off and stage separation are commanded when the IU receives a signal that the tank's fuel level has reached a predetermined point. During second stage powered flight the LVDC guides the vehicle via the best path to reach the mission objectives. During orbit, navigation and guidance information in the LVDC can be updated by data transmission from ground stations through the IU radio command system. Approximately once every two seconds, the LVDC, using iterative or "closed loop" guidance, figures vehicle position and vehicle conditions required at the end of powered flight (velocity, altitude, etc.) and generates the attitude correction signals to gimbal the engines so that the vehicle reaches its predetermined parking orbit.

IU Interior During Assembly-The large, cylindrical component simulates size and shape of the flight control computer and is used to check cable lengths and mounting arrangement. IBM-DR-17

supplied at a controlled pressure and flowrate from reservoirs within the IU.

PRELAUNCH FUNCTIONS

In addition to guidance computations, other functions are performed by the LVDC and the LVDA. During prelaunch, the units conduct test programs. After liftoff they direct engine ignition and cutoff, direct stage separations, and conduct reasonableness tests of vehicle performance. During earth orbit, the computer directs attitude control, conducts tests, isolates malfunctions, and controls transmission of data, plus the sequencing of all events.

Instrumentation

A basic requirement for vehicle performance analysis and for planning future missions is knowing what happened during all phases of flight and just how the vehicle reacted. The IU's measuring and telemetry equipment reports these facts. Measuring sensors or transducers are located throughout the vehicle monitoring environment and systems' performance.

Second stage engine cutoff comes when the IU is signaled that stage propellant has reached a predetermined level, and then the stage is separated. By this time, the vehicle has already reached its approximate orbital altitude, and the third stage burn merely gives it enough push to reach a circular parking orbit.

TRIPLE RELIABILITY

To ensure the accuracy and reliability of guidance information, critical LVDC circuits are provided in triplicate. Known as triple modular redundancy (TMR), the system corrects for failure or inaccuracy by providing three identical circuits. Each circuit produces an output which is voted upon. In case of a discrepancy, the majority rules, and a random failure or error can be ignored. In addition, the LVDC has a duplexed memory, and if an error is found in one portion of the memory, the required output is obtained from the other and correct information read hack into both memories, thus correcting the error. The ST-124-M inertial platform provides signals representing vehicle attitude. Since a signal error could produce vast changes in ultimate position, component friction must be minimized. Therefore, the platform bearings are floated in a thin film of dry nitrogen

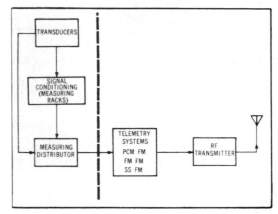

Typical Saturn Measuring System IBM-DR-1

Measuring and Telemetry

Measurements are made of mechanical movements, atmospheric pressures, sound levels, temperatures, and vibrations and are transformed into electrical signals. Measurements also are made of electrical signals, such as voltage, currents, and frequencies which are used to determine sequence of stage separation, engine cutoff, and other flight events and to determine performance of onboard equipment. In all, the IU makes several hundred measurements. A wide

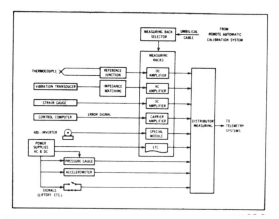

Typical Saturn Measuring System IBM-DR-1

variety of sensors are used to obtain all kinds of information required :acoustic transducers monitor sound levels: resistor or thermistor transducers monitor temperature environments; bourdon- tube or bellows transducers measure pressures; force-balance, or piezoelectric accelerometers measure force levels at critical points; flow meters determine rates of fluid flow. Various measuring devices produce a variety of outputs, and before these outputs can be effectively utilized, they must be standardized to some extent. Signal conditioning modules are employed to adapt transducer outputs to a uniform range of 0-5 VDC. Different types of data require different modes of transmission, and the telemetry portion of the system provides three such modes: SS/FM, FM/FM, and PCM/FM. Each type of information is routed to the most suitable telemetry equipment; a routing is performed by the measuring racks within the IU. To get the most out of the transmission equipment, multiplexing is employed on some telemetry channels. Information originated by various measuring devices is repeatedly sampled by multiplexers, or commutators, and successive samples from different sources are transmitted to earth. Information sent over any channel represents a series of measurements made at different vehicle points. This time-sharing permits large chunks of data to be handled with a minimum amount of equipment. The LVDC also helps in data transmission. For instance, when the vehicle is between ground receiving stations, the LVDC stores important PCM data for later transmission. Once the vehicle leaves the earth's atmosphere, sound levels requiring air for continuance no longer exist. The LVDC signals a measuring distributor to switch from unimportant measurements to those more critical to the mission. And during stage separation retro-rocket firing, when flame attenuation distorts or destroys telemetry transmissions, signals are automatically recorded by an onboard tape recorder, and transmitted later. In order to monitor vehicle performance, ground controllers must know the vehicle's precise position at all times. The RF section of the instrumentation system provides this capability, as well as linking the IU's guidance and control equipment during flight.

TRACKING SYSTEM

Several tracking systems are used to follow vehicle trajectory during ascent and orbit. Consolidation of this data not only increases data reliability, but gives the best trajectory information. Vehicle antennas and transponders, which increase ground-base tracking systems' range and accuracy, make up the IU's tracking equipment.

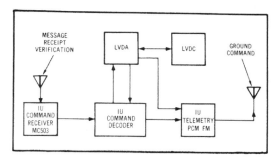

Saturn V Instrument Unit Command System IBM-DR-2

A pulse or series of pulses of RF energy sent by ground stations to the vehicle's general direction will interrogate the airborne transponder. In response, the transponder produces a pulse or series of pulses. Triangulation between precisely located ground stations determines point of origin of these reply pulses and fixes location of the vehicle. Three tracking systems are employed in conjunction with the Saturn V IU: AZUSA, C-band radar, and the S-band portion of the command and communication system (CCS). Two C-band transponders are employed to provide tracking capabilities for this system independent of vehicle attitude. A single transponder is employed with the AZUSA system. Real-time navigation, needed to update the guidance system, is received in the IU by a radio command link. But before it is sent, and before it is accepted in the IU, both ground equipment and IU instruments scrutinize update information for accuracy. The slightest error in transmission could conceivably produce a greater problem than if the original data had been left alone. The message goes from antenna to command receiver for amplification and demodulation. Then it is routed to the decoder for breakout into the original pattern of digital bits. The first validity check is here. If there is an error in a bit, or a bit is missing, the entire message is rejected. Accepted commands get further checking in the command decoder and in the LVDC. First the vehicle address is checked in the command decoder. This is important because commands for both IU instruments and the spacecraft use similar command links. If the spacecraft address is recognized, the IU ignores the message. Passing this test, the message is sent to the LVDC. Upon receipt, the LVDC tests the message to deter- mine if it is proper. If it is, then the command decoder releases a pulse via the telemetry link to the ground station verifying message acceptance. If the message fails the test, the LVDC rejects it and telemeters an error message. Depending on the mission, several types of messages can be processed. For example: commands to perform updating, commands to perform tests, commands to perform special subroutines or special modes of operation, a command to dump or clear certain sectors of the computer memory, or a command to relay a particular address in the computer memory to the ground. Provisions have been made to expand the number of types of messages if experience indicates this is necessary.

SWITCH SELECTOR

All stages, and the IU, are equipped with a switch selector. This unit has electronic and electromechanical components which decode LVDC/LVDA sequence commands and switches them to the proper circuits en each stage. This system has several advantages: reduction of stage interface lines, in- creased flexibility with respect to timing and sequencing, and conserving the discrete output circuitry in the LVDC/LVDA. Sequencing commands can come as fast as every 100 milliseconds. Stage power isolation is maintained in the switch selector by using relays as the input circuit. The relays are driven by IU power, while the decoding circuitry and driver output are powered from the parent stage. Input and output are coupled through relay contacts. These contacts drive a diode matrix used in decoding the 8-bit input code to select the output driver, producing the switch selector output. There also is a check and proceed system built into the switch selector. After the switch selector relays have been "picked," the complement of the received message is fed back to the LVDA/LVDC where it is checked. If the feedback is good, a read command is issued. If there is disagreement, a new message is sent which accomplishes the same function. (Note: For redundancy, two messages' codes are assigned for each switch selector output).

Electrical System

The electrical system powers the IU's equipment. As with most of the IU's systems, the electrical system is divided into two sectors: prelaunch and flight. Ground sources provide power through the umbilical lines before launch. At approximately 25 seconds prior to liftoff, power is transferred to the four 28 VDC IU batteries. Each battery has a 350 ampere hour capacity, and loads are equally distributed to drain. Two special power supplies are provided: a 5-volt master measuring voltage supply converts 28 VDC main supply to a highly regulated 5 VDC for reference and supply voltage to the measuring components, and a 56-volt power supply for operation of the guidance system's ST-124-M inertial platform and the platform's AC power supply. In order to get the most out of the battery stored power during flight, the LVDC and LVDA turn off unused or unimportant circuits in favor of more important applications as the mission progresses.

EMERGENCY DETECTION EQUIPMENT

The Saturn V is equipped with a myriad of equipment designed to detect malfunctions. Some of this equipment checks engine thrust, and monitors guidance computer status, attitude rates, angle of attack, and abort request. This emergency detection information is flashed to the IU where it is routed to the emergency detection distributor (EDS) in the electrical system. The EDS distributor is an interconnector and switching point and has the logic circuits which determine the emergency. In case of a malfunction, the equipment will turn on a light in front of the astronauts. If the spacecraft abort selector switch is in the automatic abort position, the abort will take place without further crew participation; the action cannot be vetoed by the astronauts. However, if the selector switch is in the manual position, the crew, consulting with NASA flight controllers, decides when to abort a mission.

FACILITIES

INTRODUCTION

Assembly, test, and launch facilities for the Saturn V consist of a combination of facilities which existed before the onset of the program as well as many specifically created for its execution. Included in these facilities are installations set up by the National Aeronautics and Space Administration to meet the greatly increased size and complexity of the Saturn program. The Marshall Space Flight Center includes installations at Huntsville, Ala., where vehicle development is the prime responsibility; Michoud Assembly Facility, New Orleans, La., where the first stage is fabricated anti assembled; and Mississippi Test Facility, Bay St. Louis, Miss., which is responsible for test operations. Launch facilities are located at the NASA Kennedy Space Center, Fla. Because of the giant size of Saturn launch vehicles and the difficulties in transporting them, fabrication and test facilities were located within easy water shipment to the launch site. At all of these NASA installations are located employees of the companies which are the prime contractors for building the various stages and components of the Saturn V. Other facilities, including the home bases of the major contractors and sub contractors, are located across the nation.

Boeing FACILITIES

The Boeing Company manufactures the Saturn V first stage at the 900-acre NASA Michoud Assembly Facility in New Orleans. The facility has about 2,000,000 square feet of manufacturing floor space and about 730,000 square feet of office space. About 60 per cent of the manufacturing area is occupied by Boeing.

Michoud - The Michoud Assembly Facility is the fabrication site of the first stage booster. Dominating the skyline is the Vertical Assembly Building. B-4793-9

The plant is arranged for logical and efficient flow of materials from the loading dock through to final assembly. Paralleling the material flow are the rework and modification area and the test and laboratory areas. There are 50,000 square feet of tooling area in the plant.

Stage Test Before leaving Michoud, the completed booster undergoes a simulated firing during which all systems function in the Stage Test Building. B-9847-7

The environmentally controlled portion of the minor assembly area contains facilities for heat treatment, chemical cleaning, conversion coating, and welding of preformed metal sections received at the loading dock. Final assembly of the propellant tanks and the joining of the major components into the complete stage occur in the Vertical Assembly Building (VAB). The VAB is a single-story structure rising the equivalent of 18 stories. A 180-ton overhead crane is used to stack the five large cylindrical segments of the first stage into a vertical assembly position. A $50 million program included the construction of three buildings -the VAB, the Stage Test Building, and the Engineering and Office Building-as well as the renovation of existing facilities at Michoud. Checkout of the stage's electrical and mechanical systems is performed in the four giant test cells of the Stage Test Building. Each of the test cells-is 83 by 191 feet with 51 feet of clear height. Each has separate test and checkout equipment. Stages leave and enter Michoud by waterways connecting to the Mississippi River or the Gulf of Mexico.

Barge Slip - First stages are loaded onto barges at Michoud and travel by waterways from New Orleans to Huntsville, Mississippi Test Facility, and Kennedy Space Center. B-8145-2

Unique Vessels - Four of six special barges used to carry Saturn rocket stages are shown moored side-by-side at the Michoud Assembly Facility. From left are the Little Lake, the Promise, the Poseidon, and the Palaemon. B-8930-10

NORTH AMERICAN SPACE DIVISION FACILITIES

The second stage of the Saturn V is manufactured and tested in facilities located from one end of the nation to the other. The main fabrication and testing facilities are located in Seal Beach, Calif., about 15 miles south of Downey, which is the headquarters of SD operations. SD subcontracts important elements of work to other North American facilities in Los Angeles and Tulsa and McAlester, Okla. The complex of buildings at Seal Beach, all built especially for the second stage, will be complemented by mid-1967 with three North American Aviation-owned buildings which will house all the second stage administrative, engineering, and support personnel who currently are located at Downey. The Seal Beach facility includes a bulkhead fabrication building, 125-foot-high vertical assembly building, 116-foot-tall vertical checkout building, pneumatic test and packaging building, and a number of other structures. The bulkhead fabrication building is a large, highly specialized structure designed solely for the construction and assembly of the second stage's three bulkheads. Among other tooling it contains an autoclave about 40 feet in diameter with a 40-foot dome for curing the large stage bulkheads.

The vertical assembly building, where the stage is assembled, contains six work stations at which successive major parts of the stage are added. After assembly, the stage is moved to the vertical checkout building, where some final installations are made and where its systems and components are given final tests. The last stop for a stage is the pneumatic test and packaging building, where stages are turned horizontally for pneumatic tests,

Over-all View - North American Seal Beach facilities include in-process storage building (left); bulkhead fabrication building (center); vertical assembly building (far right); pneumatic test and packaging building (right center); and structural test tower (right front). S-27

Night firing of Test Second Stage at Santa Susana S-28

painted, and prepared for shipping. Other buildings at Seal Beach provide for such things as processing and storage of subassemblies and machine and tool shop services. Second stage engine (J-2) testing is performed at the Rocketdyne Santa Susana facility. The Coca test area at Santa Susana operated by SD was re-built for the second stage and is where battleship test firings are conducted. SD also operates facilities at both Mississippi Test Facility and Kennedy Space Center to provide management and operational support services.

Space Truck Readied - The five engines of the Saturn V second stage dwarf technicians preparing the "battleship" vehicle for hot firing at North American's Santa Susana static test lab. S-29

Wide Load - A second stage is transported from Seal Beach to the Navy dock for shipment to Mississippi Test Facility. S-30

Aerial View of Space Systems Center, Huntington Beach, Calif. D-NRV-43

DOUGLAS Facilities

The Douglas Space Systems Center at Huntington Beach, Calif., is the master facility for engineering and production of the third stage of Saturn V. The center is headquarters for the Missile and Space Systems Division and for direction of Saturn activities in other company facilities at Santa Monica and Sacramento, Calif., and at Cape Kennedy, Fla.

Fabrication

Initial component fabrication for the Saturn V third stage is accomplished at the Santa Monica plant. It produces parts and subassemblies ranging from micro-miniature electronic components to the complete liquid oxygen tank.

Final assembly and factory checkout of the third stage takes place at the Space Systems Center. High-bay manufacturing area is provided for production of propellant tanks, skirts, and interstages. Eight tower positions are available for vertical assembly and checkout of completed vehicles. The bulk of research and development testing in the third stage program is carried out in laboratory facilities at the Santa Monica plant. There components and subassemblies are put through a complete qualification test program in some 80 different laboratories. Other test facilities include the Space Simulation Laboratory at the Space Systems Center, where major Saturn subassemblies are subjected to space-like conditions by being placed inside a 39-foot diameter vacuum chamber for extended periods of time. The chamber is capable of simulating the vacuum at an altitude of 500 miles above the earth. Structural tests on major vehicle structures such as the propellant tank, skirt sections, and interstage are conducted in the Structural Test Laboratory at the Space Systems Center. Two vertical checkout towers at the Space Systems Center provide for the final factory tests on finished third stages, prior to shipment from the plant for test firing. The vertical checkout laboratory is equipped with two complete sets of automatic check-out equipment. Actual ground test firings of the stages are accomplished at the Douglas Sacramento Test Center, where each stage is delivered following the completion of assembly and checkout at the Huntington Beach plant. Primary Saturn facilities at Sacramento include a pair of 150-foot-high steel and concrete test stands where the stages are put through the final vehicle acceptance test-a full-duration, full-power static firing, simulating actual launch operations. port of large space hardware, the plane has an inside diameter of 25 feet and a total length of 141 feet. Tail height is 46 feet-almost five stories above the ground. Cubic displacement of the fuselage is 49,790 cubic feet, approximately five times that of most present jet transports. The airplane is powered by four turbo-prop engines, producing a total of 28,000 horsepower.

IBM FACILITIES

Three IBM-owned buildings at Huntsville comprise the Space Systems Center where component testing, fabrication, assembly, and systems checkout of the instrument unit are completed. Assembly and the majority of the testing activity take place in a 130,000-square-foot building located in Huntsville's Research Park. As units are received, they are inspected and then moved to one of the testing laboratories where they are subjected to detailed quality and reliability testing. From component testing, the parts move to inventory until called out for assembly.

Assembly and Checkout Tower, Douglas Space Systems Center D-012220

IU Assembly and Test - All instrument unit assembly work and the majority of testing are done in this IBM-owned building in Huntsville's Research Park. The rear of the building is the high- bay area where assembly operations take place. IBM-DR-24

Static test Firing of Third Stage at Sacramento D-NRV-48

Automatic Checkout IBM technicians monitor systems checkout tests as another technician optically adjusts the inertial guidance platform, prior to a simulated mission. IBM-DR-25

Super Guppy H-5-29646

Following assembly operations, the IU is moved to one of two systems checkout stands-one for uprated Saturn I vehicles, the other for Saturn V.

A complete systems checkout is performed automatically. Hooked by underground cables, two digital checkout computer systems examine the IU. Each of the IU's six subsystems is tested before the IU is tested as an integrated unit. With independent computers, systems tests for two instrument units can he conducted simultaneously.

Simulation Laboratory Saturn V flight guidance and navigation programs as well as launch computer programs are tested in IBM's Engineering Building at Huntsville, Here a technician checks a computer readout of a simulated mission. IBM-DR-26

NORTH AMERICAN ROCKETDYNE FACILITIES

F-1 and J-2 engines for the Saturn V launch vehicle are manufactured at Rocketdyne's main complex in Canoga Park, Calif. F-l static testing is conducted at the Edwards Field Laboratory located at the NASA Rocket Engine Test Site, Edwards, Calif., about 125 miles northeast of Los Angeles, and the J-2 is tested at Rocketdyne's Santa Susana Field Laboratory located about 10 miles from Canoga Park. Rocketdyne operates the Neosho Facility (Missouri), which produces and tests subcomponents of the J-2 and F-1 engines.

Manufacturing of components and final assembly of both engines are carried out in eight buildings in the Canoga complex. These facilities are equipped with general purpose machine tools for precision and heavy machining as well as some 20 numerically controlled machines for performing programmed multiple machining operations. Also included are two of the largest gas-fired brazing furnaces of their type for brazing of thrust chamber tubes and injectors, eight units for ultrasonic cleaning, 21 installations for Gamma and X-ray inspection, more than 50 environmentally controlled areas for ultra- clean assembly operations, sheet metal preparation, precision cleaning, and receiving inspection.

F-l Test Stands -Three of six stands for testing F-1 rocket engines or components at full thrust are visible in this aerial view of NASA Rocket Engine Test Site. R-11

F-1 Test Firing- An F-l rocket engine developing 1,500,000 pounds of thrust is tested at NASA Rocket Engine Test Site. The stand is one of six in the complex. R-12

An Engineering Development Laboratory provides specialized facilities to support manufacturing programs. These facilities include a high-flow water test facility for checking propellant systems, 12 concrete cells for conducting hazardous tests, 28 environmental test chambers, a photo-elastic laboratory, two pneumatic flow benches, six vibration test rooms, and others for checking components as well as complete engines. Research and development testing of F-1 turbo-machinery, gas generators, heat. exchangers, seals, and splines is conducted on two test stands and three components test laboratories at Santa Susana. Six large test stands, with a total of eight test positions, and associated shops and support facilities at the Edwards Field Laboratory are used for testing complete F-1 engines as well as injectors. Six large engine test stand positions at the Santa Susana Field Laboratory are used for testing the J-2. One of these stands is equipped with a steam injection diffuser for altitude simulation testing. J-2 turbopumps, gas generators, valves seals, bearings, and other components are tested in 22 test cells in five component test laboratories in Santa Susana.

J-2 Testing-A hydrogen fueled J-2 rocket engine is tested under ambient altitude conditions at Santa Susana. R-15

F-1 Flight Engine Firing R-13

HUNTSVILLE FACILITIES

New Saturn V facilities built at the Marshall Space Flight Center at Huntsville, Ala., include the first stage static test stand, an F-1 engine test stand, the Saturn V launch vehicle dynamic test stand, a J-2 engine facility, and ground support and component test positions. The Marshall Center has completed a $39 million addition to its Test Laboratory for captive testing the Saturn V booster and F-1 engines. The Test Laboratory addition is called the West Test Area. The largest structure in the area is the first stage test stand. Completed in 1964, the stand has an overall height of 405 feet.

Growth at Huntsville —The growth of rocket testing facilities at the Marshall Space Flight Center is contrasted here by the size of the first Redstone Arsenal test stand, second from left, and stands at right built for the Saturn V program. H-5-31176

Pump Tests-Flames from gases burned during test of an F-1 engine turbopump shoot more than 150 feet in air. R-14

The nearby single engine test stand is being used for research and development tests of the 1.5 million pound thrust F-1 engine. Control and monitoring equipment for the first stage and F-1 engine stands is located in the area's central blockhouse. Water needed to cool the flame deflectors of the two stands is pumped from a nearby high-pressure industrial water station. The 365-foot tall Saturn V launch vehicle was placed in another unique Marshall Center test facility - the Dynamic Test Stand. Testing of the complete three stage vehicle and its Apollo spacecraft here was done to determine its bending and vibration characteristics. Tallest of Marshall Center's tall towers, the dynamic test stand is 98 feet square. Several tests of the liquid hydrogen-liquid oxygen powered engine have been conducted during the past year in Marshall's J-2 engine test facility. Tank- age for the facility is a battleship version of the Saturn V third stage. The J-2 engine stand is 156 feet tall and has a base of 34 by 68 feet. It is located in the MSFC's East Test Area.

Elements simulated include the first stage booster, second stage, third stage, and an instrument unit. Other Saturn V facilities at the Marshall Center include a booster checkout area, two new assembly areas and a components acceptance building.

Vibration Version - A ground test version of the Saturn V first stage moves through the West Test Area of the Marshall Space Flight Center. The large dynamic test stage was built to undergo vibration and bending tests. Test stand at right is a single F-1 engine facility. H-5-31246

Positioning - A Saturn V first stage is placed into a test stand at Marshall Space Flight Center. H-40246

A portion of the Kennedy Space Center "spaceport" has been created at the Marshall Center's ground support equipment test facility to check out giant mechanical swing arms which will be used on Launch Complex 39 to connect the Apollo/Saturn V space vehicle to the launcher tower. The 18-acre facility has eight swing arm test positions and one position for testing access arms to be used by Apollo astronauts. Also built at Marshall are an F-1 engine turbopump position in the East Test Area and a load test facility in the Propulsion and Vehicle Engineering Laboratory. A new Saturn V rocket "electrical simulator" or breadboard facility at the Marshall Center duplicates the electrical operation of the vehicle.

Static Firing -The Marshall Space Flight Center captive fired all five F-1 engines of the Saturn V S-IC-T for 16% seconds on May 6, 1965. Later they were fired for 41 seconds. H-5-23531

MISSISSIPPI TEST FACILITY

NASA has developed the Mississippi Test Facility, a field organization of the Marshall Space Flight Center, as a testing site for the Saturn V launch vehicle's two lower stages. Acceptance testing of first and second stages will be conducted at the $300 million facility. In addition, limited repair and modification of J-2 engines will be performed at MTF on behalf of all NASA operations in the Southeast.

Boeing Company is the prime contractor to NASA for developmental and acceptance testing of first stages. Stages manufactured by Boeing will be tested by the company in the first stage test complex. The U. S. Army Corps of Engineers is NASA's agent for land acquisition, design engineering, and construction.

Management, operational, and support personnel engaged at the Mississippi Test Facility after its current construction and development work is complete in 1967 will number approximately 3,000.

Checkout-Engineers and technicians of North American are shown in the second stage Test Control Center at the Mississippi Test Facility during final preparation for static firing of all systems test model of the stage. H-MTF-1432-B7

First Stage Test Stand at MTF H-MTF-67-917

Aerial View of Mississippi Test Facility H-MTF-2077

The General Electric Co., under a prime contract with NASA, operates and maintains the facility, providing site services, technical systems, and test support to NASA and to stage contractors and other tenants. North American Aviation, Inc., through its Space Division, is the prime contractor to NASA for developmental and acceptance testing of second stages. SD personnel conduct the tests within the second stage test complex. The

Stage Hoisted - The all-systems test version of the second stage is lifted into its test stand at MTF. H-MTF-4215

The MTF site was selected from 34 areas considered mainly because of its accessibility to water routes and its nearness (45 miles by water) to the Michoud Assembly Facility in New Orleans. The government-owned fee area comprises 13,424 acres and is surrounded by an acoustic buffer zone involving an additional 128,526 acres in Hancock and Pearl River counties and Saint Tammany Parish.

Static Firing-A giant plume of vapor billows skyward during the first static firing test at MTF. The Saturn second stage, built for NASA by North American Aviation, Inc., burned for 15 seconds April 23, 1966. H-MTF-66-1823A

MTF is composed of three principal complexes including approximately 60 buildings and structures. Among predominant features are the three huge test stands in the Saturn V complex. There are two separate stands for testing second stages. The first stage test stand is a dual-position structure which, with overhead crane, towers over 400 feet. The Laboratory and Engineering Complex houses engineering, administrative, and technical personnel. The Industrial Complex has facilities for equipment and personnel necessary for site and test support maintenance. The relatively small force of NASA personnel assigned to MTF has overall management and supervisory responsibilities in overseeing the work of the contractors. NASA personnel are also responsible for final evaluation of static firings and issuance of flightworthiness certificates to stage contractors.

KENNEDY SPACE CENTER

Launch Philosophy

Saturn V vehicles are assembled, checked out, and launched at Launch Complex 39 at Kennedy Space Center. Complex 39 embodies a new mobile concept of launch operations which includes superior re- liability and time savings offered by assembly and checkout in a protected environment and reduction of actual pad time as much as 80 per cent with a consequent increase in launch rate capability. The ability to adapt economically to future program requirements is another advantage. For example, the service platforms used in the Saturn/Apollo program could be used for other vehicles of similar configuration, and the area can accommodate space boosters with thrusts up to 40 million pounds.

Facilities

The major components of Launch Complex 39 include: (1) the Vehicle Assembly Building, where the space vehicle is assembled and prepared; (2) the mobile launcher, upon which the vehicle is erected for checkout, and from which, later, it is launched; (3)the crawlerway, upon which the fully assembled vehicle is carried by transporter to the launch site; (4) the mobile service structure, which provides external access to the vehicle at the launch site; (5) the transporter which carries the launch vehicle, mobile launcher, and mobile service structure to various positions at the launch complex; and (6) the launch area from which the space vehicle is launched.

THE VEHICLE ASSEMBLY BUILDING

The Vehicle Assembly Building (VAB) consists of a high bay area 525 feet tall, a low bay area 210 feet tall, and a four-story launch control center (LCC) connected to the high bay by an enclosed bridge. The VAB, with 130 million cubic feet, is the world's largest building in volume. It covers eight acres of land. There are four assembly and checkout bays in the high bay area. The low bay area contains eight stage preparation and checkout cells equipped with systems to simulate stage interface. The launch control center houses display, monitoring, and control equipment for checkout and launch operations. There are four firing rooms in the LCC, one for each high bay and checkout area. Work platforms, mounted on opposite walls in the high bay area, are designed to enclose various work areas around the launch vehicle. Platforms extend or retract in less than 10 minutes. Twenty- ton hydraulic jacks are used to align platforms. The Saturn V, after prelaunch checkout on its mobile launcher, is carried by the transporter from the VAB through a door shaped like an inverted "T". The door is 456 feet high.

The base of the door is 149 feet wide and 113 feet high; the remainder is 76 feet wide. There are four such doors in the VAB, one for each of its four high bays. In keeping with the protective environment of the building, doors were designed to withstand winds of 125 miles per hour. There are 141 lifting devices in the VAB, ranging from one-ton mechanical hoists to two 250-ton high-lift bridge cranes. Each pair of high bays shares a bridge crane. The cranes have a lifting height of 456 feet and a travel distance of 431 feet.

Checkout Vehicle-The Saturn V facilities vehicle begins its journey from the Vehicle Assembly Building to the launch pad. Its purpose was to check out facilities, train launch crews, and verify procedures at KSC. K-107-66P-237

Saturn V Facilities Vehicle Rollout at Kennedy Space Center K-107-66PC-75

LAUNCH CONTROL CENTER

Located Southeast of the VAB is the launch control center (LCC). This four-story building is the electronic brain of Launch Complex 39. Here final count-down and launch of Saturn V's will he conducted. The LCC is also the facility from which a multitude of checkout and test operations will be conducted while space vehicle assembly is taking place inside the VAB. Two separate, automated computer systems are used to check out and conduct the countdown for the Saturn V. The acceptance checkout equipment, or ACE, is used for the Apollo spacecraft. The Saturn ground computer system is used for the various stages of the vehicle. Located in the launch control center is the heart of the Saturn ground computer system. Here check-out and preflight countdown are conducted. This system has as its "brain" two RCA 110A computers. One is located in the launch control center and the other is in the mobile launcher upon which the Saturn V is erected.

Moving Tower-Personnel watch a mobile launch tower moving along the crawlerway at Kennedy Space Center. K-100-66C-813

Through the process control system, all stages are checked, and data from the engines and from the guidance, flight control, propellants, measurement, and telemetry systems is provided. The Saturn ground computer system also includes a DDP 224 display computer located in the LCC. It can drive up to 20 visual cathode ray display tubes. The RCA 110A computer is capable of transmitting 2,016 discrete signals to the vehicle where it is possible for the computer in the mobile launcher to return 1,512 discrete signals. A digital data acceptance system collects and makes available onboard analog data to the computers. A triply redundant system for discrete output information allows more reliability. There are 1,512 signals going to the mobile launcher showing "on" and "off" commands. If one signal fails or reports a wrong command and the other two signals transmit another command, the majority command is indicated in the display and transmitted to the vehicle. There are 15 display systems in each LCC firing room, with each system capable of giving digital information instantaneously. Sixty television cameras are positioned around the Saturn V transmitting pictures on 10 channels. Additionally, the LCC contains several hundred operational intercommunication channels which enable the launch team and the launch director to be in voice contact with the astronauts aboard the spacecraft. Automatic checkout of the Apollo spacecraft is accomplished through acceptance checkout equipment (ACE). Through the use of computers, display consoles, and recording equipment, ACE provides an instantaneous, accurate method of spacecraft pre-flight testing. ACE also is used at the spacecraft contractor plants and in testing at the Manned Spacecraft Center in Houston.

Vehicle Assembly Building at KSC Viewed From Across the Turning Basin of Launch Complex 39 K-100-66C-456

Computerized checkout of the Saturn stages at the launch pad and the Apollo ACE system at the Manned Spacecraft Operations Building at the Kennedy Space Center are tied together by instrumentation. The Saturn V employs completely automated computer controlled checkout systems for each of its stages. The system uses a carefully detailed computer program and associated electronic equipment to perform a complete countdown checkout of each stage and all its various systems, subsystems, and components. With electronic speed, it moves through a thorough and reliable countdown, yet permits test engineers to monitor every step of the operation and to over-ride the computer functions, if necessary. To monitor fuel and oxidizer mass for the three stages of the Saturn V vehicle, a propellant tanking computer system (PTCS) is used. This system controls propellant tank fill and replenishment. Liquid oxygen and liquid hydrogen must be replenished constantly to compensate for boil-off.

PROPELLANT STORAGE AND TRANSFER

Propellant facilities at Launch Complex 39 include a LOX system, the RP-1 system, the liquid hydrogen system, the propellant tanking computer system, the spacecraft support system, and the data transmission system. The propellant tanking computer system provides a means of monitoring amounts during the fueling operations. It also accurately controls fuel level during the final phase of tank fill and replenish. The data transmission system provides an accurate method for the transmission of propellant and environmental control system electrical signals from the launch site to the LCC. The liquid oxygen system provides oxidizer fill and drain for the three stages of the Saturn V. The system includes a storage tank, a vaporizer, two replenishing pumps, transfer lines, vent lines and drain basin, and electric circuitry for monitoring and actuating the pneumatic control system. The round liquid oxygen storage tank holds 900,000 gallons and is situated 1,450 feet from the launch site. It has a stainless steel inner wall 62 feet 9 inches in diameter. The space between this inner sphere and the outer wall is filled with gaseous nitrogen and perlite for insulation. To load liquid oxygen, a command originates in the LCC at the LOX control panel. The signal is transmitted to the mobile launcher by the data transmission system and then to the LOX storage area. The electrical signals are converted to pneumatic pressure to operate the valves, and the flow of LOX from the storage tank into the vaporizer begins. The vaporizer converts the liquid oxygen into gaseous oxygen, which then is fed back into the tank to pressurize it to the 10 psig needed to begin the flow. The pumps are started and the LOX is pumped through the transfer lines to the vehicle. The RP-1 system provides fuel fill, drain, and filtering capabilities for the first stage. The system includes three storage tanks each with a capacity of 86,000 gallons, transfer lines, a launch site facility, and electric circuitry. The liquid hydrogen system provides fueling and draining for the second and third stages. It includes a storage tank with a capacity of 850,000 gallons, a vaporizer, transfer lines, and a burn pond in which excess propellant is burned. The double walled storage tank, 1,450 feet from the launch site, has a stainless steel inner wall with a diameter of 61 feet 6 inches. The space between the inner and outer walls is filled with perlite.

FLAME DEFLECTOR

To dissipate the rocket exhaust from the F-1 engines, a flame deflector, a flame trench, and a water deluge system are used in the launch area. The inverted V-shaped steel flame deflector features a replaceable ceramic-coated leading edge. Exhaust from the outer engines strikes the point of the inverted V. At the same time, the deflector is exposed to water deluge during and after liftoff. The center engine exhaust impinges on the ceramic leading edge. The heat resistant ceramic surfaces erode slowly in the blast.

View of Pad 39A East Side at KSC and Flame Trench from North End K-100-66C-825

As they do, the great thermal energy generated is carried away in super- heated particles. All exhaust and particles are deflected through a flame trench where their energy is dissipated harmlessly into the atmosphere. The mobile deflector weighs 700,000 pounds and is moved to its position beneath the launch pedestal along a rail system. Two deflectors are available for each launch area, although only one is required per launch.

THE LAUNCH AREA

Final preparation of the space vehicle for launching, including propellant and ordnance loading, final checkout, and countdown takes place in the launch area.

Aerial of Pad 39A with VAB in background K-100-66C-5629

There are presently two launch areas on Complex 39. Each area is polygon-shaped with the linear distance from side to side at approximately 3,000 feet. The launch sites are 8,730 feet apart to allow operations on the pads to be handled independently for safety reasons. Liquid oxygen, RP-1, and liquid hydrogen are stored near the perimeter of the launch sites. Helium and nitrogen gases are stored at 10,000 psi near the center. An elevated steel and concrete hardstand is located in the center of each area. Steel support fittings for the mobile launcher and mobile service structure are anchored to the hardstand. The exhaust flame trench runs through the center of the hard- stand. Prior to the launch, the wedge-shaped flame deflector is moved along rails into the trench. The liquid oxygen system consists of a 900,000-gallon LOX storage facility and transfer system. The RP-1 system consists of a storage area containing three 86,000 gallon tanks and a transfer system. The tanks have a carbon steel shell and bonded stainless steel lining.

Gaseous nitrogen and helium are stored underground in vessels near the launch pad at pressures of 6,000 psi. Automation of vehicle prelaunch checkout is expected to uprate mission confidence and to increase launch rate capability. The heart of this automatic checkout system ia the computer complex.

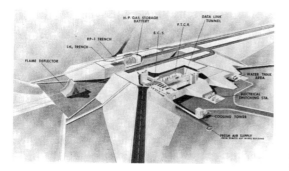

Cutaway Illustration of Pad 39A K-107-64C-2403

Facility Vehicle at Ramp of Launch Pad K-107-66PC-87

Transporter

The capacity to transport the massive mobile launcher with a fully erected Saturn V in launch-ready condition is a key to the mobile concept of Launch Complex 39. This is accomplished by a huge transporter which moves the mobile launcher and vehicle from the VAB to the launch site, approximately 3.5 miles away. The transporter moves at a maximum speed of 1 mile per hour, loaded, or 2 miles per hour, unloaded. The vehicle –131 feet long and 114 feet wide - moves on four double- tracked units, each 10 feet high and 40 feet long. Each unit is driven by an electric motor. Tractive power is provided by 16 direct current motors served by two diesel-driven direct current generators. The generators are rated at 1,000 kilowatts each and are driven by 2,750 horsepower diesel engines. Speed of the vehicle is controlled by varying the generator fields. Power for the fields is provided by two 750-kilowatt power units which also provide power for pumps, lights, instrumentation, and communications. The transporter is one of the largest land vehicles ever constructed. Yet, in transit, it must maintain a level platform within 10 minutes of arc and be capable of locating itself at its launch site and VAB positions within a 2-inch tolerance.

MOBILE SERVICE STRUCTURE

External access to the Saturn V space vehicle at the launch site is provided by the mobile service structure. The steel-truss structure rises more than 400 feet above ground level and more than 350 feet above the deck of the mobile launcher. It has five platforms which close around the vehicle. Two platforms are powered to move up and down. The remaining three are relocatable, but not self-powered. A mechanical equipment room, operations support room, communications and television equipment room, and various other equipment compartments are located in the base. The service structure is moved to and from the pad by the transporter. Once in position, either at the launch pad or in a parking area, the structure is anchored to support pedestals. The service structure remains in position at the pad until about T-7 hours when it is removed to its parking area 7,000 feet from the pad.

Arrival to Launch Pad-The facilities vehicle arrives at Launch Complex 39A. K-107-66PC-91

MOBILE LAUNCHER

The mobile launcher is a movable launch platform with an integral umbilical tower. The launcher base is a two-story steel structure covering more than half an acre. The 380-foot tower, which supports the electrical servicing and fluid lines for the vehicle, is a steel structure mounted on the base. The base and tower weigh 10.5 million pounds and stand 445 feet above ground level. Among major considerations in design of the mobile launcher were crew safety and escape provisions and protection of the platform and its equipment from blast and sonic damage. Personnel may he evacuated from upper work levels of the umbilical tower by a high speed elevator, descending at 600 feet per minute. After leaving the elevator, they can drop through a flexible metal chute into a blast and heatproof room inside the base of the pad hardstand. The mobile launcher provides physical support and is a major facility for checkout of the space vehicle from assembly at the VAB until liftoff at the launch site. The top level of the launcher base houses digital acquisition units, computer systems, controls for actuation of service arms, communications equipment, water deluge panels, and other control units. Included in the lower level are hydraulic charging units, environmental control systems, electrical measuring equipment, and a terminal room for instrumentation and communications interface. Mounted on the top deck of the base are four vehicle holddown and support arms and three tail service masts. The umbilical tower is an open steel structure providing support for nine umbilical service arms, 18 work and access platforms, and, for propellant, pneumatic, electrical, water, communications, and other service lines required to sustain the vehicle. A 250-ton capacity hammerhead crane is mounted atop the umbilical tower. The launcher restrains the vehicle for approximately 5 seconds after ignition to allow thrust buildup and verification of full thrust from all engines. The design "up-load" during the holddown period is 3 million pounds. If one or more of the engines fail to develop full thrust, the vehicle is not released, and all engines automatically are shut down.

TESTING

INTRODUCTION

The expense of the Saturn V makes it imperative that no effort be spared to assure that it will perform as expected in flight. The magnitude of the Saturn V ground test program, therefore, is unprecedented. To qualify for flight, all components and systems must meet standards deliberately set much higher than actually required. This margin of safety is built into all manrated space hardware. Compared with earlier rocket programs the ground testing on Saturn V is more extensive and the flight testing is shorter. The ground test programs con- ducted on the F-1 and J-2 engines, which power the three stages, offer an example of the thoroughness of this testing effort. The J-2 has been fired some 2,500 times on the ground, for a total running time of more than 63 hours. The F-1 has been fired more than 3,000 times for a running time of more than 43 hours. Further, in earlier rocket programs such as Redstone, Thor, and Jupiter, 30 to 40 R&D

Night Shot - A 365-foot-tall Saturn V facilities vehicle is shown in place at Launch Pad 39A. K-107-66PC-63

flight tests were standard. In the Saturn I program, where more emphasis was placed on ground testing prior to the flight phase, 10 R&D flight tests were planned. The vehicle was declared operational after the first six firings met with success. The Uprated Saturn I (Saturn IB) - an improvement on the basic Saturn I-was manrated after three flights. On the Saturn V, only two flights are planned prior to the attainment of a "manned configuration." The inspection to which flight hardware is subjected is thorough. Following are examples of many steps which are taken to inspect the Saturn V vehicle:

1. X-rays are used to scan fusion welds, 100 castings, and 5,000 transistors and diodes.
2. A quarter mile of welding and 5 miles of tubing are inspected with the use of a sound technique (ultrasonics). The same type of inspection is given to adhesive bonds, which are equivalent in area to an acre.
3. An electrical current inspection method is used on 6 miles of tubing, and dye penetrant tests are run on 2.5 miles of welding. Each contractor has his own test program that is patterned to a rather basic conservative approach. It begins with research to verify specific principles to be applied and materials to be used. After production starts each contractor puts flight hardware through qualification testing, reliability testing, development testing, acceptance testing, and flight testing.

QUALIFICATION TESTING

Qualification testing of parts, subassemblies, and assemblies is performed to assure that they are capable of meeting flight requirements. Tests under the conditions of vibration, high-intensity sound, heat, and cold are included.

RELIABILITY TESTING

Reliability analysis is conducted on rocket parts and assemblies to determine the range of failures or margins of error in each component. Reliability information is gathered and shared by the rocket industry.

DEVELOPMENT TESTING

A battleship test stage constructed more solidly than a flight stage is often used to prove major design parameters within a stage. Such a vehicle verifies propellant loading, tank and feed operation, and engine firing techniques. Battleship testing is followed by all-systems testing. For example, one of four ground test stages of the first stage completed 15 firings at Marshall Space Flight Center in Huntsville. The firings proved that the design and fabrication of the complete booster and of its subsystems were adequate. The entire Apollo/Saturn V vehicle, consisting of the three Saturn V propulsive stages, the instrument unit, and an Apollo spacecraft, was assembled in the Dynamic Test Stand at the Marshall Center. This is the only place, aside from the launch site, where the entire Saturn V vehicle has been assembled. The purpose of dynamic testing was to determine the bending and vibration characteristics of the vehicle to verify the control system design. The 364-foot assembly was placed on a hydraulic bearing or "floating platform". Electromechanical shakers caused the vehicle to vibrate, simulating the response expected from flight forces.

ACCEPTANCE TESTING

Finished work undergoes functional checkout to insure it meets operational requirements. Tests range from continuity and compatibility of wiring to all-systems ground testing. Fluid-carrying components are subjected to pressures beyond normal operating requirements, and structural components receive visual and X-ray inspections. Instruments simulate flight conditions to evaluate total performance of electrical and mechanical equipment. Rocket engines are static-fired before delivery to the stage contractor. Such tests demonstrate performance under conditions simulating flight temperatures, pressures, vibrations, etc. Each flight stage completes a series of systems tests which lead to a full-power, captive acceptance firing. Afterwards it is refurbished and given a postfiring checkout before going to Kennedy Space Center.

AUTOMATIC CHECKOUT

A fully automated, computer-controlled vehicle checkout has been designed into all major segments of the Saturn V for extensive stage test operations. Automatic checkout is used first in the final factory checkout and then throughout prefiring preparations for static tests and during the actual countdown for these firings. It is employed again throughout the postfiring checkout and finally for prelaunch checkout and launch operations at Kennedy Space Center. The system uses a carefully detailed computer program and associated electronic equipment to perform the complete countdown of each Saturn stage. With electronic speed, the automatic checkout moves through a precisely controlled and repeat-able checkout test program. The system performs a point-by-point test of each function, indicates responses to tests, and pinpoints any malfunction that occurs. The automatic checkout can also indicate ways to double check a questionable response in order to define any difficulty. It virtually eliminates the possibility of human error during a countdown.

FLIGHT TESTING

Every flight program is designed to provide a mass of vehicle performance information which is needed in planning future launches. Each stage carries a complete network of instrumentation to measure and record the performance of every system, sub-system, and vital component.

TEST DOCUMENTATION

In all Saturn V test operations, from ground development through flight, documentation of results is as important as the acquisition of data. The performance history of every part, component assembly, subsystem, and system must be accurately detailed and permanently recorded. These records give engineers a basis for making evaluations of the performance of parts and subsystems. These evaluations provide maximum confidence in every vehicle. The formidable task of record-keeping has necessitated the establishment of a test data bank for Saturn V program engineers. It can be an invaluable source of reference in the event of minor or major malfunctions in a test or flight.

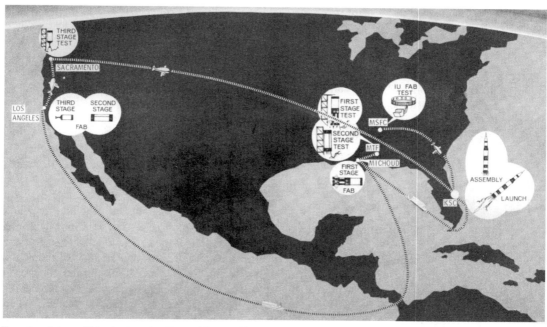

Traveling Saturn- This depicts the Saturn V assembly and test sequence and the transportation routes of rocket-carrying craft.

VEHICLE ASSEMBLY AND LAUNCH

ASSEMBLY AND CHECKOUT

Saturn V stages are shipped to the Kennedy Space Center by ocean-going vessels or by specially designed aircraft. Apollo spacecraft modules are transported by air and delivered to the Manned Spacecraft Operations Building at Kennedy Space Center for servicing and checkout before mating with the Saturn V. Saturn V stages go into the Vehicle Assembly Building low bay area where preparation and checkout begins. Receiving inspection and the low bay checkout operations are first performed before stages are erected within a high bay. After being towed into the high bay area and positioned under the 250-ton overhead bridge crane, slings tire attached to the first stage and hooked to the crane. The stage is positioned above the launch platform of the mobile launcher and lowered into place. Then it is secured to four holddown/support arms. These support the entire space vehicle during launch preparation and provide holddown during thrust

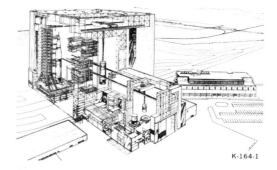

Illustration of Vehicle Assembly Building Interior at Kennedy Space Center K-164-1

buildup prior to launch. Next, engine fairings are installed on the stage and fins are moved into position and installed in line with the four outboard engines. Mobile launcher electrical ground support equipment is connected to the launch control center (LCC)via the high speed data link, and the test pro- gram is started with the actual launch control equipment. Prior to and during this time, all low bay testing is completed and the upper stages are prepared for mating. The mating operation consists of stacking the stages. Umbilical connection begins immediately and continues during the mating operation on a noninterference basis. The vertical alignment of the vehicle is performed after each stage is mated. When the launch vehicle is ready, the Apollo spacecraft is brought to the VAB and mated. Checkout of all systems is performed concurrently in the high bay. The first tests provide power and cooling capability to the vehicle, validate the connections, and establish instrumentation. When this is completed, systems testing begins. The systems tests are controlled and monitored from the LCC wherever practical and "break-in" tests are held to a minimum. Following the validation of each stage, a data review is held and the vehicle is prepared for combined systems tests.

The combined systems tests verify the flight-readiness of the overall vehicle. These tests include a malfunction sequence test, an overall test of the launch vehicle, an overall test of the spacecraft, and a simulated flight test. Prior to the simulated flight test, final ordnance installation is completed. After the test, vertical alignment is checked, a data review is held, and the vehicle is prepared for transfer to the pad. These preparations include disconnecting pneumatic, hydraulic, and electrical lines from the mobile launcher to the VAB. After the lines are disconnected, the transporter is moved into position beneath the mobile launcher. Hydraulic jacks engage the fittings on the mobile launcher and raise it approximately 3 feet so that it clears its mount mechanisms. Then the transporter moves out of the VAB, over the crawlerway, to the launch pad.

TESTING AT LAUNCH SITE

At the launch pad, the transporter moves the mobile launcher into position, lowers and locks it onto another set of mount mechanisms. The transporter then moves to the mobile service structure parking area, picks up the service tower, and positions it beside the Saturn V to provide vehicle access for pad operations. The digital data link, communications circuitry, pneumatic supply lines, propellant lines, environmental controls, and electrical power supply lines are connected. Power again is applied to the vehicle and the control and monitor links are verified. Pad testing is held to a minimum. The high bay from which the vehicle was moved remains empty during pad operations. A spacecraft systems verification test is performed, followed by a space vehicle cutoff and malfunction test. Radio frequency compatibility is established and preparations are made for a final flight readiness test, which involves sequence tests paralleling the actual countdown and inflight operations. Compatibility with the stations of the Eastern Test Range and the Integrated Mission Control Center in Houston, Tex., are verified at this time. Following an evaluation of the flight readiness test, all systems are reconfigured for launch, and all plugs reverified. A countdown-demonstration test is then performed as the final test prior to launch. The countdown-demonstration test consists of an actual launch countdown, complete with propellant loading, astronaut embarkation, etc., with the exception of actual ignition. This test exercises all systems, the launch crew, and the astronauts, and prepares the "team" for the actual operation to follow. This "dress rehearsal" is used to divulge any last minute problems and affords the mission a better chance of success. Upon completion of the countdown demonstration test, the space vehicle is recycled to pre-count status, and preparations are made for the final countdown phase of launch operations. Normal recycle time between completion of the countdown demonstration test and beginning of launch countdown is 48 to 72 hours. Propellant loading of the Apollo spacecraft is performed prior to launch day. Aerozine 50 is the fuel and nitrogen tetroxide, the oxidizer. Also prior to launch day, hypergolics for the third stage reaction control system are loaded and ordnance connected. Loading of the cryogenic propellants for the launch vehicle begins on launch day at approximately T-7 hours. (The kerosene is loaded one day before launch.) Liquid oxygen loading is begun first. The tanks are precooled before filling. Precool of one tank can be accomplished concurrently with the fill of another. Loading is started with the second stage to 40 per cent, followed by the third stage to 100 per cent. The second stage is then brought to a full 100 per cent followed by loading the first stage to 100 per cent. This procedure allows time for the liquid oxygen leak checks to be performed prior to full loading of the second stage. Liquid oxygen is pumped at a flowrate of 1,000 gallons per minute for the third stage. For the second stage, the tank rate is 5,000 gallons per minute, and the first stage tank flowrate is 10,000 gallons per minute. Liquid hydrogen loading is initiated next, beginning with the second stage to 100 per cent. Loading of the third stage liquid hydrogen is last. Liquid hydrogen is pumped to the second stage at a rate of 10,000 gallons per minute, and to the third stage at a rate of 3,000 gallons per minute. Topping of cryogenic tanks of the launch vehicle continues until launch. Total cryogenic loading time from start to finish is 4 hours and 30 minutes. At approximately T-90 minutes, after propellants are loaded, the astronauts enter the spacecraft from the mobile launcher over the swing arm walkway.

LAUNCH

During the remainder of the countdown, the final systems checks are conducted. Launch vehicle propellant tanks are then pressurized, and the first stage engines ignited. During the thrust buildup of the F-1 engines, the operation of each of these engines will be automatically check- ed. Upon confirmation of thrust OK condition, the launch commit signal is given to the holddown arms and liftoff occurs.

Assembled Vehicle-The Saturn V facilities vehicle, the 500F, arrives at Launch Complex 39A.

PROGRAM MANAGEMENT

NASA ORGANIZATION

The Saturn V development program comes under the direction of the NASA Office of Manned Space Flight, Washington, D.C. That office assigned development responsibility to the Marshall Space Flight Center, one of the three Manned Space Flight field centers. Another of those field centers, the Kennedy Space Center, has been delegated the responsibility of launching the Saturn V. (Development of the Apollo spacecraft, the first "payload" for the Saturn V, was assigned to the Manned Spacecraft Center, the other MSF field organization.)

Marshall Center Project Management Organization

More than 125,000 prime and subcontractor employees and 7,500 civil service employees are working on the Saturn program. Saturn industrial activities are scattered nationwide but there are three major areas of concentration:

1. the Northeast, with its grouping of electronic industries.
2. the Southeast, for production, test, and launch operations.
3. the West Coast, with its concentration of aerospace industries for design, production, and test work.

In addition, various research' projects by scientific institutions and subcontractor production efforts contributing to the Saturn program are spread throughout the nation.

The wide dispersion makes necessary very comprehensive and reliable management systems and control techniques to manage the program effectively. The geographic dispersion of the Saturn effort requires excellent communications. The Marshall Center must be aware of related programs carried out by other NASA centers-especially the Manned Spacecraft Center, managing the Apollo spacecraft program, and Kennedy Space Center, responsible for Saturn/Apollo launches. The Marshall Center has found that one of the more effective tools for total program visibility is especially constructed and outfitted rooms called Pro- gram Control Centers. The Saturn V launch vehicle program office and other major groups have such centers. The budget for the current fiscal year at the MSFC is about $1.7 billion. The center must have a well staffed organization responsive to the many changes which can take place in a program of this magnitude.

One of the Marshall Center's two major divisions - Industrial Operations - is responsible for the management of the Saturn launch vehicle development programs for NASA manned space flight. Dr. Arthur Rudolph, the Saturn V program manager in Industrial Operations, controls the project effort, plans, and budgets. For technical solutions to vehicle problems, the manager gets assistance from the laboratories of the Research and Development Operations - the second major and largest MSFC division reporting directly to the center director. Because of the many interfaces between the stages and with ground support equipment, program management responsibility in Industrial Operations includes establishing specifications and procedures which assure physical and functional compatibility. Formerly, the Marshall Center did the overall design of stages and major systems in-house, but more recently, particularly with subsystems and components, the program managers have concentrated on performance specifications and left the details to the contractors. This management function keeps the program engineers very much in the mainstream of technical design activity. Thus, Industrial Operations program managers are quite active in the areas of test requirements, qualification testing, product control, systems engineering, program control, and flight operations.

Marshall Center's Research and Development Operations laboratories are oriented functionally in such primary disciplines as mechanical engineering, electronics, and flight mechanics. Collectively, the laboratories provide the deep-rooted technological foundation on which the success of all Marshall projects depends. In the project offices, technical decisions are made which affect many areas. These decisions are formulated by drawing upon the full technical resources of the laboratories, which maintain a high level or professional competence. Laboratory personnel work on selected projects to keep their technical knowledge updated and their technical competence at a high pitch. This is the Marshall work bench philosophy—the "dirty hands" approach.

The Saturn V program office is headed by a program manager. There is a stage manager or project director for each of five major vehicle systems. A stage manager primarily deals with only one major contractor. In the case of the instrument unit and the ground support equipment project, there are several major contractors. The principle of a single project management focal point is the objective of each project team. Program management is vested in the program manager. Technical project management, so far as NASA is concerned, occurs at the stage or project level. The program and stage managers are fully responsible for technical adequacy, reliable performance, and for management of all related con- tractor activity. These program and project managers must be backed up and supported by technical competence in depth. This in-depth support is provided, to a degree, by a staff of competent technical and business management people in the program manager and stage manager office, and to a much larger degree, by Research and Development Operations. There is a resident manager at each of the contractor plants to act as the "official" voice for the Marshall Center. All MSFC instructions to the contractor are transmitted through the resident manager. Through the resident manager, MSFC maintains a direct contact with contractor operations and is kept informed of the status of all significant program events.

Marshall Center laboratory technical personnel are assigned to the resident managers' staffs. These technical people are assigned to each resident manager's office to provide him with assistance in resolving technical problems, and to keep the MSFC technical laboratories directly informed of field technical effort. Laboratory participation is dictated by need as determined by project management.

Many people are involved in attaining the final goal. Project management, technical, and contractor personnel are tied in a close knit group capable of managing this country's large launch vehicle program.

MANAGEMENT PERSONNEL

NASA

Dr. George E. Mueller, Associate Administrator for Manned Space Flight, NASA Headquarters.

Major Gen. Samuel C. Phillips, Director, Apollo Program, NASA Headquarters.

Dr. Wernher von Braun, Director, Marshall Space Flight Center.

Dr. Kurt H. Debus, Director of John F. Kennedy Space Center.

Dr. Arthur Rudolph, Manager, Saturn V Program Office, Marshall Space Flight Center.

Rocco A. Petrone, Director of Launch Operations, Kennedy Space Center.

MANAGEMENT PERSONNEL
BOEING

L. A. Wood, Group Vice President, Aerospace.

R. H. Nelson, General Manager, Launch Systems Branch.

G. H. Stoner, Vice President and General Manager, Space Division.

A. M. 'Tex' Johnston, Director, Boeing Atlantic Test Center.

MANAGEMENT PERSONNEL
DOUGLAS

Charles R. Able, Group Vice President, Missile and Space Systems.

Jack L. Bromberg, Vice President and Deputy General Manager, Missile and Space Systems Division.

Theodore D. Smith, Senior Director, Saturn/Apollo Programs, Assistant General Manager, Missile and Space Systems Division.

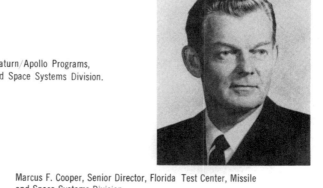

Marcus F. Cooper, Senior Director, Florida Test Center, Missile and Space Systems Division.

A. P. O'Neal, Director, Saturn Development Engineering.

MANAGEMENT PERSONNEL

IBM

McLain B. Smith, Vice President and Group Executive.

Bob O. Evans, President of the Federal Systems Division.

Arthur E. Cooper, Federal Systems Division Vice President and General Manager, Space Systems Center, Bethesda, Md.

Clinton H. Grace, Facility Manager, Space Systems Center, Huntsville, Ala.

Ammon G. Belleman, Facility Manager, Space Systems Center, Kennedy Space Center.

MANAGEMENT PERSONNEL

SPACE DIVISION — NORTH AMERICAN

William B. Bergen, Vice President, North American Aviation, Inc.; President, Space Division. Downey, Calif.

William F. Parker, Deputy Program Manager, Saturn Second Stage.

Robert E. Greer, Vice President, Space Division; Program Manager, Saturn Second Stage.

Bastian 'Buz' Hello, Vice President and General Manager, Launch Operations, Space Division, Florida.

Samuel K. Hoffman, President of Rocketdyne and Vice President of North American Aviation, Inc.

MANAGEMENT PERSONNEL
ROCKETDYNE—NORTH AMERICAN

Joseph P. McNamara, Vice President and General Manager, Liquid Rocket Division.

Paul D. Castenholz, Program Manager, J-2 Engine.

Norman C. Reuel, Assistant General Manager, Liquid Rocket Division.

David E. Aldrich, Program Manager, F-1 Engine.

FLIGHT HISTORY

AS-501 (APOLLO 4)

The first Apollo/Saturn V launch vehicle, the AS-501, performed all of its vehicle mission objectives. The vehicle was launched at 7 a.m. EST on November 9, 1967 from Launch Complex 39 at the NASA- Kennedy Space Center, Fla. The countdown proceeded smoothly and the launch came exactly on time. All vehicle systems and subsystems performed "nominally" and ground support equipment performance was satisfactory. Prime mission objectives, with respect to the rocket, included an all-up test of the vehicle with its three stages and instrument unit, the first in-orbit restart of the third (S-IVB) stage, and the first use of Launch Complex 39 and ground support equipment. Flight of the three stages was near nominal. The trajectory was near the expected and all three propulsion systems performed with no apparent anomalies. The instrument unit systems were all stable during the flight. First (S-IC) stage flight was near the expected. S-IC center F-1 engine cutoff was given by a timer at 135.5 seconds. S-IC outboard engine cutoff came by liquid oxygen depletion at 150.8 seconds with the vehicle at 38.3 miles altitude traveling at 6,024.6 miles per hour. Booster and second (S-II) stage first and second plane separations each occurred within 1.2 seconds of the predicted times. Cameras on the S-II photographed a smooth separation. Propulsion and other systems, including propellant utilization, the pressurization, and the pneumatic control pressure system, operated within expected tolerances. The S-II engines, stage propellant utilization system, pressurization system, pneumatic control pressure system, camera ejection system, and the helium injection system operated properly and within expected tolerances. All five J-2 engines operated properly during engine start and burn. Ground controllers noted that the thrust chamber jacket temperature heat-up rate was slightly higher than predicted, and the engine start bottle pressures were slightly higher than predicted but both were within required limits. S-II stage cutoff came at 519.8 seconds, 3.5 seconds later than predicted. The S-II stage's liquid hydrogen tank insulation performed satisfactorily with no defects noted during countdown or in flight.

Apollo 4 (AS-501) Launch, November 9, 1967 B-P42396

The third (S-IVB) stage first and second burns were 6.2 seconds longer and 15.2 seconds shorter, respectively, than predicted. The first burn began at 520.7 seconds. The J-2 engine was cut off by the guidance system at 665.6 seconds. This was 9.6 seconds later than expected. The vehicle was traveling 17,428.2 miles per hour and was at an altitude of 118.6 miles. The S-IVB was reignited over the eastern United States after two revolutions in earth orbit. The second burn operation was cut off by guidance and was 15.2 seconds shorter than predicted, which was attributed primarily to 37 seconds of burn time at the high thrust level operation of the J-2 engine during second burn. A low liquid hydrogen ullage pressure reading was recorded at the Kennedy Space Center immediately before J-2 engine restart. The reading was 28 pounds and the expected minimum pressure was 31 pounds. This had no effect on engine operation. The pressure in the helium repressurization spheres was apparently lower than expected during S-IVB restart preparations but reignition was achieved without difficulty. Hydraulic systems on all three stages performed without evidence of out-of-tolerance conditions. Maximum engine deflection was 0.6 degree on the S-IC and 0.8 degree on the

S-II. Structurally, the Saturn V vehicle performed with no problems. Maximum bending occurred between 70 and 80 seconds. Longitudinal loads were near nominal throughout flight, and longitudinal acceleration at S-IC center engine cutoff was 4.15 G, which was very near the expected value. The instrument unit on this flight was the first to be flight tested since an external structural stiffener was added to reduce vibration effects on the inertial platform. Vibrations in the area were lower than those on previous flights of the Saturn IB. The instrument units on the Saturn IB and Saturn V are essentially the same. Telemetry taken during the first 560 seconds of powered flight showed guidance and control to be nominal. The emergency detection system was flown "open loop" on this flight. All indications were that the system operated satisfactorily. The EDS was developed for the manned Apollo flights so that astronauts and ground controllers could know of impending troubles in the rocket in time to take corrective action. Apollo 4 experienced only a few measurement failures. Two known measurement failures and 40 questionable measurements were identified out of the approximately 2,862 taken on the flight. This is a loss of less than two per cent. Both onboard cameras viewing first and second stage separation recorded excellent quality pictures. The cameras were recovered shortly after being ejected into the Atlantic Ocean.

AS-502 (APOLLO 6)

The second Saturn V launch vehicle, AS-502, was not totally successful although it achieved most of its objectives and placed more than 264,000 pounds into earth orbit. The vehicle was launched from Complex 39 at the NASA-Kennedy Space Center on April 4, 1968. The launch occurred on schedule at 7 a.m. EST after a smooth countdown. The first (S-IC) stage performed as planned and hydraulic system performance was satisfactory. Stage thrust was essentially the same as predicted during the first portion of the flight. However, a longitudinal oscillation ("Pogo" effect), measured at five cycles per second, was experienced during the latter portion of first stage burn. The phenomenon was also noted on the first Saturn V flight, AS-501, but on AS-502 it was much greater. Second stage engines numbered 2 and 3 cut off prematurely at 408.7 and 410 seconds after liftoff, respectively, causing a 58-second longer than normal second stage burn and larger than expected deviations from second stage flight end conditions. S-II performance was satisfactory through first stage boost, S-II ignition and the early portion of S-II powered flight. The earliest observed deviations were decreasing temperatures on the main oxidizer valve and its control line on engine number 5 and a steady increase in engine number 2 yaw actuator pressure, occurring at 278.4 seconds. A sudden 5,000 pound thrust decrease and other deviations at 318 seconds preceded a cutoff signal to engine number 2. That cutoff signal also caused engine number 3 to shut down, because the wires carrying cutoff commands to engines numbered 2 and 3 were interchanged. Hydraulic system performance was satisfactory on the second stage until about 140 seconds before premature shutdown of the two engines. At this time the increase in the yaw and pitch actuator differential pressures occurred. First burn of the third (S-IVB) stage was 29.2 seconds longer than planned to compensate for the early cutoff of the two second stage engines. The result was a high cutoff velocity and an elliptical parking orbit. The attainment of this orbit was a demonstration of the unusual flexibility designed into the Saturn V.

Apollo 6 (AS-502) Launch, April 4, 1968

All engine and stage restart conditions appeared normal but the S-IVB's J-2 engine did not restart in orbit. The restart was to have propelled the S-IVB and Apollo spacecraft into a simulated translunar trajectory. The third stage performed satisfactorily through first burn and orbital coast. Shortly after orbit insertion a cold helium supply leak was observed but bottle pressure was sufficient to meet second burn requirements. Even though normal engine and stage prestart conditions were observed, the engine received the start signal and the engine valves opened properly, the engine did not reignite. Study of data relating to the S-IVB reignition problem indicated a leak in one of the two propellant lines leading to the J-2 engine's augmented spark igniter (ASI). In such a case, propellants reaching the spark plugs were insufficient, or inadequate in mixture, to achieve the proper start conditions. Third stage hydraulic system performance was normal through first burn. Shortly before space- craft separation, a programmed command to initiate the auxiliary hydraulic pump was given but the pump failed to operate. Ground commands after spacecraft separation also failed to start the system. Pump operation was not a requirement for engine restart. Guidance and other instrument unit functions were satisfactory. Flight profile was nominal up to the loss of engine number 2 on the second stage.

At second stage cutoff the altitude was high and velocity low. This led to a longer burn of the third stage and a velocity slightly higher than normal, causing the third stage and spacecraft to go into an elliptical orbit. Prior to launch 29 measurements were waived. During flight there were nine known failures and 19 questionable measurements of the approximately 2,800 measurements planned originally. Telemetry performance was good on all links. Onboard television cameras gave good data. Only two of the six on-board film cameras were recovered. Both these cameras viewed the separation of the first and second stages. A study of data relating to the failure of the number 2 J-2 engine on the second stage and the single J-2 on the third stage indicated that in each case a propellant line leading to the engine's augmented spark igniter (ASI) ruptured. Those lines have been redesigned to remove the flexible sections where the breaks occurred. The new lines have been tested and proven adequate with a sufficient safety margin. All J-2's in the future will use the new lines. The oscillations in the first stage also prompted an extensive investigation which led to the decision to create "shock absorbers" in the large liquid oxygen (LOX) lines leading to four of the five F-1 engines. This was done by injecting helium into cavities in the existing LOX prevalves to damp out LOX surges.

APPENDIX - GLOSSARY

The following list defines acronyms, abbreviations, nomenclature, and other terminology used in the Saturn V News Reference.

TERM	Description
APS	Auxiliary propulsion system
Bulkhead	A dome-shaped segment which encloses the end of a propellant tank
Burnout	Point at which engines shut down due to lack of fuel or oxidant
Burst Diaphragm	A disc designed to rupture at a predetermined pressure differential
Bus	A main circuit for transfer of electrical current
Cavitation	The formation of bubbles in a liquid, occurring whenever the static pressure at any point in the fluid flow becomes less than the fluid vapor pressure
Convection	Mass motions within a fluid
Cryogenic	Ultra-low temperature
DDAS	Digital data acquisition system
Exhaust Nozzle	The lower section of the thrust chamber of a liquid rocket engine
Expansion Area Ratio	The ratio of the measurements of an engine nozzle exit section to that of the nozzle throat area
Exploding Bridgewire	Wire which explodes when subjected to a high voltage, high energy pulse
Fusion Weld	To join two pieces of metal together by bringing the surfaces to a molten state by electric arc or gas flame controlled to produce a localized union through fusion or recrystallization across the interface
Gimbal	A device on which a reaction engine may be mounted and which allows for angular movement in two directions
GOX	Gaseous oxygen
GSE	Ground support equipment
Hydrostatic Test	Use of water for pressure test of propellant containers
Hypergolic Liquids	Liquids that ignite spontaneously when mixed with each other
Impeller	A device that imparts motion to a fluid or air
Inducer	A pump which increases the pressure and motion of a fluid
KSC	Kennedy Space Center
LH2	Liquid hydrogen
LOX	Liquid oxygen
LVDA	Launch vehicle data adapter
LVDC	Launch vehicle digital computer
Monocoque	A structure in which all or most of the stresses are carried by the skin
MSC	Manned Spacecraft Center
MSFC	Marshall Space Flight Center
Multiplexer	A mechanical or electrical device for time sharing of a circuit
NASA	National Aeronautics and Space Administration
ODOP	Offset Doppler System
Pitch	Movement of the vehicle from its lateral axis
PSI	Pounds per square inch
PSIA	Pounds per square inch absolute
PSIG	Pounds per square inch gage
Purge	To remove residual fluid or gas
Retrorocket	A rocket fitted to a stage to produce thrust opposed to the stages forward motion
RF	Radio frequency
RJ-I	A grade of kerosene which is used in the hydraulic system prior to lift-off
Roll	The rotation of a vehicle about its axis
RP-I	A rocket fuel consisting essentially of kerosene

Squib	An explosive device used in the ignition of a rocket engine. Usually called an igniter
Stator	A mechanical part that remains stationary with respect to a rotating or moving part or assembly
Thermocouple	A device which converts thermal energy directly into electrical energy
Thrust	The force developed by a rocket engine
Thrust Vectoring	An attitude control for rockets wherein one or more engines are gimbal-mounted so that the direction of the thrust force may be changed in relation to the center of gravity of the vehicle to produce a turning movement
Torus	A circular duct (manifold) used to collect fluid or gases
Ullage	The amount that a container, such as a fuel tank, lacks of being full
Umbilical	Any of the servicing lines between the ground or tower and a launch vehicle
Volute	A flow passage that collects and redirects fluids
Yaw	Movement of a vehicle from its longitudinal axis

APPENDIX A-SATURN V SUBCONTRACTORS

The following are lists of subcontractors who have played a major role in the development and production of the Saturn V launch vehicle. It should be recognized that many more subcontractors contributed to the total vehicle and program; however, it is not practical to list all in this document.

BOEING MAJOR SUBCONTRACTORS

SUBCONTRACTOR	LOCATION	PRODUCT
Aeroquip Corp.	Jackson, Mich.	Couplings, pneumatic, and hydraulic hoses
Aircraft Products	Dallas, Tex.	Machined parts
AiResearch Manufacturing Co.	Phoenix, Ariz.	Valves
Applied Dynamics, Inc.	Ann Arbor, Mich	Analog computers
Arrowhead Products, Div. of Federal-Mogul Corp.	Los Alamitos, Calif.	Ducts
The Bendix Corp., Pioneer-Central Div.	Davenport, Iowa	loading systems and cutoff sensors
Bourns, Inc. Instrument Div.	Riverside, Calif.	Pressure transducers
Brown Engineering Co., Inc.	Lewisburg, Tenn.	Multiplexer equipment
The J. C. Carter Co.	Costa Mesa, Calif.	Solenoid valves
Consolidated Controls Corp.	Bethel, Conn. Los Angeles, Calif.	Pressure switches, transducers, and valves
The Eagle-Picher Co., Chemical & Metals Div.	Joplin, Mo.	Batteries
Electro Development Corp.	Seattle, Wash.	AC and DC amplifiers
Flexible Tubing Corp.	Anaheim, Calif.	Ducts
Flexonics, Div. of Calumet and Hecla, Inc.	Bartlett, Ill.	Ducts
General Precision, Inc. Link Ordnance Div.	Sunnyvale, Calif	Propellant dispersion systems
Gulton Industries, C. G. Electronics Div.	Albuquerque, N. M.	Wiring boards
Hayes International Corp.	Birmingham, Ala.	Auxiliary nitrogen supply units
Hydraulic Research & Manufacturing Co.	Burbank, Calif.	Servoactuators and filter manifolds
Johns-Manville Sales Corp.	Manville, N.J.	Insulation
Kinetics Corporation of California	Solano Beach, Calif.	Power transfer switches
Ling-Temco-Vought, Inc.	Dallas. Tex.	Skins, emergency drains, and heat shield curtains
Marotta Valve Corp.	Boonton, N. J.	Valves
Martin Marietta Corp.	Baltimore, Md.	Helium bottles
Moog, Inc.	East Aurora, N. Y.	Servoactuators
Navan Products, Inc.	El Segundo, Calif.	Seals
Parker Aircraft Co.	Los Angeles, Calif.	Valves
Parker Seal Co.	Culver City, Calif.	Seals
Parsons Corp.	Traverse City, Mich.	Tunnel assemblies
Precision Sheet Metal, Inc.,	Los Angeles, Calif.	Filter screens, anti-vortex, and adapter assemblies
Purolator Products, Inc., Western Div.	Newbury Park, Calif.	Umbilical couplings

Subcontractor	Location	Product
Kandall Engineering Co.	Los Angeles, Calif.	Valves
Raytheon Co.	Waltham, Mass.	Cathode ray tube display system
Rohr Corp.	Chula Vista, Calif.	Heat shields
Servonic Instruments, Inc.	Costa Mesa, Calif.	Pressure transducers
Solar, Division of International Harvester	San Diego, Calif.	Ducts
Southwestern Industries, Inc.	Los Angeles, Calif.	Calips pressure switches
Space Craft, Inc.	Huntsville, Ala.	Converters
Stainless Steel Products, Inc.	Burbank, Calif.	Ducts
Standard Controls, Inc.	Seattle, Wash.	Pressure transducers
Statham Instruments, Inc.	Los Angeles, Calif.	Pressure transducers
Sterer Engineering and Manufacturing Co.	Los Angeles, Calif.	Valves
Stresskin Products Co.	Costa Mesa, Calif.	Insulation
Systron-Donner Corp.	Concord, Calif.	Servo accelerometers
Thiokol Chemical Corp., Elkton Div.	Elkton, Md.	Retrorockets
Trans-Sonics. Inc.	Burlington, Mass.	Measuring systems and thermometers
Unidynamics/St. Louis, A Division of UMC Industries, Inc.	St. Louis, Mo.	Spools, harnesses and ducts
United Control Corp.	Redmond, Wash.	Ordnance devices and control assemblies
Vacco Industries	South El Monte, Calif.	Filters, relief valves, and regulators
Whittaker Corp.	Chatsworth, Calif.	Valves and gyros
Fred D. Wright Co., Inc.	Nashville, Tenn.	Support assemblies and measuring racks

DOUGLAS MAJOR SUBCONTRACTORS

SUBCONTRACTOR	LOCATION	PRODUCT
Accessory Products Co., Div. of Textron, Inc.	Whittier, Calif.	Valves heaters
Aeroquip Corp., Marman Div.	Los Angeles, Calif.	Clamps
Airdrome Parts Co.	Inglewood, Calif.	Fittings
AiResearch Mfg. of Ariz.	Phoenix, Ariz.	Valves
Airtex Dynamics, Inc.	Compton, Calif.	Tank assemblies
Amco Engineering Co.	Chicago, Ill.	Cabinets
American Electronics, Inc.	Fullerton, Calif.	Batteries
Amp, Inc.	Harrisburg, Pa.	Electrical panels
Ampex, Corp.	Los Angeles, Calif.	Tape recorders
Arnphenol Borg Electronics Corp.	Chicago, Ill.	Connectors
Anaconda Metal Hose Div.	Waterbury, Conn.	Metal hose
Astrodata, Inc.	Anaheim, Calif.	Telemetry
Avnet Corp.	Westbury, L.I., N.Y	Electrical connectors
Barry Controls	Watertown, Mass.	Electronic controls
Bertea Products	Pasadena, Calif.	Transmitter, Fabricated assemblies
Brown Engineering Co., Inc.	Huntsville, Ala.	Telemetry equipment
Calmec Mfg. Co.	Los Angeles, Calif.	Valves
Capital Westward, Inc.	Paramount, Calif.	Filters
Christie Electric Corp.	Los Angeles, Calif.	Meters
Consolidated Electrodynamics	New York, N.Y.	Electronic equipment
Data Sensors	Gardena, Calif.	Transducers
Deutsch	Los Angeles, Calif.	Fittings
Dynatronics, Inc.	Orlando, Fla.	Telemetry equipment
Eagle-Picher Co.	Joplin, Mo.	Batteries
Electra Scientific Corp.	Fullerton, Calif.	Transducers
Electrada Corp.	Culver City, Calif.	Electrical components
Fairchild Camera Inst. Corp.	Plainview, L.I., N.Y.	Cameras, oscilloscopes
Fairchild Controls Div.	Montebello, Calif.	Valves
Fairchild Hiller Corp.	Bayshore, L.I., N.Y.	Valves
Fairchild Semiconductor	Hollywood, Calif.	Semiconductors
Fairchild Stratos	Bayshore, L.I., N.Y.	Valves
Flexible Metal Hose Mfg. Co.	Costa Mesa, Calif. Northridge, Calif.	Metal hose
Flomatics, Inc.	Natoma, Calif.	Valves
Frebank Co.	Glendale, Calif.	Switches
Control Data Corp.	Minneapolis, Minn.	Computers
General Electric Co.	Waterford, N.Y.	Electrical components
Giannini Controls Corp.	Durate Pasadena, Calif.	Transducers
Grove Valve Regulator Co.	El Segundo Oakland, Calif.	Valves

Hadley, E. H. Co.	Pomona, Calif.	Relays
Hewlett Packard Co.	Pasadena, Calif.	Oscilloscopes recorder
Hexcel Products, Inc.	Berkeley, Calif.	Honeycomb panels
Honeywell, Inc.	Minneapolis, Minn.	Temperature controls, gyros, sensors
ITT Wire and Cable	Clinton, Mass.	Wire and cable
K-Tronics	Los Angeles, Calif.	Semiconductors
Kaiser Aluminum Chemical Sales, Inc.	Spokane, Wash. Halelhrope, Md.	Raw material, cable
Kinetics, Corp.	Solano Beach, Calif.	Electronic equipment
Ladewig Valve	Los Angeles, Calif.	Valves
Lanagan, W. M. Co., Inc.	Costa Mesa, Calif.	Valves
Leonard, Wallace O., Inc.	Pasadena, Calif.	Valves
Linde Co., Div. Union Carbide	El Monte, Calif.	Batteries
Litton Industries of Calif.	Beverly Hills, Calif.	Transducers
Magnesium Alloy Prods. Co.	Compton, Calif.	Castings
Magnetika, Inc.	Venice, Calif.	Batteries
Marotta Valve Corp.	Santa Ana, Calif.	Valves
Marshall, G. S. Co. San Marino, Calif.	Electronic hardware	
Mason Electric Div. Ansul.	Los Angeles, Calif.	Switches
Master Specialties Co.	Los Angeles, Calif.	Meters
Menasco Mfg. Co. Burbank, Calif.	Fabricated assemblies	
Military Products Div., Clary	San Gabriel, Calif.	Elec. fittings
Moog Servo Controls, Inc.	Aurora, N.Y.	Valves
Motorola Semiconductor Products	Hollywood, Calif.	Semiconductors
Pacific Valve, Inc.	Long Beach, Calif.	Valves
Parker Aircraft Co.	Los Angeles, Calif.	Fittings/valves
Pesco Products Div., Borg Warner	Bedford, Ohio	Pumps
Philco Corp.	Philadelphia, Pa.	Radio equipment
Planautics Corp.	Solano Beach, Calif.	Switches
Potter Brumfield	Princeton, Ind.	Switches, relays
Purolator Products, Inc., Western Div.	Van Nuys, Calif.	Filters
Reynolds Metals Co.	Birmingham, Ala.	Raw material (alum.)
Rosemount Eng. Co.	Minneapolis, Minn.	Temperature controls
Sandorn Co.	Waltham, Mass.	Recorders
Scintilla Div. Bendix	Sidney, N.Y.	Electrical connectors
Sealol Corp.	Providence, R.I.	Valves
Servonic Instruments, Inc.	Costa Mesa, Calif.	Transducers
Signet Scientific	Burbank, Calif.	Ovens
Snap Tite	Union City, Pa.	Connectors
Sperry Gyro Co. Div., Sperry Rand	Great Neck, L.I.,N.Y.	Instruments
Stainless Steel Products	Burbank, Calif.	Flexible ducts
Statham Instruments, Inc.	Los Angeles, Calif.	Transducers
TRW, Inc.	Cleveland, Ohio	Attitude control rocket engines
Trans-Sonics, Inc.	Lexington, Mass.	Temperature controls
Technology Instruments Corp.	Newbury Park, Calif.	Potentiometers
Telemetrics, Inc. Subsy of Arnoux Corp.	Santa Ana, Calif.	Ground support electronics
Texas Instruments, Inc.	Dallas, Tex.	Resistors/transistors
U.S. Steel Corp. U.S. Steel, Supply Div.	Seattle, Wash.	Raw material
Vacco Valve Company	El Monte, Calif.	Valves
Vickers	Detroit, Mich.	Pumps
Vinson Mfg. Co., Inc.	Van Nuys, Calif.	Valves
W & S Industries	El Monte, Calif.	Connectors
Winsco Instruments Control	Santa Monica, Calif.	Temp. control units
Wyman Gordon Co.	Worcester, Mass.	Forgings

IBM MAJOR SUBCONTRACTORS

SUBCONTRACTOR	LOCATION	PRODUCT
Aerodyne Controls Corp.	Farmingdale, N. Y.	First stage regulator, Gas bearing pressure regulator
Airtek Div., Fansteel Metallurgical Corp.	Compton, Calif.	2-cubic-foot bottle 165-cubic-inch sphere
Applied Microwave Laboratory, Inc.	Andover, Mass.	T M power divider
Astro Space Laboratories, Inc.	Huntsville, Ala.	Bleeder assembly, Union orifice assembly
Automatic Metal Products Corp.	Brooklyn, N. Y.	Coaxial switch
AVCO Corp.	Nashville, Tenn.	Cable trays

	Huntsville, Ala.	Gas bearing mounting panel
		Measuring range cards
		Measuring range cards software
		Thermal conditioning panels
Avion Electronics, Inc.	Paramus, N. J.	Telemetry directional coupler
		VSWR measuring assembly
Bourns, Inc.	Riverside, Calif.	Transducer pressure gauges
Brown Engineering Co., Inc.	Huntsville, Ala.	Acquisition system
		Alternating current amplifier assembly
		Measuring rack selector
		Measuring rack assembly
		Mod 145 multiplexers
		Mod 245 multiplexers
		Mod 270 multiplexers
		Pulse code modulated
		digital data
		Radio frequency analog
		Radio frequency pulse code modulated
		Telemetry assembly A3
		Telemetry assembly B1
		Telemetry assembly SSB
		Telemetry calibrators
		Telemetry pulse code
Chrysler Corp.	Huntsville, Ala.	Q-ball, Vehicle plate assembly
Conic Corp.	San Diego, Calif.	UHF transmitter
Cox Instruments Div., Lynch Corp.	Detroit, Mich.	Flowmeters
Eagle-Picher Industries, Inc.	Joplin, Mo.	Primary battery
Electro Development Corp.	Seattle, Wash.	Amplifier direct current
		Channel selector A
		Channel selector B
Electronic Communications, Inc.	St. Petersburg, Fla.	Control computer
EIMAC Div., Varian	San Carlos, Calif.	UHF transmitter
The Foxboro Co.	Foxboro, Mass.	Gas temperature probe
Fenwall Electronics, Inc.	Framingham, Mass.	Temperature gauges
Flodyne Controls, Inc.	Linden, N. J.	Shut-off ball valve
Gulton Industries, Inc.	Hawthorne, Calif.	Electrical cables
		Vibration accelerometers
		5-volt power supply
Hamilton-Standard Div., United Aircraft Corp.	Windsor Locks, Conn.	Preflight heat exchanger
Hayes International Corp.	Huntsville, Ala.	Network cables
Hydro-Aire Div., Crane	Burbank, Calif.	Coolant pump, R&D
ITT Wire and Cable Div., International Telephone and Telegraph Corp.	Clinton, Mass.	Wire
		Cables
		Coaxial cables
Marotta Valve Corp.	Boonton, N. J.	Shut-off valve
		Solenoid valve
Martin Co.	Orlando, Fla.	Control signal processor
Melpar, Inc.	Falls Church, Va.	C-band antenna
		Command antenna
		Command directional coupler
		Command power divider
		Telemetry antenna
		Telemetry power divider
Motorola, Inc.	Scottsdale, Ariz.	C-band transponder
		Command receiver
North American Aviation, Inc.	Tulsa, Okla.	Structure segments
Nortronics Div., Northrop Corp.	Norwood, Mass.	Rate gyro package
Ralph M. Parsons Electronics Corp.	Pasadena, Calif.	Tape recorder
Perkin-Elmer Corp.	Norwalk, Conn.	Retroreflector
Potter Aeronautical Corp.	Union, N. J.	2 flowmeter
Purolator Products, Inc.	Los Angeles, Calif.	Gas bearing pressure regulator
		Quick disconnect coupling
Rantec Corp.	Calabasas, Calif.	Telemetry RF coupler
Raytheon Co.	Bristol, Tenn.	AZUSA antenna
Resistoflex Corp.	Roseland, N. J.	Flex hose assembly

Subcontractor	Location	Product
Rosemount Engineering Co.	Minneapolis, Minn.	2 temperature gauge
Servonic Instruments, Inc.	Costa Mesa, Calif.	2 transducer pressure
Sierra Electronics Div., Philco-Ford Corp.	Menlo Park, Calif.	Coaxial terminal
Solar Div., International Harvester Co.	San Diego, Calif.	Gas bearing heat exchanger
		Manifold assembly
		Water methanol accumulator
Space Craft, Inc.	Huntsville, Ala.	C1U 501
		Command decode
		Frequency DC converter
		2 multiplexers 410
		Servoaccelerometer unit
Spaco, Inc.	Huntsville, Ala.	2 Servoaccelerometer unit
Statham Instruments, Inc.	Los Angeles, Calif.	Control accelerometer
Systron - Donner Corp.	Concord, Calif.	2 Force balance accelerometers
Tavco, Inc.	Santa Monica, Calif.	Pressure switch
Teledyne Precision, Inc.	Hawthorne, Calif.	Thermal probes
Transco Products, Inc.	Venice, Calif.	CCS coaxial switch
TRW, Inc.	Cleveland, OH.	Coolant pump
United Control Corp.	Redmond, Wash.	Temperature control assembly
Vacco Industries	South El Monte, Calif.	Filter
		Quality test filter
Watkins-Johnson Co.	Palo Alto, Calif.	Power amplifier
Wyle Laboratories	Huntsville, Ala.	Component testing
	Structure testing	

NORTH AMERICAN SPACE DIVISION MAJOR SUBCONTRACTORS

SUBCONTRACTOR	LOCATION	PRODUCT
Acoustica Associates	Los Angeles, Calif	Transducers
American Brake Shoe Co.	Oxnard, Calif.	Hydraulic pumps
Amp, inc.	Hawthorne, Calif.	Patch panels
Amphenol Space & Missile Systems Div., Amphenol-Borg Electronics Corp.	Chatsworth, Calif.	Patch panels
Astrodata, Inc.	Anaheim. Calif.	Time code generator, Digital multimeter
Babcock Relay, Div of Babcock Electronics	Costa Mesa, Calif.	Relays Generator
Barry Controls	Glendale. Calif.	Shock mounts
Boonshaft and Fuchs, Division of Weston	Monterey Park, Calif.	Servoanalyzer
Computer Measurements	San Fernando, Calif.	Counters
Consolidated Electrodynamics Corp.	Pasadena, Calif.	Tape recorder
Deutsch Co., Electronic Components Division	Banning, Calif.	Connectors
Electrada Corp.	Culver City, Calif.	Test conductor console
Electronic Specialty Co.	Los Angeles, Calif.	Hybrid junction, band pass, filter, low pass filter
Electroplex, Subsidiary Borg-Warner Corp.	Santa Ana, Calif.	Logic modules
		Power supplies
Fairchild Precision Metal Products, Division of Fairchild Camera and Instrument	El Cajon, Calif.	Cryogenic lines
B. H. Hadley (Royal Industries)	Pomona, Calif.	Disconnects
Hallicrafters Pacific Division	Santa Ana, Calif.	Power supply
W. O. Leonard, Inc.	Pasadena, Calif.	Vent valves
Micro - Radionics, I n c.	Van Nuys, Calif.	RF couplers
Non-Linear Systems, Inc.	So. Pasadena, Calif.	Scanners
Parker Aircraft Co.	Los Angeles, Calif.	Hydraulic systems
Rantec Corp.	Calabasas. Calif.	Mistram antenna, Mistram coupler
		Multiplexer
Rocketdyne Div., North American Aviation	McGregor, Tex.	Ullage rocket motors
Servonic Instruments, Inc.	Costa Mesa. Calif	Transducers
Solar Division, International Harvester Co.	San Diego, Calif.	Cryogenic lines
Stainless Steel Products	Burbank, Calif.	Cryogenic lines
Transco Products, Inc.	Venice, Calif.	Power dividers
United Electrodynamics, Inc.	Pomona, Calif.	PCM RF assemblies

NORTH AMERICAN ROCKETDYNE MAJOR SUBCONTRACTORS

SUBCONTRACTOR	LOCATION	PRODUCT
A & M Castings Inc.	South Gate, Calif.	Aluminum castings
Ace Industries	Santa Fe Springs, Calif.	Machined assemblies, nozzles, rotors, stators, and brg. supports
Adept Mfg. Co.	Los Angeles, Calif.	Inducers
Amphenol Corp.	Broadview, Ill.	Connectors
Anaconda Metal Hose	Los Angeles, Calif.	Flex. hoses
Anaconda Amer. Brass Co.	Detroit, Mich.	OFHC copper
Arcee Foundry	Norwalk, Calif.	Castings
Arcturus Mfg. Co.	Oxnard, Calif.	Forgings
Bendix Corp., Scintilla Div.	Sidney, N. Y.	Connectors
Beuhler Corp., Indiana Gear Wks.	Indianapolis, Ind.	Gears and shafts
Borg Warner Corp., Borg Warner Mech. Seals Div.	Vernon, Calif.	Seals
Calif. Doran Heat Treating Co.	Los Angeles, Calif.	Thermal processing of various major engine components
Cam Car Company	Rockford, Ill.	RD bolts
Chicago Rawhide Co.	Chicago, Ill.	Seals
Cleveland Graphite Bronze Div, Clevite Co.	Cleveland, Oh.	Seals
Consolidated Electrodynamics Corp.	Pasadena, Calif.	Connectors and transducers
Aerospace Div., DK Mfg. Co.	Batavia, Ill.	Flex lines, bellows and gimbals
Fairchild Metal Products Div. Fairchild Camera & Instru.	El Cajon, Calif.	Bellows, ducts, gimbals, and line assemblies
General Labs, Inc.	Norwich, N. Y.	Exciters and igniters
Globe Aerospace	North Hollywood, Calif.	Machined metal parts, fittings, and elbows
Herlo Engineering Corp.	Hawthorne, Calif.	Major machined components
Hollywood Plastics, Inc.	Los Angeles, Calif.	ABS Royalite closures, covers, and other protective devices
Howmet Corp., Austenal Div.	Dover, N. J.	Castings
Huntington Alloys, International Nickel Co.	Huntington, W. Va.	Inco sheet and plate
Industrial Tectonics, Inc. General Controls, Inc.	Burbank, Calif.	Actuators, brgs.
Kentucky Metals	Louisville, Ky.	Honeycomb
L. A. Gauge Co., Inc.	Sun Valley, Calif.	Valves
Langley Corp.	San Diego, Calif.	Seals, gimbals, machined assemblies
LeFiell Mfg. Co.	Santa Fe Springs, Calif.	Tubing-thrust chamber
McWilliams Forge Co.	Rockaway, N. J.	Forgings
Orbit Machine Corp.	Gardena, Calif.	Seals
Parker Seal Company	Culver City, Calif.	0-Rings, seals, and orifice plates
Paragon Die Tool & Engr.	Pacoima, Calif.	Stators, brg. supports
Precision Sheet Metal, Inc.	Los Angeles, Calif.	Major sheet metal subassemblies, jackets
Quadrant Engr. Company	Gardena, Calif.	Valves and components
Reisner Metals, Inc.	South Gate, Calif.	Forgings
Rohr Corp.	Chula Vista, Calif.	Major sheet metal subassemblies
Rosemount Engr. Co.	Minneapolis, Minn.	Pressure and temp transducers
Scientific Data Systems	Pomona, Calif.	Circuit boards
Southwestern Industries	Los Angeles, Calif.	Switches
Solar, Div. of International Harvester Co.	San Diego, Calif.	Valves
Statham Instruments	Los Angeles, Calif.	Transducers, connectors, electronic assemblies
Standard Pressed Steel	Santa Ana, Calif. & Jenkinstown, Penna.	RD bolts
Texas Instruments	Dallas, Tex.	Transistors
Turbo Cast Inc.	Los Angeles, Calif.	Castings
Union Carbide Corp., Haynes Stellite Div.	Kokomo, Ind.	Castings, Hastelloy C, Rene 41
Viking Forge & Steel Co.	Albany, Calif.	Forgings
Western Arc Welding Co.	Los Angeles, Calif.	Welded assemblies
Western Way Inc.	Van Nuys, Calif.	Ducts, line assemblies, heat exchangers
Winsco Instruments & Controls	Santa Monica, Calif.	Transducer and receptacles, temp. and pressure transducers
Wyman-Gordon Company	N. Grafton, Mass.	Forgings

APPENDIX B-INDEX

A

Acceptance testing	100, 107-108, 132
Aft Interstage	12, 49-50, 61, 63-66, 75-76
Aft LOX Bulkhead	49, 52, 55, 129, 131
Aft Skirt	49-53, 55, 60-65, 72, 75-76
ARPA	15, 17
Assembly and Checkout	67, 95-96, 101, 109
Automatic Checkout	75, 96, 103, 105, 108

B

Battleship	19, 94-95, 99, 108
Bearing Coolant Control	42-43
Boeing	9, 11, 18, 21, 26, 93, 100, 123, 129

C

Canoga Park	11, 97
Coca	94
Common Bulkhead	49-50, 52-55, 62-65, 71
Complex	39 14, 19, 99, 101-105, 119-120, 131
Control Pressure System	27, 34-35, 119, 130

D

Development testing	95, 97, 107-108
Douglas	9, 11, 18, 67, 95-96, 124
Downey	94

E

Edwards Field Laboratory	97-98
Electrical System	32-33, 39, 45, 61, 85, 92
Emergency Detection	27, 92, 120
Engine Cutoff	28, 30, 48, 59, 61, 70, 76, 84, 90, 119-120
Engine Description	39, 78
Engine Interface	44-45
Engine Operation	36, 40, 44, 46-47, 65, 68-69, 71, 81-84, 119
Environmental Control System	26, 35, 67, 71, 75, 85, 87-88, 103
Exhaust System	78, 82

F

F-1	13-14, 17-20, 22, 26-30, 34, 37-41, 47, 50, 97-99, 104, 106, 111
Fabrication	21, 53, 64, 86-87, 93-96, 108
Fairings	24-25, 76, 109
Fins	21, 24, 53, 109
First Stage	12-14, 16-19, 21-27, 31-37, 39, 50-51, 57, 59, 61, 80, 83, 89, 93, 98-101, 104, 108-111, 120-121, 125
Flame Deflector	104
Flight Control System	60, 67, 71-72
Flight Testing	103, 106-108
Fluid Power System	31-32
Forward Skirt	20-21, 24-26, 33, 35-36, 49, 51-52, 54-55, 60, 63-65, 70, 74-76
Fuel Pump	28, 41-42, 47, 71, 82
Fuel System	27, 32, 39-40, 45, 68-69, 73
Fuel tank	20-23, 25-30, 32, 37, 42, 45, 69-71, 123

G

Gas Generator System	39, 42, 68
Gimbal	22, 39-40, 60, 72, 78, 89, 122-123
Gimbal Bearing	39, 78

Ground Support	28, 31, 33, 58, 62, 67, 87, 98-99, 109, 112, 119, 122, 125
Guidance and Control	17, 85, 88-89, 91, 120

H

Houston	14, 103, 110
Huntington Beach	11, 66-67, 95
Huntsville	11, 15, 21, 86-87, 93, 96-98, 108, 124-127
Hydraulic Control System	39, 42, 45
Hypergol Cartridge	40, 45, 47

I

IBM	11, 16, 19, 86-91, 96-97, 125
Instrumentation System	27, 33, 39, 46-47, 60, 68, 74, 78, 83, 85, 91
Interstage	12, 49-52, 55, 59-66, 75-76, 95
Intertank	20, 23-26

J

J-2	12-15, 17-18, 49-52, 56, 58, 60-65, 67-74, 77-80, 98, 100, 106, 121

K

Kennedy Space Center	9, 11, 13-15, 18-19, 21, 50, 86, 93-94, 99, 101-103, 108-109, 111, 119-120, 122
KSC	15, 19, 102-104, 122

L

Launch Control	14, 101-102, 109
Launch Site	93, 101, 103-106, 108, 110
LH2	12, 49, 54-55, 57-59, 63, 66-71, 73, 84, 122
Liquid Hydrogen Tank	50-54, 56-58, 62, 119
Liquid Oxygen Tank	23, 50-53, 55-56, 58, 62, 95
Los Alamitos	123
Los Angeles	94, 97, 122-128
LOX	12-13, 20-32, 34, 36-37, 39, 41, 43-49, 52-53, 55, 57, 59, 63, 65-69, 71, 73, 83-84, 103-104, 121-122
LOX System	28, 30, 69, 103

M

Marshall	9, 11, 13, 15, 18-19, 21, 93, 98-100, 108, 111-112, 122, 125
Marshall Space Flight Center	9, 11, 13, 15, 18, 21, 93, 98-100, 108, 111, 122
Measurement System	32-33, 60
Michoud Assembly Facility	18, 21, 25, 93-94, 101
Mississippi Test Facility	18, 21, 93-95, 100
Mobile Launcher	101-106, 109-111
MSFC	18-19, 86, 99, 112, 122

N

NASA	11, 13-18, 25, 86, 92-93, 97, 100-101, 111-112, 119-120, 122
New Orleans	11, 18, 21, 93, 101
North American	9, 11, 17-18, 50, 94-95, 97, 100-101, 126-128
North American Aviation	9, 11, 17-18, 50, 94, 100-101, 126-127

O

Ordnance System	61
Oxidizer Dome	39, 43-45, 78
Oxidizer Pump	41-42, 46, 71
Oxidizer tank	41, 68, 82

P

Pearl River	101
Poseidon	94
Pressurization System	27-29, 31, 39, 44, 58, 68, 70, 73, 119
Propellant Feed Control System	43
Propellant System	34, 56, 67-68
Propulsion System	27, 32, 47, 65-67, 72-73, 78, 83, 122
Pyrotechnic Igniter	40

Q

Qualification testing	107, 112

R

Range Safety System	32, 34, 76
Reliability testing	96, 107
Rocketdyne	9, 11, 16-18, 78, 94, 97, 127-128
RP-1	12-13, 20, 27, 31-32, 37, 39, 103-104

S

Sacramento	19, 67, 95-96
Santa Monica	95, 125, 127-128
Santa Susana Field Laboratory	97-98
Seal Beach	11, 51, 54-55, 94-95
Second Stage	12-14, 16-19, 24, 34, 37, 49-56, 59-62, 65, 76, 78, 80-83, 89-90, 94-95, 99-101, 110, 120-121
Silverstein Committee	15, 17
Start Tank	78, 83-84
Super Guppy	96
Switch Selector	32, 34, 71, 74, 92
Systems Tunnel	51, 54

T

Thermal Control System	60
Thermal Insulation	39-40
Third Stage	12-15, 18-19, 49-51, 54, 60-61, 64-68, 71-76, 78, 81-85, 87, 90, 95-96, 99, 110-111, 121
Thrust Chamber Assembly	39, 78, 81
Thrust Chamber Body	39-40, 78
Thrust Chamber Injector	39, 45, 78, 83
Thrust Chamber Nozzle	39-40
Tracking System	33, 91
Turbine	31, 40-47, 71, 80-84,
Turbopump	37, 39, 41-48, 59-60, 69-72, 78, 80-84, 98-99

U

Ullage Motors	49, 51, 59-61

V

VAB	18-19, 93, 101-102, 104-106, 110
Vehicle Assembly Building	17-18, 101-103, 109
Vertical Assembly Building	22, 25-26, 54-55, 93-94
Visual Instrumentation	35-36
Von Braun	15

The Complete Manufacturing and Test Records

by Alan Lawrie

F-1 ENGINE POSITION MATRIX FOR S-1C STAGES

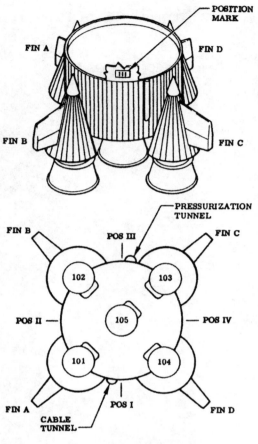

F1-1-5

J-2 ENGINE LOCATION - SATURN S-II

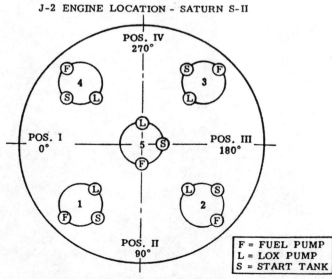

F = FUEL PUMP
L = LOX PUMP
S = START TANK

ORIENTATION VIEW LOOKING AFT

Facilities and transportation

NASA, George C Marshall Space Flight Center (MSFC)

The George C Marshal Space Flight Center (MSFC) was activated on 1 July 1960 with the transfer of personnel and facilities from the US Army Ballistic Missile Agency. The Saturn V program as a whole was managed from NASA's Marshall Space Flight Center in Huntsville, Alabama under the direction of Wernher Von Braun.

New Saturn V facilities built at MSFC included the first stage static test stand, an F-1 engine test stand, the Saturn V launch vehicle dynamic test stand, a J-2 engine/battleship test stand, and ground support and component test positions.

The new area containing the Saturn V booster test stand and the F-1 test stand was designated the West Test Area and supplemented the existing East Test Area. The East Area included the two-position Saturn I/IB test stand (facility 4572) that was modified to allow static firings of F-1 engines. Nearby the J-2/battleship test stand (facility 4520) was constructed.

In addition to the static firing test stands, the largest test facility at MSFC is the Dynamic Test Stand (facility 4550), constructed in 1964 to perform vibration tests on the complete Saturn V vehicle in a vertical orientation. It stands 360 feet high and is 122 by 98 feet at the base. Vibration loads can be applied in 6 degrees of freedom as well as yaw, pitch and roll directions. It was made a National Historic Landmark on 22 January 1986.

The S-IC booster test stand (facility 4670), constructed in 1964, had an overall height of 405 feet and contains 12 million pounds of concrete in its base legs. Modifications to the test stand began in 1974 to accommodate LH2 for space shuttle external tank verification testing. Further modifications in 1986 were made to accommodate the Technology Test Bed engine, a derivative of the SSME. At this time its name was changed to the Advanced Engine Test Facility. It was designated a National Historic Engineering Landmark on 28 October 1993.

As well as the center for manufacture and testing ground versions of the S-IC stage, MSFC was responsible for manufacturing the first two S-IC flight stages, and static firing the first three S-IC flight stages.

NASA/Boeing, Michoud Assembly Facility (MAF)

The 832-acre NASA Michoud Assembly Facility (MAF) is located in New Orleans, Louisiana, some 15 miles East of downtown New Orleans. It is a NASA-owned facility that has been operated by several contractors over the years. During the Apollo program the facility was used by the Chrysler Corporation for building Saturn I/IB stages and by Boeing for building the Saturn V booster stage, the S-IC.

The plant gets its name from Antoine Michoud who operated a sugar cane plantation and refinery in the area in the middle of the 19th century. Two brick smokestacks from the original refinery still stand before the Michoud facility.

68-64889 Michoud dock with plant in background 4.68.

In 1940 the Government purchased a 1,000 acre tract and constructed the world's largest building at the time – 43 acres under one roof – for the manufacture of cargo planes and landing craft. During the Korean War the facility produced tank engines.

On 7 September 1961 NASA took over the facility for the design and manufacture of Saturn first stages. The facility includes a port with deep-water access for the transportation of large space structures. The facility includes a high bay area, the Vertical Assembly Building, where Saturn V S-IC stages were assembled vertically. The VAB is a single story structure rising the equivalent of 18 stories.

Checkout of the stage's electrical and mechanical systems was performed in the four giant test cells of the Stage Test Building. Each of the Cells is 83 by 191 feet with 51 feet of clear height. Each had separate test and checkout equipment.

Stages left and entered Michoud by waterways connected to the Mississippi River or the Gulf of Mexico.

In 1973, with the Apollo program winding down Chrysler and Boeing made way for Lockheed Martin Michoud Space Systems who designed and manufacture the Space Shuttle External tanks at Michoud. The same barges as used to transport the Saturn rockets are still used today to move the External tanks.

The S-IC-15 stage, manufactured at Michoud, now rests horizontally within the perimeter fence but visible to passing motorists along the Old Gentilly Road.

North American Aviation, Seal Beach

The main fabrication and testing facilities for the second stage were located at the Space and Information Division of North American Aviation at Seal Beach, southeast of downtown Los Angeles and close to the Douglas facilities at Huntington Beach.

The complex at Seal Beach comprised NASA-owned facilities operated by NAA for the Saturn second stage activities and offices owned by NAA. The Seal Beach facilities included a bulkhead fabrication building, 125-foot-high vertical assembly building, 116-foot-tall vertical checkout building, remote pneumatic test area and a number of other structures.

The vertical assembly building, where the second stage was assembled, contained six work stations at which successive major parts of the stage were added. After assembly the stage was moved to the vertical checkout building, where some final installations were made and where its systems and components were given final tests.

After checkout the stages were transported by road to the nearby Naval docks for travel by sea to the static firing facility at MTF. From there the stages went directly to KSC.

The Space and Information Division became the Space Division during the Saturn era, and also in this time the company merged and became North American Rockwell. Today the Seal Beach facility is owned by Boeing

Douglas Aircraft Company, Huntington Beach

The Douglas Aircraft Company, based at Huntington Beach in southern Los Angeles, designed and manufactured the S-IVB third stage at that facility. The center was the headquarters for the Missile and Space Systems Division.

Initial component fabrication for the third stage was accomplished at Douglas' Santa Monica plant. Final assembly and factory checkout of the third stage took place at the Space Systems Center at Huntington Beach, just down the coast from Santa Monica. High bay manufacturing area was provided for the production of propellant tanks, skirts and inter-stages. Eight tower positions were available for vertical assembly and checkout of completed vehicles. Structural tests on major vehicle structures such as the propellant tank, skirt sections, and inter-stage were conducted in the Structural Test Laboratory at the Space Systems Center.

Two vertical towers at Huntington Beach provided for the final factory tests on finished third stages, prior to shipment from the plant for test firing. Completed stages were transferred by road to the nearby Los Alamitos Naval Air Station and airlifted on a Super Guppy aircraft to Douglas' Sacramento Test Facility for static firing. The stages were then transported directly from Sacramento to KSC.

In 1967 Douglas became McDonnell Douglas and the Huntington Beach plant is now owned by Boeing.

Rocketdyne, Canoga Park

Rocketdyne designed, manufactured and tested all the main engines (F-1 and J-2) for the Saturn V vehicles. During the active phase of the Saturn V program Rocketdyne was a division of North American Aviation. Their main engineering and manufacturing facility, then as now, is located at Canoga Park, to the northwest of Los Angeles. Rocketdyne opened the 51-acre facility in 1955 and have designed and manufactured many of the nation's rocket engines.

Manufacturing of components and final assembly of both Saturn engines was carried out in eight buildings in the Canoga Park complex. As well as machining facilities there were two large brazing furnaces for the brazing of the thrust chamber tubes and injectors. Special areas for precision cleaning and assembly and for various types of non-destructive inspection existed. An Engineering Development Laboratory provided test facilities to support manufacturing production.

Test firing of the engines was performed at Rocketdyne's Santa Susana Test Facility, not far from Canoga Park, and at Edwards Field Laboratory 125 miles away in the desert.

Rocketdyne Propulsion & Power is currently part of the Boeing Company. Rocketdyne has other local facilities at West Hills and DeSoto in Los Angeles. Visitors to Canoga Park will notice a full-size F-1 engine displayed outside the main entrance. This is in fact a mock up on loan from the Smithsonian.

NASA, Mississippi Test Facility (MTF)

Located in Hancock County in southwest Mississippi, 55 miles from New Orleans, is the site of the NASA's rocket propulsion test center. On 25 October 1961 the federal government announced that it had selected this area to be the site of a static test facility for launch vehicles to be used in the Apollo program. It was the largest construction program in the state of Mississippi and the second largest in the USA at that time.

The site was selected for logical and practical reasons. Excellent water access ensured ease of transportation of the large Saturn stages to and from the manufacturing and launch sites. The area also provided the 13,428-acre test facility with a sound buffer of 125,442 acres.

The Army Corps of Engineers opened a real estate project office on 18 April 1962 to begin land acquisition negotiations. The following year, on 17 May, workmen cut the first tree to start clearing the test area for construction. An eight-mile canal, complete with lock system, was dug to connect the test stands to the East Pearl river.

The B test stand at SSC in 2004

The A-1 (right) and A-2 (left) test stands at SSC in 2004

The facility was initially known as the Mississippi Test Operations (MTO), but was changed to the Mississippi Test Facility (MTF) on 1 July 1965.

Three large test stands were built in 1965 for the Saturn V rockets. The dual position B1/B2 test stand (building 4220) was used for static firings of Saturn S-IC first stages. The B2 position was used in every case. This test stand is 407 feet tall and is constructed from steel and concrete resting on 1600 steel pilings each 98 feet long. A test control center with necessary facilities forms part of the B complex.

Close by the individual A1 (building 4120) and A2 (building 4122) test stands, each 154 feet tall, were used for static firings of the S-II Saturn V second stages. The first Saturn firing was when the S-II-T stage was tested in the A2 stand on 23 April 1966. The final S-II firing was on 30 October 1970 in the A2 stand. The final S-IC firing was on 30 September 1970 in the B2 stand. The A2 test stand has since been modified for simulated altitude firings with the addition of a diffuser into which the engine on the stand fires. The pressure conditions within the diffuser simulate an altitude of 65,000 feet. Barges storing propellants are used to re-supply run tanks. These days flight engines are fired in the A2 and B1 positions. The test stands are now listed as National Historic Landmarks.

During Saturn V rocket testing at MTF NASA logged 160 noise complaints, of which 18 resulted in financial settlements totaling $39,405.

On 14 June 1974 MTF was renamed the National Space Technology Laboratories (NSTL) and space shuttle main engine testing started on 24 June 1975. On 20 May 1988 the center was renamed the NASA John C Stennis Space Center and as recently as 21 January 2004 the center achieved the one million seconds of SSME test and flight operations during a firing at SSC. It is a NASA field center and was previously an outpost of the Marshall Space Flight Center.

Douglas, Sacramento Test Operations (SACTO)

In the 1960s Douglas operated a test facility known as the Sacramento Test Facility (SACTO), located just outside Sacramento, California, in Rancho Cordova. This facility was used for numerous tests, mostly associated with static firing of S-IV and S-IVB stages.

The stages were manufactured and functionally checked at the Huntington Beach facility. From there the stages were driven on the back of a trailer the short distance to the Los Alamitos Naval Air Station, on the Los Angeles coast. The stages were flown from there aboard a Super Guppy aircraft to Mather Air Force Base, located just outside of SACTO. Final transportation to the test facility was along public roads by truck.

68-66896 S-IVB APS module ready for firing at the Gamma 3 cell at SACTO 5.68.

68-66304 Firing a S-IVB O2/H2 burner in the Alpha II-B test site at SACTO 4.68.

At SACTO there were a number of test sites. Those that had a relevance to the Saturn V program are listed below;

- Alpha 1 site 3 – S-IVB APS vibration testing
- Alpha 2-B – O2/H2 burner test firing
- Alpha site 4 – Cryogenic tank testing
- Beta I – S-IVB stage static firing
- Beta III – S-IVB stage static firing
- Gamma cell 3 – S-IVB APS test firing
- Kappa cell A – Fuel pre-valve testing

In addition to the test firing sites there was a building complex for processing, testing and storing the stages. The main building was the large Vertical Checkout Laboratory (VCL).

The site is now owned by Aerojet and is used for a variety of propulsion activities.

Rocketdyne, Santa Susana Field Laboratory (SSFL)

The Santa Susana Field Laboratory (SSFL) is a rocket testing facility outside of Los Angeles that was used extensively during the development and production phases of the Saturn V rocket. It was used for qualification and acceptance testing of all J-2 engines and for the S-II battleship stage testing.

The site is located in the Simi Hills area of Ventura County, 30 miles northwest of Los Angeles. The site is situated on a ridge, between 1640 and 2250 feet above sea level, overlooking Simi Valley to the north and the San Fernando Valley to the southeast. Close to the Rocketdyne manufacturing facility at Canoga Park, "The Hill" as it is affectionately known by those who have worked there, has been used for testing rocket engines, components and complete stages since 1948. The beginning of the official testing at SSFL was on 15 November 1950 when the first successful main stage test of a US-designed-and-built large liquid propellant rocket engine took place. Because of its stark, remote mountainous appearance the site was used by many Hollywood film makers as the set for films and TV series.

The SSFL facility is currently jointly owned by Boeing and NASA and is operated by the Rocketdyne Propulsion and Power Division of Boeing. During the Saturn development ownership was also in the hands of Rocketdyne and NASA, although Rocketdyne at that time was a division of North American Aviation. The SSFL site consists of four administrative areas used for research, development and test operations and buffer areas on the southern and northwestern boundaries of the facility.

Area 1, to the east, consists of 641 acres owned by Rocketdyne and 42 acres owned by NASA (formerly owned by the US Air Force). Area 1 includes three former rocket engine test areas, two of which were used for the Saturn and Saturn-era programs;

> **Bowl**. As its name suggests this is a natural bowl feature with test positions around the rim. The Bowl was the first test area to be activated and there were originally five test stands built. This area was used for testing Atlas, Navaho, Redstone and Saturn J-2 engines. J-2 engines were fired in Vertical Test Stand 2 (VTS-2), Vertical Test Stand 3A (VTS-3A) and Vertical Test Stand 3B (VTS-3B). J-2 Thrust Chamber Assemblies (TCA) were fired in the Horizontal Test Stand (HTS). Test stand 3A (for altitude) was unique in that a special vacuum diffuser allowed firings to take place at simulated altitude conditions, which were appropriate for the J-2 mission. The original Navaho and Redstone firings were performed in the Vertical Test Stand 1 (VTS-1). Work in the Bowl area has now stopped although it is not shut completely.
> **Canyon**. Similar to Bowl, this was another natural feature containing several test stands. Canyon was activated after the Bowl Area, in the mid-to-late 1950s. Canyon has been used to test Jupiter, Thor and Saturn H-1 engines. For H-1 testing there was a two-position stand (horizontal and vertical) and a one-position stand (vertical). The Canyon area is now closed.

In addition, Area 1 contained laboratory facilities where Saturn components were tested;

Component Test Laboratory III (CTL III). F-1 GG/HX, J-2 Turbo-pumps, J-2 ASI and J-2 GG.
Component Test Laboratory V (CTL V). J-2 Turbo-pumps.

Area 2, consists of 410 acres in the north-central position of the site and is owned by NASA and operated by Rocketdyne. Area 2 contains four test areas, Alfa, Bravo, Coca and Delta, two of which are still active today.

> **Alfa**. Used for RS-27 Delta, Atlas, Navaho, Jupiter and Thor engine tests.
> **Bravo**. Used for Atlas, Navaho, Thor, Saturn E-1 (pre- F-1), and Saturn V F-1 components (TCA and Turbo-pump). A gas generator exploded during heat exchanger acceptance testing at Bravo 1C test stand on 7 October 1965. The Bravo test area is now closed.
> **Coca**. Used for Atlas, SSME, S-II-TS-B and Saturn S-II battleship stage testing. The S-II battleship testing was performed at the Coca I site, which was activated in November 1964. Coca IV was constructed for the proposed testing of the S-II-T All-Systems Test stage. However, it was ultimately decided to test that stage at MTF. However Coca IV was used for testing of the structural test stage, S-II-TS-B. The Coca test site is now closed.
> **Delta**. Used for Atlas, Jupiter, Thor, Saturn V J-2 and Saturn E-1 (pre- F-1). Test stands Delta 2A and Delta 2B were used for J-2 single engine development, qualification and acceptance firings from December 1963. The Delta stands are now closed.

The Saturn V F-1 component testing at the Bravo test stand ran from 1959-1971.

The Saturn V J-2 testing at the Coca and Delta test stands took place between 1960 and 1971.

Area 3, consists of 114 acres in the northwest portion of the site and is owned and operated by Rocketdyne. No Saturn-related work was performed in this area.

Area 4, consists of 290 acres owned and operated by Rocketdyne and 90 acres leased by the Department of Energy. The DOE operated several nuclear reactor facilities in this area and in recent years has been the subject of close monitoring to ensure that the ground and surrounding area is free from contamination. It was in this area that a Sodium Reactor Experiment suffered a fuel damage "meltdown" on 26 July 1959.

Buffer zones to the northwest and south consist of 175 and 1140 acres respectively of undeveloped land.

With the growing expansion of greater Los Angeles and the nearby residential areas of Simi Valley the usefulness of SSFL as a remote test site for rocket engine testing with hazardous propellants has diminished. In recent years there have been several investigations into the ground contamination in surrounding areas, looking particularly for perchlorate samples. In the long term SSFL will likely be closed with engine testing moving to purpose-built facilities such as NASA Stennis Space Center.

Air Force Research Laboratory, Edwards Air Force Base

Situated out in the Mojave Desert, 125 miles northeast of Los Angeles, Edwards Air Force Base (EAFB) is famous because of its many links to historical aviation and space feats. Lesser known is its role in the Saturn V engine test program. But it was here that the development, and all acceptance test firings of the F-1 engines took place.

Because of the high thrust level of the F-1 engine testing at SSFL was precluded due to the proximity to inhabited areas. A more remote site was needed and Edwards fitted the bill perfectly.

The Air Force Research Laboratory (AFRL) is located in the eastern side of Edwards Air Force Base, occupying about 65 square miles of the Base. Construction of the rocket testing facilities at AFRL began in late 1949 and the first test stands were activated in 1952. The test sites are laid out at 1-mile intervals in a pattern perpendicular to the prevailing wind.

Between the years 1963 and 1987 the facility was known as the Air Force Rocket Propulsion Laboratory (AFRPL). It was also known as the Rocket Engine Test Site (RETS). During this time many different rocket engines were tested there. Between 1963 and 1969 all F-1 engines were acceptance test fired there in stands specifically built for high thrust engines. Testing of F-1 engines continued there up until April 1973 with F-1A development and flight support work.

As the F-1 engine was only used on the Saturn V first stage that operated from the ground the engine was never required to be fired in the vacuum of space. For that reason all the F-1 testing was performed at sea-level conditions. In any event there are no facilities that could support vacuum firings of such a thrust level. Consequently the test stands used for the F-1 at AFRPL were relatively simple constructions that allowed the engines to fire vertically downwards out over natural ridges.

There were five test stands with six positions constructed and used for F-1 firings. These were split into two test areas located on the northeast end of Luehman Ridge.

> **Area 1-120**, now known as the Advanced Launch System Complex. This area consists of three liquid rocket stands, with five firing positions, a control center and various support facilities. Vertical stand 1-A is a single position stand designed for 1,500,000 lbs thrust rocket systems. Vertical stand

1-B is a dual position stand now capable of 6,000,000 lbs thrust rocket systems. All three positions (1-A, 1-B1 and 1-B2) were used for F-1 engine testing. The 1-A (for Atlas) and 1-B (for Thor) test stands had been used since the 1950s, but were adapted for the first Saturn F-1 firings in 1961. Stand 1-A underwent further modification in the spring of 1964 as the flame deflector and eroded ground adjacent to this needed rework. A third stand in this complex, stand 2-A, is a two-position horizontal stand designed for 750,000 lbs thrust component testing. This stand was never used for F-1 engine firings although it was used to test F-1 Thrust Chamber Assemblies. Stand 2-A underwent an $18.5 million, 18 month, modernization and refurbishment activity that was completed as recently as January 2004.

Area 1-125, now known as the Large Systems Complex. This area consists of three large rocket stands and necessary support structures and equipment. This complex was built specifically for F-1 engine testing and was activated in mid 1964. Vertical stand 1-C, activated on 10 June 1964, today can support solid rocket motors with thrusts up to 1,600,000 lbs. Vertical stand 1-D is capable of firing engines up to 1,500,000 lbs thrust. It was built in the first half of 1964 and first fired an F-1 engine on 1 July 1964. Vertical stand 1-E originally had the same characteristics as stand 1-D, but is now the location of the National Hover Test Facility (NHTF). This stand was activated in September 1964. NASA formally accepted the complex on 9 October 1964 with a firing ceremony witnessed by Wernher Von Braun. All three stands (1-C, 1-D and 1-E) were used for F-1 engine firings. In addition to the F-1, some Saturn H-1 engine firings were performed at this complex.

In total more than 5,000 F-1 and F-1-related firings were conducted at AFRPL. Apart from a few verification firings performed on two stands at the NASA Marshall Space Flight Center in Huntsville, all single engine F-1 firings took place at AFRPL.

Arnold Engineering Development Center (AEDC)

The Arnold Engineering Development Center (AEDC) is the world's most diverse complex of aerospace ground test facilities. Located in Middle Tennessee, not far from the NASA Marshall Space Flight Center, it is operated by the US Air Force Materiel Command. The main attribute of AEDC is its unique capability for rocket engine firings at simulated altitude. In order to maintain the necessary vacuum conditions powerful steam diffuser systems are needed to draw the rocket engine exhaust gasses away from the firing chamber. AEDC operates some of the most powerful of these systems thus allowing altitude firings of very high thrust engines.

Because of its unique facilities AEDC was selected for a series of test firing programs with the Saturn J-2 engine, an engine that in flight was fired only in vacuum and near-vacuum ambient conditions. Production J-2 engines were acceptance tested in the ambient and semi-vacuum test stands at Rocketdyne's Santa Susana Field Laboratory, close to the production facility at Canoga Park. However for testing in the highest vacuum firings were performed at AEDC that would allow correlation of the SSFL firings to full vacuum conditions. AEDC was also used for development testing of the second-generation J-2S engine.

For altitude rocket engine testing AEDC currently has four main test cells, J-3, J-4, J-5 and J-6. Only test cell J-4 was used for Saturn J-2 engine testing, although the smaller test cell J-3 was used for testing the AJ10-137, Apollo service module SPS.

Rocket Development Test Cell J-4. This is a vertically oriented test cell designed for testing large liquid propellant engines at simulated altitudes of up to 100,000 feet. The J-4 test cell became operational in 1964 and has since tested over 400 rocket engines. It has a 48 feet diameter with an available capsule height of 82 feet. It is capable of firing engines of up to 1,500,000 lbs thrust although is currently limited to one third of this level. The other unique characteristic of this facility is its temperature-conditioning system that allows testing anywhere in the range 50F to 110F. Cryogenic rocket engines, last tested in 1976, recently returned to J-4 following a $9.7 million upgrade.

Barges and ships

Barges and ships were used to transport all three Saturn V stages, F-1 engines and various components between the various manufacturing and testing sites across the USA. For the S-IVB third stage air transportation was the primary method for moving the stage with water as the occasionally used back up. F-1 engines were also previously transported by air and by road.

The following barges were used for transportation of Saturn V hardware.

Poseidon

Covered barge. This is a modified Navy YFNB. River and Ocean barge used for S-IC and S-II stage transportation from MTF and MSFC to KSC. Still used today to transport External Tanks from Michoud to KSC.

Orion

Covered barge. This is a modified Navy YFNB. Orion was mostly used to ferry Saturn V first and second stages from Michoud and MTF to KSC. Still used today to transport External Tanks from Michoud to KSC.

Little Lake

Open-decked shuttle barge. This is a modified Navy YFNB used for transporting S-IC and S-II stages between MAF and MTF.

Pearl River

Open-decked shuttle barge. This is a modified Navy YFNB used for transporting S-IC and S-II stages between MAF and MTF. Still used today to transport External Tanks.

The following ship was used to transport Saturn V hardware:

Point Barrow

Built at the Maryland Shipbuilding and Drydock Company, Baltimore, MD, the Point Barrow was launched on 25 May 1957. She was placed in service by the Military Sea Transportation Service (MSTS) for Arctic service, as Point Barrow (T-AKD-1) on 28 May 1958. Later the Point Barrow was used primarily to transport S-II stages and F-1 engines between Seal Beach Naval docks, Michoud Assembly Facility and Kennedy Space Center, although it did make other trips in support of the Saturn program. The Point Barrow provided a stable, covered transport facility for the vital rocket hardware.

Around 1974 it was reclassified as a Deep Submergence Support Ship and renamed Point Loma (T-AGDS-2). She was struck from the Naval Register on 28 September 1993 and later scrapped.

Aircraft

On the Saturn I/IB program the Pregnant Guppy aircraft had been used for the transportation of S-IV and S-IVB stages from Huntington Beach to the tests site at SACTO and then on to KSC. The same plane was used initially in the Saturn V program, but this time to airlift F-1 engines from Rocketdyne to the S-IC manufacturing facilities at MSFC and MAF. Later on the engines were transported by more economical means. S-IVB-500 stages for the Saturn V were airlifted by the second transportation plane to come into service, the Super Guppy.

The F-1 engines were flown from LAX freight terminal to New Orleans Naval Air Station (for MAF) or Redstone airfield (for MSFC). The S-IVB-500 stages were flown from Los Alamitos Naval Air Station to Mather Air Force Base for testing at SACTO. Onward transport was to the KSC skid strip.

Aero Spacelines 377PG Pregnant Guppy

The Pregnant Guppy was constructed from parts of a Stratocruiser, Boeing B-377 NX1024V that had made its initial flight on 7 October 1947. Both the length and the width of the original plane were enlarged. The Pregnant Guppy first flew on 19 September 1962. After serving the Saturn V program transporting F-1 engines and the occasional test stage the Guppy carried more conventional cargo until it was retired in 1979 and scrapped.

Aero Spacelines 377SG Super Guppy

As the Pregnant Guppy had been so successful a second, larger aircraft was built. The Super Guppy (originally Clipper Constitution N1038V built in 1949) made its first flight on 31 August 1965 from Van Nuys Airport. The aircraft was first used to transport an S-IVB stage to MSFC in March 1966.

After service the Super Guppy has been placed in storage at Pima AFB in Arizona where it sits today.

F-1 Engine

Introduction

The F-1 engine was developed to provide the thrust for the Saturn V first stage. Five engines were utilized in each stage, with each engine having a nominal thrust of 1,500,000 lbs and specific impulse of 265 seconds. Fuel was RP-1 kerosene and the oxidizer was liquid oxygen. The engine was designed by Rocketdyne at its Canoga Park facility in northwestern Los Angeles.

Background development history

The engine evolved from the smaller E-1 engine designed in the early 1950s and which had a thrust of about 400,000 lbs. That concept was first tested with a development firing at Santa Susana Field Laboratory on 10 January 1956. Research and development work on the F-1 started on 23 June 1958 and by 6 August 1958 the Rocketdyne Division of North American had secured the award of the preliminary F-1 contract. The award of the full Research and Development contract, NASw 16, was announced on 19 January 1959. The thrust chamber of the F-1 was successfully static fired at SSFL on 6 March 1959. On 10 February 1961 the first engine thrust chamber main-stage test had taken place at Edwards Air Force Base. On 11 July 1961 a complete F-1 prototype engine began firings at EAFB. The first complete system firing took place on 16 August 1961 to be followed a year later by the first rated thrust and duration test. This test, on 26 May 1962, lasted 2.5 minutes at EAFB.

Flight standard engines

The first production engine, F-1001, was delivered to NASA on 29 October 1963. This engine is now the property of the Smithsonian Institute. The contract for the first 76 production engines, NAS8-5604, was awarded to Rocketdyne on 30 March 1964. Component development continued and involved some spectacular failures such as turbo pump explosions in February and April 1964 that delayed testing. Flight Readiness (or Rating) Testing (FRT) was completed during November and December 1964 and involved engines F-2004 and F-2006. This testing demonstrated the chosen design concepts in complete engines and paved the way for full qualification testing.

The first engine qualification was completed in mid 1965 probably involving engine F-2009. A qualification reliability program ran from October 1965 to February 1966. Component qualification was completed in July 1966 and the second engine qualification, to up-rate the thrust to 1,522,000 lbs, was completed on 6 September 1966, using engines, F-5037 and F-5039. The initial contract, NASw 16 was completed in May 1967. A further follow-on contract, 18734A comprised extended life validation and further production deliveries and flight support.

An F-1 engine on display at ASRC in Huntsville in 2004

63-35341 36 second test firing of the F-1001 engine in the S-IB static test stand in the East test area at MSFC 17.12.63.

64-01638 An F-1 engine is test fired at Edwards Air Force Base 7.6.62.

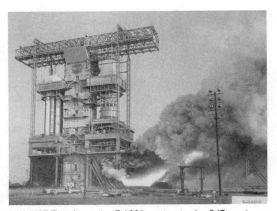

64-04597 Test firing the F-1001 engine in the S-IB static test stand in the East test area at MSFC on either 20 or 24 March 1964 3.64.

65-21185 Nine F-1 engines in the Engine Preparation Shop at MSFC, including F-2010 about to be fired in 7 days time 19.3.65.

68-66986 F-1 engine F-5038-1 firing in the F-1 test stand in the MSFC West Test area. 5.68.

68-62846 Acceptance firing of the F-104-4 engine at Edwards Air Force Base. 8.67

69-75821 F-1 engine final assembly line at Canoga Park 4.69.

98-08563 Ten second test firing of the F-1001 engine in the S-IB static test stand in the East test area at MSFC 5.12.63.

68-64127 The F-6073 (S-IC-10) engine arrives at MAF on the Point Barrow. It was the first engine to be transported by sea 16.2.68.

In total Rocketdyne manufactured and delivered 98 production F-1 engines. A breakdown of the engine disposition is as follows.

    65 engines were launched on S-IC stages
    10 engines are on flight stages not launched
    16 engines are either in storage or on display in a museum
    2 engines are on the S-IC-D museum stage
    2 engines are on the S-IC-T museum stage
    3 engines were scrapped or consumed

In addition there were 2 engines (F-3T1 and F-4T2) that were manufactured by NASA as a learning exercise using spare parts. These engines are on the S-IC-T museum stage. Finally there are 7 serialized mock-ups. Three of these engines are on the S-IC-D museum stage, one is on the S-IC-T museum stage, one was consumed and the other two are on display/in storage.

Engine serialization

Two serialization systems were in operation – those generated by Rocketdyne and NASA. The number that appears on the identification plate on the engine is the NASA number not the Rocketdyne one that was used for in-house purposes only. Listed below is a cross reference of Rocketdyne and NASA serial numbers for F-1 engines.

	Rocketdyne s/no.	NASA s/no.	First flight
1	F-1001 to F-1002	F-1001 to F-1002	N/A
2	F-1003	F-3T1*	N/A
3	F-2003 to F-2010	F-2003 to F-2010	N/A
4	F-2011 to F-2016	F-3011 to F-3016	Apollo 4
5	F-2017 to F-2028	F-4017 to F-4028	Apollo 6
6	F-2029 to F-2042	F-5029 to F-5042	Apollo 9
7	F-2043 to F-2098	F-6043 to F-6098	Apollo 11

*Manufactured by NASA from spare parts.

Engine configuration changes

There were six major F-1 configurations. Summarized below are the main changes between configurations;

Configuration	Rocketdyne s/no.
1	Pre-production R&D
2	F-1001 to F-1003
3	F-2003 to F-2016
4	F-2017 to F-2028
5	F-2029 to F-2042
6	F-2043 to F-2098

Changes from configuration 1 to 2

Externally tied flexible high pressure ducts were replaced with internally tied high pressure ducts. This simplified the fabrication and reduced the number of parts from eight to three.
Electrical harness no longer loosely attached in bundles as it was prone to vibration and thermal damage. An armored harness with protective insulation was introduced.
Hydraulic control system plumbing no longer connected via flared tubes, "B" nuts and AN fittings which were prone to leaks. Replaced with welded joints and manifolded tubing with bolted flange joints and seal plates.
Constant area turbine exhaust manifold was replaced with constant pressure turbine exhaust manifold in order to provide a better distribution of the gases to the thrust chamber nozzle extension resulting in the elimination of local hot spots.
No protective thermal insulation provided for these three engines as they were planned to be fired only as single engines.
No interface panel was provided for these three engines as the panel served as a thermal and electrical interface with the stage, and these engines were not to be part of a stage.

Changes from configuration 2 to 3

The gas generator spark ignition system was replaced by pyrotechnic igniters because of vibration-induced problems with the spark ignition system.
The LOX dome inlet was redesigned to improve flow characteristics and reduce turbo pump speed requirements.
Redundant flight instrumentation was added to ensure that, during the early flights, engine performance data was obtained successfully.
Thermal insulation was added as these engines would be part of stage testing and would require protection from nearby engines.
Wrap-around lines were added across the gimbal plane to provide the necessary hydraulic and gas supply and returns between the vehicle and the engines.
A cast interface panel was added in order to provide electrical interface connections between the stage and the engines and to serve as a heat shield between the stage and the engines.

Changes from configuration 3 to 4

The internally tied high pressure ducts were replaced with rigid high pressure ducts as the former ducts proved difficult to fabricate due to electron beam welding problems.
The cast interface panel was replaced with a fabricated panel because of fabrication problems resulting from the large size of the panel.
Stage static firing instrumentation was added in order to provide additional stage static test data.

Changes from configuration 4 to 5 (these changes were all minor)

A thrust "OK" pressure switch was added to provide flight onboard engine monitoring capability.
A number of minor detail component design improvements were incorporated in order to increase the reliability of the engine.

Changes from configuration 5 to 6 (these changes were all minor)

The redundant flight instrumentation was deleted because sufficient data had already been obtained from earlier flights, and the deletion would improve engine reliability by reducing the number of access ports with their associated seals.

Apart from engine configuration there was one further change in the operating parameters of the engines. The first 28 engines delivered had a nominal thrust level of 1,500,000 lbs. From serial number F-5029 onwards the thrust level was increased to 1,522,000 lbs.

F-1 engine position matrix for S-IC stages

The four outer engines were identified as positions 101 to 104. These engines could be canted. The center engine, position 105, was in a fixed position. The location of the engines in the stage is identified in the figure on page 134.

Engine usage

The first R&D engines were not serialized and were used to test the F-1 concept during tests at EAFB. Engines F-1001 to F-2010 were used in single engine development testing, flight readiness testing and qualification testing. In addition two of the engines in this group were installed in the S-IC-T all systems test stage. Engines from this batch were acceptance tested at the EAFB test stands and subject to development and qualification testing at both Edwards and at MSFC. Engines from serial number F-3011 onwards were used in flight vehicles. All these engines were acceptance tested at Edwards, although some also performed additional firings at MSFC (in two test stands there) if further verification of performance was deemed necessary.

Engine logistics

All F-1 engines were manufactured and assembled at Rocketdyne's design and manufacturing plant at Canoga Park. Each engine was trucked into the Mohave Desert to be acceptance tested on one of the AFRPL stands within Edwards Air Force Base. The engines returned to Canoga Park by truck for final delivery readiness checks and evaluation of test results. Following acceptance by NASA each engine was transported to its next destination. Both the destination and means of transport varied.

Ground test engines that were not intended to be used in stages either remained at Canoga Park or were shipped by Pregnant Guppy aircraft to the Redstone airfield at MSFC in Huntsville for further single engine testing. Two of these engines, flown to MSFC, were used in the S-IC-T ground test stage. From F-3011 to F-4021 the engines were flown to MSFC where they were integrated into the first two flight boosters that were manufactured at Marshall. Later engines were transported to the Michoud Assembly Facility in New Orleans where the remaining Saturn S-IC booster stages were assembled. These engines were airlifted to the New Orleans Naval Air Station, just south of downtown New Orleans, and transported the short distance to MAF by truck.

In an effort to reduce costs the means of transportation was changed two times. Initial engines were flown to either MSFC or MAF using Guppy aircraft. These aircraft usually flew from the LAX freight depot. Later, engines were transported by truck as this was cheaper and the vibration loads en-route had been shown to be low. Finally, as there were ships transporting S-II stages from the place of manufacture, Seal Beach, to MAF it was realized that the F-1 engines could hop-a-lift onboard. All that was needed was for the engines to be trucked across Los Angeles to the Naval docks at Seal Beach in the south of the City. On arrival at MAF the engines were offloaded and the S-II stage was transferred to a barge for the short trip to the test site at MTF (MTF was not capable of taking the larger Ocean-going ships).

Generally each S-IC stage was kitted with 6 engines, which included one spare. For various reasons the spare engine was frequently called into use and a new spare was allocated.

J-2 Engine

Introduction

The J-2 engine was developed to provide the thrust for the Saturn IB second stage, the Saturn V second stage and the Saturn V third stage. Five J-2 engines were used in the Saturn V second stage and one engine in each of the other applications. In the Saturn V configuration a restart capability was required in order to propel the Apollo spacecraft on Translunar Injection (TLI). Each engine had a thrust of 200,000 lbs, later upgraded to 230,000 lbs. Fuel was liquid hydrogen and the oxidizer was liquid oxygen. The engine was designed by Rocketdyne at its Canoga Park facility in northwestern Los Angeles.

Background development history

The R&D contract for the J-2 engine, NAS8-19, was awarded to Rocketdyne in 1960. A contract for production of 55 flight engines, NAS8-5603, was awarded on 24 June 1964. On 24 August 1964 NASA announced that it would purchase another 102 J-2 engines. Development engine firings took place at Rocketdyne's test facility at Santa Susana Field Laboratory (SSFL), not far from their design and manufacturing headquarters at Canoga Park in Los Angeles.

Design improvements during the development phase included the addition of a rigid T-ring device to reinforce the thrust chamber to prevent distortion that occurred during the stresses of repeated firings. During May 1964 engine systems tests concentrated on elimination of fuel pump stall at flight stage inlet conditions. The major component problem to resolve in this time period concerned performance of the fuel turbo-pump. Various thrust chamber injectors were tested in order to obtain a stable injector for the FRT engine.

Other engine components tested and improved at SSFL included the LOX pump inducer, fuel pump diaphragm, augmented spark igniter, and heat exchanger.

Flight standard engines

It was planned that Rocketdyne initiate PFRT testing in June 1964 using engine J-2004, or J-2005 as a back up. However both engines were damaged during acceptance testing and had to be returned to Rocketdyne.

64-17049 A J-2 engine being prepared to be fired in the VTS-3A test stand at the Bowl area at SSFL 2.63.

68-62867 J-2 engine J-2118 (S-II-11) ready for shipment from Rocketdyne 29.8.67.

A J-2 engine (white-painted version) on display at ASRC in Huntsville in 2004

01-00986 A J-2 engine being fired in the VTS-3A test stand at the Bowl area at SSFL 1963

68-62882 J-2 injector tests at the Vertical Test Stand at SSFL 10.67.

70-04803 Non-flight J-2 engine J-2148 being test fired in the Delta 2 stand at SSFL 8.69.

69-75832 J-2S engine J-113A during test firing at SSFL 4.69.

70-04809 S-II-15 J-2 engine J-2152 undergoing the final ever J-2 acceptance hot fire test at SSFL 2.1.70.

70-04802 The J-2 manufacturing production engine assembly area at Canoga Park 8.69.

70-06065 The final J-2 production engine, J-2152, being turned over to NASA at Canoga Park 28.1.70.

PFRT testing was accomplished during November 1964 using engine J-2008. On 11 December 1964 a major milestone in the J-2 program was achieved when a ground test engine demonstrated restart capability. After a firing of 165 seconds, and a simulated coast of 75 minutes the engine was restarted for 7 seconds, followed by a further six-minute shutdown and a final restart firing of 310 seconds. FRT testing was performed with engines, J-2023 and J-2022 in mid-1965. Qualification 1 was completed on 13 December 1965 using engine J-2032. The engine was fired a total of 30 times with a cumulative duration of 3774 seconds. On 14 March 1966 Rocketdyne successfully completed the reliability demonstration program for the 230,000 lb configuration of the J-2 engine. Qualification 2 was completed on 22 August 1966 using engine J-2072.

In total Rocketdyne manufactured and delivered 152 production J-2 engines. A breakdown of the engine disposition is as follows.

- 65 engines were launched on S-II stages
- 9 engines were launched on S-IVB-200 stages
- 12 engines were launched on S-IVB-500 stages
- 10 engines are on S-II flight stages not launched
- 2 engines are on S-IVB-500 stages not launched
- 1 engine is on an S-IVB-200 stage not launched
- 37 engines are either in storage or on display in a museum or unaccounted
- 5 engines are on the S-II-F/D museum stage
- 11 engines were scrapped or consumed or stripped for spare parts

Engine serialization

J-2 engines were serialized sequentially from J-2001 to J-2152, with no distinction being made between engines for Saturn IB or Saturn V applications.

Engine configuration changes

There were six major J-2 configurations. Summarized below are the main changes between configurations;

Configuration	Rocketdyne s/no.
1	Pre-production R&D
2	J-2001 to J-2011
3	J-2012 to J-2019
4	J-2020 to J-2059
5	J-2060 to J-2139
6	J-2140 to J-2152

Configuration 1

Early R&D engine design was for a tank head start. Stall problems and subsequent overheating necessitated a change to an engine mounted start tank.
Helium control gases had to be kept at a low temperature to prevent venting off of gas. A helium tank integral with the hydrogen start tank eliminated this problem.
An oxidizer turbine bypass valve was added to better regulate the rate of increase in oxidizer pump speed.

Configuration 2

These engines were ground test engines.

Configuration 3

First production-type engines.
Bleed system was increased to 1.5" diameter.
Gas generator was made integral with the fuel pump and incorporated an improved film coolant combustor.
The spark de-energize timer was reduced from 5 sec to 3 sec and the start tank discharge delay timer changed from 0.50 to 0.64 sec.
Added accumulator to the pneumatic control system.

Configuration 4

Insulated start tank.
Incorporated high performance T/C injector.
Electrical control assembly gold plated circuit boards replaced with solder plated boards.
Thrust chamber painted white – deleted requirement at J-2038 and subsequent, retrofit stripping of paint from fuel inlet manifold to exit end of chamber.
Added stage static instrumentation.
Added redundant main stage OK pressure switch.
Incorporated armored harness.

Configuration 5

Engine thrust up-rated to 230,000 lbs.
Incorporated thermostatic control orifice in closing side of main oxidizer valve actuator, rotated sequence valve.
Up-rated oxidizer and fuel turbo pumps.

Configuration 6

Eliminated excessive and redundant instrumentation.

J-2 engine position matrix for S-II stages

The four outer engines were identified as positions 201 to 204. These engines could be canted. The center engine, position 205, was in a fixed position. The location of the engines in the stage is identified in the figure on page 134.

Engine usage

The first R&D engines were not serialized and were used to test the J-2 concept during tests at SSFL. Engines J-2001 to J-2014 were used for ground testing at engine level and in ground test stages. Engine J-2015 was the first engine to be launched, on S-IVB-201 on 22 February 1966. All other engines were used on a combination of S-II, S-IVB-200 and S-IVB-500 stages as well as for ground test and as spares.

Engine logistics

All J-2 engines were manufactured and assembled at Rocketdyne's design and manufacturing plant at Canoga Park. Each engine was trucked to Rocketdyne's nearby test center at Santa Susana Field Laboratory (SSFL) for acceptance testing. SSFL included a test stand for simulated altitude testing by means of a steam ejector system. The engines returned to Canoga Park by truck for final delivery readiness checks and evaluation of test results. Following acceptance by NASA each engine was transported to its next destination that was generally Seal Beach for S-II stage integration, or Huntington Beach for S-IVB stage integration. Both of these locations are just a few hours drive from Canoga Park in greater Los Angeles. Some engines were installed in the Battleship stages that were tested at SSFL, MSFC, SACTO and AEDC. Unlike the F-1 engines, there was no need to ship J-2 engines great distances by air, sea or truck as the locations of the stage manufacturers were close to the Rocketdyne facilities.

S-IC-S

Summary

Structural test elements used to qualify the S-IC at MSFC. Included a structural test failure in 1966.

Stage activities

The structural test stage, S-IC-S, was one of two ground test stages built at MSFC. The stage was constructed in sections for use in various structural tests.

The test fuel tank, S-IC-C, was the first structural test component in the Saturn V development program at MSFC. In early 1964 technicians were completing assembly of the tank in the MSFC Vertical Assembly Facility. The tank was then cleaned and the aft adapter and special half-inter-stage were attached. On 6 March 1964 the tank was moved to the Load Test Annex of the P&VE Laboratory for load proof testing by Boeing.

On 4 August 1964 MSFC hydrostatically tested the S-IC test fuel tank. Partial collapse of the tank's upper bulkhead occurred during the test. After rework further hydrostatic testing occurred on 28 September and was successful. Inertial loads testing began in October. Various other structural tests continued through 1965.

The ME Laboratory at MSFC began welding bulkheads for the S-IC-S stage in March 1964. Boeing's difficulties in manufacturing acceptable fuel tank fittings and gore segments delayed assembly of this stage. The fuel tank for S-IC-S was 10 weeks behind schedule by June 1964, due to weld defects and to defective and late parts from Boeing.

Boeing at Michoud delivered the S-IC-S inter-tank to MSFC by September 1964. On 20 November ME Laboratory at MSFC completed fuel tank welding in the vertical assembly facility. MSFC technicians hydrostatically tested the tank on 1 December 1964. The LOX tank lower assembly, completed at MSFC in December, moved into the vertical assembly facility for installation of baffles and helium bottles.

The aft section, consisting of the thrust structure and fuel tank, went from the ME Lab to the Propulsion and Vehicle Engineering (P&VE) Lab's Load Test Annex on 19 February 1965 and was turned over to Boeing for test preparations. Workmen installed the stage on simulated hold-down arms and load tests simulating various static firing conditions were performed during April, May and June 1965. These tests simulated maximum thrust and gimbal loads. The last captive-firing condition test occurred on 21 June 1965 when the structure withstood 140% of the design load.

Personnel then added the inter-tank to the aft section in preparation for flight simulation tests. In parallel, fixtures were prepared for testing the fin and fairing assemblies and apex gore specimens.

During October 1965 Boeing successfully performed the launch pad standby load test on the aft section, achieving the required 140% limit load. On 28 October, with the fuel tank empty and un-pressurized, personnel conducted to 140% limit the ground wind/launch pad condition test. On 19 November 1965 a launch rebound test demonstrated the section's integrity to 135% of design limit load, with the fuel tank filled with water.

The S-IC-S forward assembly, comprising the LOX tank, inter-tank and forward skirt, that was delivered to P&VE at MSFC in March 1965, was stored upright in an outside holding area to await future tests planned for the following year.

During January 1966 test personnel moved the S-IC-S oxidizer tank assembly into the Vertical Assembly Building (VAB) at MSFC for installation of instrumentation in preparation for tests in the Load Test Annex.

Between February and March 1966 structural tests were performed on the S-IC-S fuel tank/thrust structure inter-tank assembly. The first and second test condition of Max Q Alpha were completed in February. A rerun of the Flight Cutoff condition was successfully conducted on 1 March. The test was run to a loading of 140% of the design limit. The rerun was necessary because a prior test had been terminated at 130% when high stresses were noted. Both the Slow Release Fitting and Heat Shield Loading test conditions were completed on 17 March. On 13 April 1966 personnel at the MSFC Load Test Annex removed the S-IC-S fuel tank from the facility following the teardown of the fuel tank/thrust structure/inter-tank setup.

Fins and fairings structural loading tests took place in the first half of 1966.

S-IC-5 Oxidizer tank assembly testing took place in the second half of 1966. The cut-off empty load test was performed on 12 July to a load of 140%. The second and third tests, cut-off full and launch rebound were completed by August.

The S-IC-S oxidizer tank assembly was moved to the Load Test Annex on 17 August from the Vertical Assembly Building. The first 7 of 9 test conditions were completed between 5 and 7 October 1966. Inertial load and burst testing followed. Burst testing of the fuel tank was delayed from late 1966 to early 1967.

The center engine alignment strut test was carried out to 140% of design limits on 29 September 1966. Testing began in July 1966 on the first of five Apex Gore specimens. Testing was completed on 20 July on the Upper LOX Bulkhead Gore. Testing of the other specimens ran through to January 1967.

Testing of the fairing to stage fittings was successfully completed on 21 July 1966 when failure occurred at 253% of design limit load.

The S-IC/S-II interface test was conducted to simulate loads during S-IC burnout. Stacking of the components began on 23 September 1966 and the completed specimen was delivered to the Load Test Annex at MSFC on 12 October. The five-part test began on 13 December. However, during the third test, the S-II interstage portion of the S-IC/S-II interface test specimen failed during loading at 127% design limit load on 15 December 1966. The specimen was compressed about 4 inches.

On 10 March 1967 MSFC test personnel installed the S-IC-S fuel tank in the S-IC Test Stand and began preparations for completing a series of hydrostatic pressure tests. The test data would be used by Boeing for tank design evaluation.

On 28 April 1967 the S-IC-S fuel tank hydrostatic pressure tests were completed in the S-IC Test Stand at MSFC.

During the first half of 1967 the final test (ultimate pressure of short LOX tank) of the 16 major tests required by the S-IC-S structural test program was completed satisfactorily.

However, one test was re-run. The Apex Gore Test, previously interrupted by a flash fire that destroyed the tarpaulin roof and much of the wiring in the testing facility, was resumed in early July 1967. Testing of the simulated hold repair was satisfactorily completed on 10 July 1967 at 225% of design load. On 31 July the testing of the plug-nut configuration hole repair was successfully accomplished at 225% of design load. No additional tests were required.

S-IC-F

Summary

Facilities checkout vehicle built at Michoud. First stage of the AS-500F vehicle, the first Saturn V at pad-39 in 1966.

Engines

The stage had four simple dummies and mock-up engine FM-105 installed.

Stage manufacture

The facilities checkout vehicle, S-IC-F, was built by Boeing at the Michoud Assembly Facility (MAF). Initiation of activities on the S-IC-F stage was delayed in 1964 due to S-IC-D bulkhead welding problems at Michoud. Major assembly work started on the S-IC-F stage in early 1965. Boeing welded the bulkhead and skin assemblies for the fuel and LOX tanks during the first quarter of 1965. On 14 June 1965 the fuel tank was completed and hydrostatic testing followed shortly afterwards. The fuel tank then moved to the final assembly station on 16 July 1965. LOX tank assembly was completed in early July followed by the hydrostatic test on 11 August 1965. Work on the forward skirt was completed on 30 June 1965 and the thrust structure was completed in early July.

Vertical assembly operations started on 15 July 1965 and ran through to 25 August. Following this the stage was transferred to the horizontal assembly station where it was mounted on to stationary support fixtures having been removed from the transporter that was needed for S-IC-2 horizontal assembly activities at MSFC.

68-66961 In order to assist in tests for the elimination of POGO the LOX duct was removed from the S-IC-F stage at MAF 5.68.

Erection of the S-IC-F stage on the mobile launch platform in the VAB at KSC. Note single mock-up engine and four dummies 14.3.66. (Courtesy Kipp Teague)

During systems installation in September three days were lost and minor damage sustained (rain spots on the LOX tank) when hurricane Betsy struck the area. Work on the stage was also affected by a labor strike in mid-September.

During October 1965 Boeing attempted to install F-1 mock-up engine FM-105 into position on the S-IC-F stage. Several practice runs were conducted using the engine installer tool and the engine simulator without incident. However, after lifting FM-105 from the dolly and positioning it on the stage, the installer tool could not be removed. Boeing then resorted to unwritten procedures to remove the tool. After removal of the tool, extensive thrust chamber tube dents were observed in the areas where the installer tool contacts the engine. Investigations were performed and both the installer tool and the procedure were improved.

On 16 December 1965 engine FM-105 was successfully temporarily installed in the stage in a trial run using the revised tool and procedure.

Following horizontal assembly and systems installation, S-IC-F moved to a systems test cell for post-manufacturing checkout on 19 November 1965.

Stage testing

Boeing workmen at MAF completed integrated testing of the S-IC-F stage, including final integration of mock-up engine FM-105 and four simple dummies.

Despite parts shortages post-manufacturing checkout proceeded in January 1966. Integrated testing was completed on 6 January and the stage was transferred from Test Cell I to Test Cell II for weighing and preparation for shipment. Post-manufacturing checkout was completed on 13 January 1966 at MAF. The stage was moved to the MAF dock, but it was found that the barge's GN2 pressurization system was contaminated

69-02620 The S-IC-F stage being loaded onto the Poseidon at MAF for the trip to MSFC 27.9.69.

so an alternative system had to be fabricated and installed. On 14 January 1966 the S-IC-F stage was loaded on board the barge Poseidon for its journey to KSC. It left MAF dock on the following day, arriving at KSC on 19 January 1966.

On 7 February 1966 the S-IC-F stage was moved into the low bay of the VAB in preparation for stacking. The stage was installed on LUT-1 on 15 March.

On 25 March the S-II-F stage was mated with the S-IC-F stage in the VAB. Between 28 and 29 March 1966 technicians at KSC stacked the S-IVB-F atop the S-IC-F and S-II-F. The following day the S-IU-500F was also added to the rocket.

Following completion of the AS-500F, facility checkout vehicle, power was first applied on 13 May 1966. Systems tests were completed on 24 May 1966 and the complete vehicle was rolled out of the VAB towards Launch Complex 39A on the following day. The vehicle traveled on crawler-transporter No. 1 and the journey took most of the day.

On 8 June 1966 AS-500F processing and test activities at LC-39A were interrupted because of the approach of Hurricane Alma. The vehicle was rolled back into the VAB as a precaution. Two days later, with the threat of the hurricane past, the AS-500-F vehicle was again moved out to pad 39A. The journey took about eight hours.

During the S-IC-F LOX loading test at LC-39A at KSC on 19 August 1966 the facility LOX system failed, causing delay in the S-II-F tanking schedule. An 18-inch flex line at the LOX storage tank ruptured resulting in the loss of the entire LOX load in the storage tank and some damage to the tank itself. Subsequent NASA inspection revealed that approximately 800,000 gallons of LOX were lost and that the inner shell of the LOX storage tank had collapsed inward about 16 feet in one area. Repairs to the tank were achieved by hydraulic pressurization of the tank to return it to its former shape.

The automatic loading of fuel, that had began on 12 August 1966, was completed successfully on 24 August 1966. Manual LOX loading tests on the S-IC-F stage were completed on 20 September 1966. The automatic LOX loading ended on 3 October 1966. The complete AS-500F vehicle automatic LOX and LH2 loading was satisfactorily accomplished on 12 October 1966 at LC-39A, following an aborted attempt on 8 October. The launch vehicle drain was satisfactorily accomplished, the sequence being, LOX drain preparations, simultaneous manual S-IVB LH2 and S-II LH2 drain, simultaneous automatic S-IVB and S-II LOX drain and S-IC LOX drain. At this point, AS-500F–1 wet tests were considered complete.

Following completion of testing at the pad, AS-500F was rolled back to the VAB from LC-39A on 14 October 1966. A minor bearing overheating problem on the Crawler Transporter was encountered during the move.

On 21 October 1966 KSC technicians completed de-erection of the AS-500F vehicle in the VAB. The de-erection process had began on 15 October with the CSM and Instrument Unit, continued on 16 October with the S-IVB-F and S-II-F stages and finished with the S-IC-F stage. Following de-erection, the S-IC-F stage was prepared for shipment to MAF for storage.

The S-IC-F stage arrived at MAF aboard the barge Poseidon on 10 December 1966 and was placed in storage.

As part of the effort to eliminate POGO vibrations in the S-IC stages, the LOX duct was removed from the S-IC-F stage at MAF during May 1968.

On 29 September 1969 Boeing delivered the S-IC-F stage to MSFC. The stage traveled from Michoud aboard the covered barge Poseidon.

S-IC-D

Summary

First S-IC stage built at Michoud. Used for AS-500D dynamic testing of the Saturn V stack in 1967. On display at ASRC in Huntsville.

Engines

The initial and final engine configuration was:

Position 101: Dummy engine
Position 102: Mass simulator
Position 103: Mass simulator
Position 104: Mass simulator
Position 105: Mass simulator

The engine configuration in the S-IC-D stage displayed at the Space and Rocket Center, Huntsville is:

Position 101: F-2003
Position 102: F-2007
Position 103: FM-102
Position 104: FM-104
Position 105: FM-105

Note: although the serial numbers are correct the positions cannot be verified.

Stage manufacture

The dynamic test stage, S-IC-D, was the first S-IC stage built and assembled by Boeing at the NASA Michoud Assembly Facility (MAF).
Boeing started fabrication of the S-IC-D stage in April 1964. Late in 1964 there were delays in tank bulkhead assembly at Michoud. At the end of the year personnel proceeded with assembly of the S-IC-D LOX and fuel tanks.

Early in 1965 Boeing completed the thrust structure and on 24 March 1965 it was installed in the Vertical Assembly Building at MAF. The fuel tank was installed in the hydrostatic test tower and the test was accomplished on 6 May 1965. Final welding and assembly of the LOX tank was completed and the tank underwent hydrostatic testing on 26 May 1965.

Final assembly of the forward and aft sections of the stage occurred in the VAB from 28 May to 16 June 1965, and the stage emerged from the assembly tower on 27 June 1965 after a delay for rework of GSE fittings.

The S-IC-D stage was transferred from the vertical assembly tower to the horizontal assembly station at the end of June. Activities here were completed on 25 August 1965, following which the stage underwent post-manufacturing testing.

Stage testing

Hurricane Betsy struck the area on 9-10 September, delaying work on S-IC-D, but otherwise the stage was undamaged. Damage to the stage test facility was sustained and required that S-IC-D be moved on 13 September to permit repairs to the checkout building. The dynamic booster stage was returned to the checkout building on 20 September for completion of tests, weighing, and preparation for shipment to MSFC where it was to undergo dynamic testing.

The S-IC-D stage on display at ASRC in Huntsville in 2004

67-61031 The S-IC-D stage being installed in the B-2 test stand at MTF 10.11.67.

The S-IC-D stage on display at ASRC in Huntsville in 2004

The S-IC-D stage on display at ASRC in Huntsville in 2004

99-03035 The S-IC-D stage is removed from the Saturn V dynamic test stand at MSFC following Configuration I testing 4.4.67.

68-64317 Transfer of the S-IC-D stage to Poseidon at MTF for the trip to MSFC 19.4.68.

S-IC-D left the MAF dock aboard the barge Poseidon at 1417 on 6 October 1965, arriving at MSFC docks at 0545 on 14 October 1965. This was Poseidon's first trip. Removal from the barge began at 1230 on the day of arrival and by 1615 the stage had been moved to the ME Lab, building 4755, for storage until needed. On 23 November 1965 a facility inspection cleared the way for the start of preparation for installation of S-IC-D into the Dynamic Test Stand at MSFC. All simulated engines were installed in S-IC-D on 7 December 1965. Mass dummies were used in four locations and a dummy engine in the fifth.

On 13 January 1966 workmen moved the S-IC-D stage from the ME Laboratory building and hoisted it into the Dynamic Test Stand. Damping tests began on 24 January 1966. Between 11 and 18 February 1966 S-IC-D suspension system and damping tests were performed. During March technicians began installing vertical fins, fairings and retrorockets on the stage. The initial phase of the S-IC-D vehicle-site integration testing was completed on 24 June 1966.

On 11 August 1966 S-IC-D thrust vector control tests began in the Dynamic Test Stand. The S-IC-D was ballasted with a full fuel tank, and the LOX tank was filled to the lower "Y" ring level. The TVC tests could not be completed due to the discovery of two defective actuators. While actuator problems were being resolved, Engine Lateral Modes (ELM) tests were initiated. Testing of the shaker was further delayed due to a burned out water pump in the cooling system. ELM testing continued using the live engine with engine gimballing locks not operating. This test was completed on 9 September. An ELM test with the gimballing locks operating was attempted on 12 September, but was stopped when the four lead weights used to simulate standard engine weight came loose.

Shaker testing and evaluation was re-instituted on 13 September and continued until 27 September. ELM testing was resumed on 27 September with mechanical locking of the engine gimballing actuators and was completed on 28 September. On 3 October 1966 S-IC-D fin and fairing tests that had started on 29 September were completed. Prior to the start of the Longitudinal test, cracked brackets on the upper thrust ring splice fitting were discovered and had to be replaced. On 17 October 1966 stage ballasting was completed and longitudinal testing started in the Dynamic Test Stand. Testing was suspended twice for various problems including the need to repeat ELM testing with a simulated engine. This testing was completed on 12 November 1966.

On 22 November 1966 Boeing personnel at Huntsville stacked the S-II-F/D stage on the S-IC-D stage in the Saturn V Dynamic Test Facility. The S-IVB-D Aft Inter-stage stacking began on 23 November and was completed on 28 November. The S-IVB-D stage was added on 30 November 1966, and the Apollo CSM and LES were added on 3 December 1966, completing the assembly of the AS-500D dynamic test vehicle.

The dynamic test campaign was classified as the AS-500D Configuration I test series. Configuration I was the complete Saturn V vehicle. Testing started in early January 1967 with the roll test being completed on 7 January. However a difference in the hardware configuration compared with the flight vehicle was noticed and it was decided to repeat the test with an improved configuration.

The Configuration I test program included roll testing, completed on 16 January 1967, pitch testing, from 20 to 23 January 1967, yaw testing, completed on 15 February 1967, and longitudinal testing, completed on 26 February 1967. A LOX vent line ruptured during the final longitudinal test. On 6 March MSFC supplied a spare vent line and authorized additional Configuration I tests to verify the Flight Control System.

Testing continued until 11 March 1967 when the final test was performed to verify the flight control system. De-stacking of the AS-500D vehicle began during the second half of March. By 30 March the LES, CSM, IU, S-IVB-D, and S-II-F/D stages had been removed.

On 4 April 1967 technicians removed the S-IC-D vehicle from the Dynamic Test Tower and began preparations for shipping the S-IC-D booster to MTF for storage in the Booster Storage Building.

The S-IC-D stage arrived at MTF from MSFC on 15 April 1967. It was placed in storage in the Booster Storage Building before being installed in the B-2 test stand on 10 November 1967. It underwent a propellant loading test on 14 December 1967 followed by a series of fuel tank drain tests beginning on 19 December 1967.

On 29 January 1968 Boeing personnel at MTF completed validation of anti-vortex modifications in the fuel tank of the S-IC-D stage. They had conducted four fuel tank drain tests during the test program.

The stage was removed from the B-2 test stand at MTF on 1 February 1968.

On 19 April 1968 the stage left MTF aboard the barge Poseidon en route for MSFC.

Ground breaking for the Alabama Space and Rocket Center in Huntsville began in July 1968. The following year the S-IC-D stage was moved to the outdoor exhibit area. The stage was equipped with five engines for display purposes. On 26 June 1969 it was moved to a position near the Astronautics Lab at MSFC. Two days later, on Saturday 28 June at 0500 CDT, the stage was transported along Rideout Road to the museum. A number of power lines had to be disconnected, road signs taken down and some poles moved. The museum opened its doors the following year and the stage has been there ever since. On 15 July 1987 the MSFC Saturn V was designated as a National Historic Landmark. Over the following 34 years weathering took its toll and there is now a campaign to restore the complete Huntsville Saturn V and place it under cover.

S-IC-T

Summary

"Thunderbird" stage used for static firing tests of the S-IC stage and F-1 engines at both MSFC and MTF. First Saturn V booster to be built at MSFC. First Saturn V first stage to be fired.

Engines

There were a number of different engine configurations in the S-IC-T stage, so for clarity each different build configuration is listed here.

Initial build and test numbers S-IC-01 to S-IC-03:

Position 101: F-2005 fitted but not fired
Position 102: F-2007 fitted but not fired
Position 103: F-2008 fitted but not fired
Position 104: F-2010 fitted but not fired
Position 105: F-2003

Configuration for test numbers S-IC-04 to S-IC-08:

Position 101: F-2005
Position 102: F-2007
Position 103: F-2008
Position 104: F-2010
Position 105: F-2003

Configuration for test numbers S-IC-09 to S-IC-10:

Position 101: F-2005
Position 102: F-2010
Position 103: F-2008
Position 104: F-2007
Position 105: F-2003

Configuration for test numbers S-IC-11 to S-IC-15:

Position 101: F-4T2
Position 102: F-2003
Position 103: F-2008
Position 104: F-2007
Position 105: F-3T1

Configuration for interface tests

Position 101: F-2003
Position 102: F-2007
Position 103: F-2008
Position 104: No engine
Position 105: F-3T1

Configuration for test numbers S-IC-T1 to S-IC-T2 (at MTF) and S-IC-20 to S-IC-22 (at MSFC):

Position 101: F-2003
Position 102: F-2007
Position 103: F-2008
Position 104: F-2010
Position 105: F-3T1

Current configuration on display at the Saturn Center at the Kennedy Space Center:

Position 101: *FM-106*
Position 102: *FM-106*
Position 103: F-2008
Position 104: F-2010
Position 105: F-3T1

Note: it is not known whether FM-106 is in the 101 or 102 position. Also the final engine is not known.

Stage manufacture

The all-systems test stage (S-IC-T) was built by MSFC in Huntsville. It was the first Saturn V booster stage built at MSFC and one of two ground test stages built at MSFC.

Assembly of the S-IC-T stage was several months behind schedule at MSFC in early 1964. Welding problems slowed progress. By mid-March the baffles in the fuel tank's upper half had been installed and the fuel exclusion riser had been placed in the tank's lower section. Assembly of the LOX tank also began at MSFC in March 1964. At Michoud Boeing proceeded with manufacture of the stage's forward skirt and inter-tank assemblies. A breakdown of automatic welding equipment in the MSFC Vertical Assembly Facility delayed completion of the fuel tank until late April. The stage's thrust structure moved into the assembly stand on 18 June. By June the fuel tank had completed hydrostatic, calibration, and leak checks; some internal components had been installed in the tank; and the completed LOX tank was in the vertical stand awaiting functional pressure tests.

The S-IC-T stage on display in the open at KSC in 1991

64-14299 The forward skirt being attached to the LOX tank of the S-IC-T stage at MSFC 18.10.64.

The S-IC-T stage on display in the Saturn Center at KSC in 2001

64-14376 The fuel and LOX tanls for the S-IC-T stage at MSFC 20.10.64.

64-14979 The assembled LOX tank for the S-IC-T stage at MSFC 3.11.64.

65-20225 The S-IC-T stage being hoisted into the S-IC Static Test stand at the MSFC West Test area 1.3.65.

65-18750 Trial fitting of an F-1 engine in the S-IC-T stage at MSFC 29.1.65.

64-16485 The S-IC-T fuel and LOX tanks being mated in building 4705 at MSFC 18.12.64.

64-16952 The assembled S-IC-T stage at MSFC 21.12.64.

64-16493 The fuel tank assembly for the S-IC-T stage being transported to the MSFC building 4705 for mating with the LOX tank 9.12.64.

65-17717 Rear view of the S-IC-T stage without engines at MSFC 11.1.65.

64-16507 The fuel tank assembly for the S-IC-T stage being transported to the MSFC building 4705 for mating with the LOX tank 9.12.64.

65-21629 An F-1 engine is installed in the S-IC-T stage at the S-IC Static Test stand in the MSFC West Test area 27.3.65.

65-20216 The S-IC-T stage en route from the MSFC manufacturing Engineering Laboratory to the S-IC Test Stand in the West Test area 1.3.65.

65-21876 F-1 engines being installed in the S-IC-T stage showing one engine on the vertical engine positioner 30.3.65.

65-20222 The S-IC-T stage being hoisted into the S-IC Static Test stand at the MSFC West Test area 1.3.65.

65-22474 First five engine firing of the S-IC-T stage at MSFC for a duration of 6.5 seconds 16.4.65.

65-23540 Five engine firing of the S-IC-T stage at MSFC; either 6 or 20 May 1965.

Technicians discovered and repaired numerous minute weld cracks in both the LOX and fuel tank bulkheads. LOX tank hydrostatic testing ended in August 1964, and fuel tank hydrostatic testing ended in November. The S-IC-T forward skirt and inter-tank, made at Michoud, arrived at Huntsville in July for rework and assembly.

The forward skirt was joined to the LOX tank on 18 October 1964. The fuel tank was joined to the thrust

65-21634 F-1 engine installation on the S-IC-T stage showing the skirt adapter on the second engine 27.3.65.

structure on 24 November. LOX delivery lines were completed on 3 December. The fuel and LOX tanks were joined to the inter-tank on 18 December 1964.

In early 1965 the final horizontal assembly of the forward and aft sections of the stage took place in the Manufacturing Engineering Laboratory. On 27-28 February 1965 the stage was prepared for transfer to the S-IC static test stand in the MSFC West Test Area.

Stage testing

The S-IC-T stage without its engines arrived at the S-IC West test stand on 1 March 1965. Installation in the test stand took approximately 2.5 hours. The booster was aligned, positioned, and pre-clamped in the test stand. On 8 March installation and checkout of the manual GSE was completed, nearly 11 weeks ahead of schedule. Between 8 and 10 March 1965 load tests of the structure were performed. Liquid oxygen was loaded to 15% on 24 March 1965. On 25 March the tanks were pressurized and leak checks were performed. All five F-1 engines were installed between 27 and 30 March 1965. The initial hot fire test was planned using the center engine, F-2003.

Test firing S-IC-01

Single engine firing performed at 1620 CDT on 9 April 1965. The test duration was 3 seconds (planned duration 7 seconds). The test was unintentionally terminated by operating personnel when the thrust had only reached the 40% level.

Test firing S-IC-02

Single engine firing performed at 1845 CDT on 9 April 1965. The test duration was 2.5 seconds (planned duration 7 seconds). This test performed just over two hours after the first test was automatically terminated by a safety circuit when it failed to obtain verification that no. 1 MFV was open. The cause was a broken wire in a cannon connector.

Test firing S-IC-03

Single engine firing performed at 1710 CDT on 10 April 1965. The test duration was 16.73 seconds (planned duration 15 seconds). The first successful single engine stage firing was terminated as planned with a console cutoff.

Test firing S-IC-04

Five engine firing performed at 1458 CDT on 16 April 1965. The test duration was 6.5 seconds (planned duration 7 seconds). A major milestone in the Saturn program was this first five engine stage test of the S-IC booster that was performed two months ahead of schedule. Performance was as expected and cutoff was initiated by the stage timer.

Test firing S-IC-05

Five engine firing performed at 1510 CDT on 6 May 1965. The test duration was 15.55 seconds (planned duration 15 seconds). During this test the 101 engine (F-2005) was gimbaled. The performance was again nominal and the firing was terminated by the firing sequencer as planned.

Test firing S-IC-06

Five engine test performed at 1458 CDT on 20 May 1965. The test duration was 40.84 seconds (planned duration 40 seconds). The flak curtains were removed prior to this test. Console cutoff was achieved as planned.

Test firing S-IC-07

Five engine test performed at 1608 CDT on 8 June 1965. The test duration was 41.1 seconds (planned duration 90 seconds). The test was prematurely terminated by the pillbox observer because of an aspirator flow-back.

Test firing S-IC-08

Five engine test performed at 1459 CDT on 11 June 1965. The test duration was 90.9 seconds (planned duration 90 seconds). The firing was terminated by a console initiated cutoff as planned. The inspection following this test revealed cracks in the main injectors of the engines in positions 101, 102, 104 and 105. This necessitated the removal of the engines for injector replacement. Following repairs the engines were reinstalled in the stage that remained on the test stand. The engines in the 102 and 104 positions were transposed.

Test firing S-IC-09

Five engine firing performed at 1756 CDT on 29 July 1965. The test duration was 17.6 seconds (planned duration 40 seconds). For this test the fuel and LOX tank onboard pressurization system was installed. Engine cutoff was initiated by an observer monitoring the LOX pump inlet pressure. He detected a low pressure on engine 103. The GFCV and auxiliary LOX pressurization had failed to operate properly.

Test firing S-IC-10

Five engine test performed at 1602 CDT on 5 August 1965. The test duration was 143.6 seconds (inboard engine), 147.6 seconds (outboard engines). The test was planned to run until LOX depletion. This was the first full duration firing and was completely successful. The test was terminated when LOX depletion was reached and was commanded by an observer prompted by the LOX pump inlet temperature. Cutoff engine sequence was 1-2-2. This test concluded the S-IC-T manual configuration firings.

The stage was removed from the test tower and returned to the ME Lab on 12 August where the stage configuration was modified for automatic firings.

Technicians completed electrical modifications on 7 September and mechanical modifications early in October. The stage was then returned to the West Area static test tower.

Test firing S-IC-11

Five engine automatic test performed at 1641 CDT on 8 October 1965. The test duration was 42.38 seconds (inboard engine), 47.80 seconds (outboard engines). The test was planned to run until LOX depletion that was predicted to be at 45 seconds. This was the first automatic configuration firing and was completely successful. Prior to this test there was a rearrangement of the engines on S-IC-T including the addition of two new engines. These two new engines had been assembled from spare parts inventory as part of a NASA learning exercise. The cutoff was initiated by the observer of the LOX pump inlet temperature noting LOX depletion. Cutoff engine sequence was 1-2-2.

Test firing S-IC-12

Five engine automatic test performed at 1640 CST on 3 November 1965. The test duration was 90.5 seconds. The test was planned to run until LOX depletion that was predicted to be at 145 seconds. However the test was inadvertently terminated by the LOX tank ullage pressure redline observer after 90.5 seconds. This was the first test that was run by Boeing personnel, with MSFC Test Laboratory supervision.

Test firing S-IC-13

Five engine automatic test performed at 1307 CST on 24 November 1965. The test duration was 148.4 seconds (inboard engine), 153.4 seconds (outboard engines). The test was intended to run to LOX depletion at about 145 seconds. Unusually the propellants had been held on board the night of 23 November through to the test day. The test was very successful and was terminated as planned by the LOX depletion observer. Engine cutoff sequence was 1-2-2.

Test firing S-IC-14

Five engine automatic test performed at 1609 CST on 9 December 1965. The test duration was 146.07 seconds (inboard engine), 150.02 seconds (outboard engines). The scheduled duration was 150 seconds. The successful test was terminated as planned by the console operator.

Test firing S-IC-15

Five engine automatic test performed at 1500 CST on 16 December 1965. The test duration was 40.96 seconds (inboard engine), 45.96 (outboard engines). The planned duration was 40 seconds. The successful test was terminated as planned by the console operator. This concluded the automatic program of firings on the S-IC-T stage. Cumulative firing time on the stage was 867.1 seconds. Of the 67 propulsion test objectives, during the testing campaign, 57 had been fully completed and 5 had been partially completed. Three objectives had been cancelled, one was incomplete and one was not performed because of possible hazard to stage and facility.

Subsequent to the final firing pre-valves and flow-meters were inspected, automatic procedures were given final checkouts and four engines were removed for the installation of gold plated injector plates. Preparations began for the removal of the stage from the S-IC West test stand. The S-IC-T stage was removed from the stand on 19 January 1966 to make room for the acceptance firings on the first three S-IC flight stages. The stage was transferred to the Manufacturing Engineering Hanger building for storage prior to its conversion to the S-IC-4 configuration. S-IC-T was to return to the MSFC S-IC test stand on 7 July 1966 for a short interface check before being removed on 29 July 1966. There was consideration of performing a 40 second firing and some fuel and LOX loading tests but due to the urgency to checkout the MTF stand further firings at MSFC were deferred. During this time S-IC-T had only 4 engines attached and the positions of some of the engines had changed since the final automatic test some 7 months earlier. After the configuration tests engine F-2010 was installed in the 104 position in preparation for static tests at MTF.

Transfer of S-IC-T to MTF

Following updating of the stage to the S-IC-4 configuration the S-IC-T stage was transferred by the Poseidon from MSFC to MTF, leaving on 17 October 1966, for a short series of verification firings in the new B-2 test stand at MTF. These tests would qualify the test stand in terms of booster interfaces and performance measurement when compared with the results obtained on the same stage at MSFC.

The stage arrived at MTF on 23 October 1966 and was stored pending completion of the B-2 test stand. The stage configuration for the MTF testing was identified as S-IC-T/4. On 16 December 1966 the stage was removed from storage and on the following day placed in the B-2 test stand in preparation for the static firings.

Prior to the static firings a series of pre-static firing tests were performed.

Tests included electrical tests

- bus resistance
- bus voltage
- bus isolation
- power transfer
- electrical heaters
- thrust vector control
- sequencing
- range safety

mechanical tests

- external leak checks
- valve seat leakage
- primary seal leakage with flow meters at bolted

flanges
- relief valve settings
- regulator settings
- pressure switch settings
- valve timing

instrumentation tests

- liquid level initialization
- measurement power
- measurement profile

Test firing S-IC-T1

In preparation for the first firing at MTF two propellant loading trials were performed, on 14 and 25 February 1967. Fuel for the firing was loaded on 2 March 1967 and the LOX was loaded on the firing day. The firing itself was performed at 1721 CST on 3 March 1967 and marked the first use of the B-2 stand at MTF. The duration of test was 15.2 seconds as measured from simulated liftoff to cutoff complete (planned duration 15 seconds). Engine system performance was satisfied with a successful 1-2-2 engine start sequence and a 3-2 engine shut down sequence. There were various minor problems encountered in the test:

Engine 105 indicated a low thrust level
A small fire was present on engine 101 after cutoff
Pre-pressurization and umbilical pressures were below required limits
The Parker vent and relief valve operated extremely slowly
The GSE LOX dome purge lockup pressure was above the required limit
A number of taper pins in GSE and facility electrical terminal boards vibrated loose during the firing, causing loss of some electrical signals between the stage and ground equipment
A number of fuel leaks were detected on engine 101 during post test inspection.

On a positive note, there were no stage structural deficiencies during the tests. The stage electrical systems were satisfactory as were the stage instrumentation and telemetry systems.

The engine data recorded during the static firing was used to demonstrate that the engine thrust, specific impulse, engine manifold torus temperature, fuel pump balance cavity pressure and heat exchanger performance were within specification.

The stage data recorded during static firing was used to demonstrate that the stage thrust vector control, pressurization, electrical power, control pressure, GN2 purge, flight measurements, range safety, propellant loading, sequencing and pogo suppression systems were performing within specified limits.

Limited post firing checks were conducted before the stage was readied for a final, longer, firing at MTF.

Test firing S-IC-T2

Again the stage was in the S-IC-4 functional configuration for this test. The firing took place at 1516 CST on 17 March 1967. Firing duration was 60.184 seconds as measured from simulated liftoff to cutoff complete (planned duration 60 seconds). The 1-2-2 engine start and 3-2 engine cutoff sequences were achieved.

The test was considered successful with the following specific comments being made

the engine systems met the static firing objectives
stage propellant loading and delivery systems were satisfactory
the onboard purge and control pressure systems were satisfactory
the thrust vector control system was satisfactory
the stage electrical systems were satisfactory
there were no indications of stage structural deficiencies

The stage instrumentation and telemetry systems were active but were not evaluated. The GSE and facility systems were satisfactory for this firing, but corrective action was required prior to flight stage firings.

Post firing checks, similar to those performed before the firings, were performed in order to check that the firings had not had an adverse effect on the stage systems.

Overall the tests were extremely successful and validated the use of the B-2 test stand at MTF for use with S-IC flight stages. In particular this test validated the flame bucket water flow pattern of the B-2 test stand.

The S-IC-T stage was removed from the B-2 test stand on 24 March 1967. Removal from the test stand was delayed due to a failure in the electrical controls of the MTF test stand main derrick. The derrick control system was repaired and proof load tested, subsequently allowing the S-IC-T stage to be removed from the test stand.

The stage was transported from MTF to MAF aboard the barge Pearl River on 31 March 1967. At Michoud dockhands transferred the stage aboard the covered barge Poseidon for shipment to Huntsville. The stage left MAF on 1 April 1967 arriving back at the MSFC on 9 April 1967. At MSFC it entered storage in the ME Laboratory to await use in research and development tests.

S-IC-T was reinstalled in the MSFC S-IC test stand on 1 June 1967 for a planned final 40 second static firing test to verify continuity of performance after the series of tests at MTF and to demonstrate the operational readiness of the S-IC test complex, the S-IC-T stage, and provide training for KSC launch crews. Intended to be a single test it was expanded to three due to problems with the first two attempts when the firings were inadvertently terminated by ground observers. The engine configuration remained the same as for the mid-1966 configuration tests.

Test firing S-IC-20

Five engine test performed at 1500 CDT on 1 August 1967. The test duration was 2.15 seconds (planned duration 40 seconds). The test was inadvertently terminated by the observer who was checking the 102 engine turbine temperature. To correct this problem and prevent its recurrence 5 of the 10 chamber pressure channels were changed completely for the next test.

Test firing S-IC-21

Five engine test performed at 1500 CDT on 3 August 1967. The test duration was 3.60 seconds (planned duration 40 seconds). Again the test was inadvertently terminated this time by the observer on the 105 engine fuel pump inlet pressure. A check of the transducers after the test revealed that the original 5 chamber measurements were discrepant. The 5 transducers changed after test number 21 were in their normal range. To further isolate the problem, one of the original chamber pressure transducers was changed prior to test 22.

Test firing S-IC-22

Five engine test performed at 1923 CDT on 3 August 1967. This was a second attempt during the day to achieve a 40 second firing. The test was successful with a main-stage duration of 41.74 seconds being achieved. There was an automatic cutoff by the terminal countdown sequencer as planned. This test marked the final static firing of the S-IC-T stage. In total it was fired 18 times at MSFC and 2 times at MTF.

MSFC conducted a series of LOX tank fill and drain tests on the S-IC-T stage in order to evaluate the LOX transfer and bubbling procedures used at KSC. The first two tests were run in January 1968 and the third and final test was performed in February 1968.

A new LOX loading test was run at MSFC in May 1968. As a result of this test and the earlier ones performed two changes were recommended for S-IC flight vehicles;

- to incorporate a one-hour standby with LOX in the suction lines prior to tanking to eliminate negative LOX tank pressures
- to provide an additional safety margin against geysering by increasing the helium bubbling rates.

On 10 July 1968 the MSFC materials laboratory submitted a sample of the S-IC-T LOX ring baffle to evaluation of stress corrosion cracking.

Very little is known about the whereabouts of S-IC-T between 1968 and the mid 1970s when it was shipped to KSC to form a Saturn V display outside the VAB. It was on display in an un-mated horizontal configuration from April 1976. In 1979 the title was turned over to the Smithsonian. The stage remained there until 1996, when, on 11 May, it was transferred 2 miles to its new home. It was refurbished, configured as the S-IC-6 stage and located in the new indoor Saturn V center at KSC that opened its doors to the public on 5 December 1996.

S-IC-1

Summary

This was the first flight S-IC stage produced. It included an engine allocation change before engine installation in the stage. Two of the engines were subject to single engine hot fire calibration checks at the NASA Marshall Space Flight Center (MSFC) prior to installation in the stage. Another engine was removed from the stage after the stage firings and subjected to a single engine verification prior to being reinstalled in the stage. This was one of only two S-IC flight stages (S-IC-11 being the other) that were subject to two stage static firings.

Engines

The initial engine configuration was:

Position 101: F-3013
Position 102: F-3015
Position 103: F-3014 – allocation changed and replaced by F-3016
Position 104: F-3012
Position 105: F-3011

Spare engine: F-3016 – replaced F-3014

The stage hot fire and final flown engine configuration was:

Position 101: F-3013
Position 102: F-3015
Position 103: F-3016
Position 104: F-3012
Position 105: F-3011

Stage manufacturing

The S-IC-1 booster began to take form at MSFC towards the middle of 1964. By that time part of the fuel tank's two bulkheads had been welded together. By the end of 1964 the upper bulkhead of the fuel tank was complete and baffle installation was in progress.

At MSFC the fuel tank bulkhead subassembly was moved into the MSFC Vertical Assembly Facility on 3 February 1965. After joining of the two assemblies to form the tank, workmen installed baffles on 5 March 1965. Welding of LOX tunnels into the tank followed. The thrust structure, which had arrived from Boeing at MAF on 27 January 1965, underwent several major engineering changes.

Joining of the fuel tank and thrust structure started on 27 April 1965. The completed assembly moved to the horizontal assembly area of the ME Lab on 10 May 1965. LOX tank assembly was completed on 26 April

1965 and hydrostatic testing and cleaning of this tank was completed on 1 June 1965. The forward skirt was mated between 1-9 June 1965 and the inter-tank attached between 9-25 June 1965. The LOX tank was then moved to MSFC building 4705 for mating with the fuel tank thrust structure assembly.

During the second quarter of 1965 MSFC rejected all electrical cables fabricated by Boeing in Huntsville for the S-IC-1 thrust structure area because of improper length. New cables fabricated at MAF by Boeing were checked out on the S-IC mockup at Huntsville before use. The shortage of acceptable cables lasted through to the end of 1965.

Failure of the destruct ordnance cowling on the S-IC-T stage during static firing dictated the redesign of the cowling. Engineers reworked the S-IC-1 cowling by installing additional bonded brackets to the propellant tank skin.

Horizontal assembly proceeded through July, August and most of September. During this time the five F-1 engines were installed in the stage between 18 August and 1 September. MSFC personnel completed horizontal assembly and transferred the stage to the Quality Laboratory, building 4708, on time on 27 September 1965. Late delivered parts and systems, and modification kits were installed out of sequence during quality checkout of the stage.

Stage testing

The S-IC-1 stage was subjected to post-manufacturing checkout (PMC).

The first S-IC flight stage, S-IC-1, completed post manufacturing checkout (PMC) at MSFC on 16 January 1966.

Following the validation of the S-IC test stand at MSFC with the final firing of the S-IC-T stage on 16 December 1965 the path was clear for the static firing testing of flight booster stages. S-IC-1 stage was hoisted atop the test stand on 24 January 1966. After verification of all systems, including pressure testing and engine alignment checks, the fuel and then the LOX were loaded in preparation for the two planned firings (as this was the first flight stage it was the only flight stage to have two firings planned).

Test S-IC-16

The firing was performed at 1518 CST on 17 February 1966. The firing duration was 40.79 seconds (planned duration 40 seconds) and all main objectives were met. The logbook records that the engines were lowered to remove the suction line screens. Firing cutoff was initiated by the firing panel operator as planned.

Test S-IC-17

The firing was performed at 1459 CST on 25 February 1966. The firing duration was 83.2 seconds. It had been planned that the test would run for about 125 seconds with cutoff occurring after depletion of LOX as witnessed by the LOX pump inlet pressure redline observer. However there was a combustion chamber pressure measurement failure on both the primary and redundant systems on engine 101 which caused an early shutdown. As the cumulative firing time on the stage was 123.99 seconds this total was considered sufficient and no further firings took place.

The engine data recorded during the static firings was used to demonstrate that the engine thrust, specific impulse, engine manifold torus temperature, fuel pump balance cavity pressure and heat exchanger performance were within specification.

The stage data recorded during static firings was used to demonstrate that the stage thrust vector control, pressurization, electrical power, control pressure, GN2 purge, flight measurements, range safety, propellant loading, sequencing and pogo suppression systems were performing within specified limits.

Following a post firing evaluation checkout the S-IC-1 stage was removed from the test stand on 14 March 1966 and placed in MSFC's new ME Lab (building 4755) for refurbishment prior to start of post-static checkout. Because of a doubt about the 101 engine, F-3013, it was removed from the stage and underwent a single engine test at MSFC on 7 April 1966 before being declared acceptable and reinstalled in the stage. Post-static refurbishment was completed on 12 May 1966 and the stage was moved to the Quality Lab for post static checkout. This was completed on 9 August 1966 and the stage was turned over to ME Lab on 10 August to prepare for shipment to KSC. The stage was loaded on to the barge Poseidon on 26 August 1966 to start its first journey away from MSFC. The ship was delayed in New Orleans due to a strike between the Towing subcontractor and the Union. It left there on 7 September for the remainder of the journey to KSC, arriving on 12 September 1966.

The S-IC-1 stage eventually provided the first stage of AS-501 that launched Apollo 4 on 9 November 1967.

S-IC-2

Summary

This was the final S-IC stage to be built and hot fire tested at the NASA Marshall Space Flight Center (MSFC). Although the following stage was also tested at MSFC it was to be built at the NASA Michoud Assembly Facility (MAF). S-IC-2 included an engine that was removed following the stage static firing test, retested at engine level, and re-installed in the same location on this stage.

Engines

The initial and final flown engine configuration was:

Position 101: F-4017
Position 102: F-4018
Position 103: F-4019
Position 104: F-4021
Position 105: F-4020

Spare engine: F-4028 – not used

Stage manufacturing

Welding of the fuel and LOX tank gore segments for the S-IC-2 booster stage began in early 1965. Fuel tank closeout welding was completed on 28 May 1965. LOX tank assembly was completed and baffle installation started in June 1965. In the ME Lab's VAB at MSFC, mating of the fuel tank to the thrust structure was completed on 18 September 1965. The following day horizontal assembly was started in building 4755. It was the first flight hardware to occupy building 4755, the Multi-purpose Vehicle Assembly building.

Assembly of the thrust structure to the fuel tank ended on 8 November. LOX tank hydrostatic testing and cleaning occurred from 28 September to 18 October 1965. The forward skirt and LOX tank assemblies were joined together on 25 October 1965. The inter-tank was completed on 18 October 1965 and then joined to the forward skirt/LOX tank assembly between 25 October and 8 November 1965.

This forward section moved from the vertical assembly tower to the horizontal assembly area at MSFC (building 4705) on 4 December 1965. Following this transfer, the forward and aft sections of the stage were mated.

The five F-1 engines were installed on the stage between 6 and 11 December 1965. They were subsequently removed, because they were contaminated with hydraulic fluid, cleaned and reinstalled between 7 and 14 April 1966.

Horizontal assembly of the S-IC-2 stage was completed in the MSFC Manufacturing Engineering Laboratory on 17 January 1966.

Stage testing

Following completion of manufacture and assembly the stage was moved to the Quality and Reliability Assurance Laboratory for post-manufacturing checkout (PMC).

Stage status checks were completed on 1 February 1966 and power was applied to the S-IC-2 stage two days later. The engines were reinstalled in April as described above and the stage returned to the Laboratory on 25 April for continuation of post-manufacturing checkout, which was completed on 12 May 1966.

The second flight stage, S-IC-2, was moved to the static test stand and was installed in the test stand on 17 May 1966. Propellant loading trials were performed on 26-27 May 1966. After verification of all systems the fuel and then the LOX was loaded in preparation for the planned firing.

Test S-IC-18

The firing was performed at 1843 CDT on 7 June 1966. The firing duration was 126.3 seconds (planned duration 125 seconds). For this test the engines were canted to remove the suction line screens. Firing cutoff was initiated by the firing panel operator as planned.

The engine data recorded during the static firing was used to demonstrate that the engine thrust, specific impulse, engine manifold torus temperature, fuel pump balance cavity pressure and heat exchanger performance were within specification.

The stage data recorded during static firing was used to demonstrate that the stage thrust vector control, pressurization, electrical power, control pressure, GN2 purge, flight measurements, range safety, propellant loading, sequencing and pogo suppression systems were performing within specified limits.

Following a post firing evaluation checkout the S-IC-2 stage was removed from the test stand on 16 June 1966 and transferred to ME Lab for refurbishment. Because there was a high thrust problem with the 101 engine, F-4017, it was removed from the stage and subjected to three single engine static firings at MSFC on 25 July, 23 August and 24 August 1966. With its problem corrected and performance verified it was reinstalled in the S-IC-2 stage on 27 August 1966.

On 10 August ME Laboratory personnel at MSFC completed the refurbishment of the S-IC-2 stage and moved it to the Quality Lab checkout station for commencement of post-static checkout. The late availability of change hardware and documentation prolonged the scheduled completion of checkout during October and November. The stage distributors were rejected after being pulled for inspection. Post-Static checkout testing was held up until 21 November in order to allow for repair and reinstallation of the distributors. The stage simulated flight tests began on 30 November and were completed on 2 December.

The long-delayed Post Static Checkout was completed on 12 December 1966 and the stage was disconnected from power and transferred to ME Laboratory for incorporation of changes and modifications. On 6 February 1967 the stage entered final checkout in the MSFC Quality Laboratory. On 21 February 1967 S-IC-2 was returned from the Quality and Reliability Assurance Laboratory to the ME Laboratory following rerun of certain system tests, completed on 13 February

1967, and which were necessary to complete post-static checkout. The tests became necessary following major rework of the stage distributors during post-static modification. The stage was loaded on to the barge Poseidon to start its first journey away from MSFC. The ship left port on 3 March 1967 for the journey to KSC, arriving on 13 March 1967.

The S-IC-2 stage eventually provided the first stage of AS-502 that launched Apollo 6 on 4 April 1968.

S-IC-3

Summary

This was the last S-IC flight stage to be hot fire tested at the NASA Marshall Space Flight Center (MSFC). It was the only S-IC stage, subsequently used for a manned launch, to be hot fire tested at MSFC. It was the first S-IC flight stage to be built at MAF. Also it was the only S-IC stage built at the NASA Michoud Assembly Facility (MAF) but hot fire tested at MSFC. It was the only S-IC stage to include a particular engine that was installed in two different locations on the same stage at different times. This stage had two engines replaced due to defects. Finally, this stage had an engine replaced after the stage static firing test which meant that one engine in the final configuration had not been exposed to a stage-level static firing test.

Engines

The initial engine configuration was:

Position 101: F-4023
Position 102: F-4022
Position 103: F-4025
Position 104: F-4026
Position 105: F-4024 – replaced by F-4027

Spare engine: F-4027 – ultimately utilized in place of F-4023

The engine configuration for the static firing test was:

Position 101: F-4023 – replaced by F-4024
Position 102: F-4022
Position 103: F-4025
Position 104: F-4026
Position 105: F-4027

The final flown engine configuration was:

Position 101: F-4024
Position 102: F-4022
Position 103: F-4025
Position 104: F-4026
Position 105: F-4027

Stage manufacturing

Assembly of the third S-IC flight stage, and first one to be built at MAF, started in late 1964, and progressed early in 1965. Welding of the LOX tank and fuel tank subassemblies ended in June. Boeing workmen completed the thrust structure on 30 August 1965 and the inter-tank on 10 September 1965. Completion of the forward skirt final assembly was delayed until 9 October 1965 because of the effects of Hurricane Betsy (that struck on 9 September) and a labor strike. Both the fuel tank and the LOX tank sustained some corrosion and pitting damage from water and falling gravel when the VAB roof was severely damaged by the hurricane.

Hydrostatic testing of the fuel tank was completed on 4 October 1965, and the similar test on the LOX tank was finished on 11 October 1965. Vertical assembly of the stage was completed on 8 December 1965 after which the stage was moved to the horizontal assembly area.

67-59502 The S-IC-3 stage being moved from storage in the Manufacturing Building at MAF 12.9.67.

The five F-1 engines were installed in the stage during January and February 1966. Two of these engines subsequently would be removed from the stage for different reasons.

Stage testing

On 9 March 1966 the stage was placed in the Stage Checkout Building at MAF for post-manufacturing checkout. Hardware shortages and design changes were delaying assembly and on 19 March the incomplete stage was moved to the systems test area with 162 installations remaining to be completed. Power was applied on 23 March. On 15 May 1966 post-manufacturing checkout was interrupted when about 50% complete to accommodate a planned period of stage modification.

On 6 June 1966 post-manufacturing checkout of the S-IC-3 stage resumed at MAF following incorporation of

21 change orders. Finally, post-manufacturing checkout was completed successfully on 24 August 1966. S-IC-3 was the first S-IC stage to undergo this testing at MAF.

This third flight stage, S-IC-3, was shipped from MAF to MSFC for the static firing test. It was the only S-IC stage to be built at MAF and fired at MSFC. The stage left MAF on 22 September 1966 on board a barge, arriving at MSFC on 2 October 1966. It was installed in the test stand on 3 October 1966. Power was applied to the stage on 10 October, permitting the start-up of electromechanical tests. The GN2 purge system test was accomplished by 20 October. The S-IC-3 propellant load test was started on 25 October 1966 and completed successfully the following day. Before the static test could take place a problem was discovered in engine F-4024 in the 105 position. There was suspected contamination of the engine fuel system. This engine was removed on 4 November 1966 and replaced with the spare, engine F-4027 that had been brought to MSFC separately, leaving on 1 November 1966 and arriving at MSFC on 6 November 1966. The spare engine was installed in the stage on 7 November. After verification of all systems the fuel and then the LOX was loaded in preparation for the planned firing.

Test S-IC-19

The firing was performed at 1538 CST on 15 November 1966, just a week after the 105 engine had been replaced. The firing duration was 121.7 seconds (planned duration 125 seconds). For this test the engines were canted to remove the suction line screens. Firing cutoff was initiated by the firing panel operator as planned. It was determined that the incentive "burn time" was 127.3 seconds. This is the time from open command to the first "four-way control valve" start solenoid, until chamber pressure decay in the last engine shutdown reached 10% of main-stage value.

The engine data recorded during the static firing was used to demonstrate that the engine thrust, specific impulse, engine manifold torus temperature, fuel pump balance cavity pressure and heat exchanger performance were within specification.

The stage data recorded during static firing was used to demonstrate that the stage thrust vector control, pressurization, electrical power, control pressure, GN2 purge, flight measurements, range safety, propellant loading, sequencing and pogo suppression systems were performing within specified limits.

Following a post firing evaluation checkout the S-IC-3 stage was removed from the test stand on 21 November 1966. The following day the stage was shipped back to MAF on the barge Poseidon, arriving on 27 November 1966. It would be the first S-IC stage to undergo refurbishment and post static checkout (PSC) at MAF. The stage was moved to the Booster Test Building for refurbishment. Refurbishment was completed and Post Captive Checkout started on 17 December 1966.

On 3 February 1967 Boeing completed the post-static checkout of the S-IC-3 stage at MAF. On 16 February the stage was moved from the Stage Test Building to the Vertical Assembly Building for incorporation of a splice angle change. The change was completed on 27 February, and the stage was moved back to the Stage Test Building, where it was weighed and relays and distributors were installed.

The stage was accepted by NASA on 15 March 1967. On 28 March 1967 NASA placed the stage in storage in the Manufacturing Building at Michoud to await shipment to KSC. Boeing would complete the post-static modifications to the stage during this period of storage.

Special cleanliness controls were put in place for the change out of the helium bottles in the LOX tank. It was deemed necessary to replace the bottles as they were found to be susceptible to stress corrosion cracking. On 17 July the first bottle was winched from the LOX tank. Change out of the bottles was completed during the first week of August. On 22 August 1967 Boeing personnel at Michoud moved the S-IC-3 stage from the Stage Test Building to the Manufacturing Building for temporary storage. On 12 September 1967 the stage was moved from storage to the Stage Test Area at MAF for post-storage modifications and tests.

On 13 November 1967 Boeing began a retest of S-IC-3 at MAF to check out the numerous engineering change procedures incorporated since the acceptance of the stage by NASA. A successful simulated flight test of the S-IC-3 stage on 5 December 1967 was the last of these retests required because of the long post-acceptance storage period in which workmen had completed modifications originally scheduled for installation at KSC.

All testing was completed on 8 December 1967 and NASA accepted the stage as "ready for shipment" on 11 December 1967. Final weighing took place in Test Cell number 1 of the Stage Test Building on 14 December 1967. The weight was recorded as 311,970 pounds.

The stage was loaded on to the barge Poseidon on 21 December 1967 to start its journey to KSC. The ship left port on 23 December 1967, arriving at KSC on 27 December 1967. That was not quite the end of the story for the S-IC-3 stage as a further engine change was required, and uniquely in the Saturn program this occurred at the launch site. Problems with the 101 engine, F-4023, meant that it was removed and replaced with F-4024 on 31 May 1968.

The S-IC-3 stage eventually provided the first stage of AS-503 that launched Apollo 8 on 21 December 1968.

S-IC-4

Summary

Manufactured at MAF and static fired at MTF. Launched Apollo 9.

Engines

The initial and final flown engine configuration was:

Position 101: F-5029
Position 102: F-5032
Position 103: F-5031
Position 104: F-5033
Position 105: F-5030

Stage manufacture

Welding operations on the S-IC-4 stage began in the second quarter of 1965 at Michoud. Thrust structure final assembly began on schedule on 13 August 1965, however the completion was delayed by a labor strike. Prior to the arrival of hurricane Betsy on 9 September the fuel tank was almost complete. There was some corrosion and pitting damage to the fuel tank's upper bulkhead. By the end of September 1965 the LOX tank was nearly complete. By the end of 1965 the fuel tank was being installed in the hydrostatic test tower and the LOX tank subassemblies were being joined and welded.

67-58549 S-IC-4 stage being weighed at MAF test cell 1 of the stage test building 21.8.67.

Following completion of the thrust structure of the stage on 10 January 1966 the inter-tank assembly was completed on 20 January 1966. The forward skirt assembly was completed on 8 February and the hydrostatic test of the fuel tank took place on 10 February. Vertical assembly of the S-IC-4 stage began at MAF on 28 February 1966.

The five F-1 engines were installed in the S-IC-4 stage between 16 and 27 June 1966.

Stage testing

On 3 August 1966 the stage was moved into the Stage Checkout Building at MAF for the start of post-manufacturing checkout.

Checkout was completed successfully on 1 December 1966 after which the stage was put into storage at MAF on 16 December 1966.

On 11 January 1967 Boeing moved the stage from its storage area in the MAF Manufacturing Building to the VAB to incorporate an engineering change necessitated by stress corrosion. The change would involve replacement of 32 splice plates of the upper and lower thrust rings. The modifications were completed by 15 February, but the stage had to remain at Michoud until the S-IC test stand at MTF could be successfully checked out.

The S-IC-4 stage was shipped by the barge Pearl River from MAF to MTF on 4 April 1967. The stage was installed in the B-2 position of the B test stand on the following day and pre-static firing tests were performed. These tests essentially repeated the post-manufacturing checkout (PMC) already performed at MAF prior to shipment.

Test firing S-IC-4-1

S-IC-4 was the first booster flight stage to be tested at MTF following the verification of the test stand using the S-IC-T stage two months earlier. Power-on was achieved on 14 April. A propellant loading test was performed on 5 May 1967. The fuel for the firing itself was loaded on 15 May 1967 and the LOX loaded on the firing day. The firing took place at 1520 CDT on 16 May 1967. The firing duration was 125.096 seconds as measured from simulated liftoff to cutoff complete (planned duration 125 +/- 2 seconds).

The engine systems were satisfactory. The 3-2 engine cutoff sequence was successful, but the 1-2-2 start sequence was not achieved.

There were no indications of stage structural deficiencies.

Suspect pressure data meant that the GG flow rate and turbine manifold torus temperature performance could not be evaluated.

The following systems were satisfactory:

    the thrust vector control system
    fuel and LOX systems were satisfactory
    the onboard purge and control pressure systems
    stage environmental control systems
    electrical systems
    the range safety system
    stage instrumentation/telemetry systems

A number of parameters exceeded their limits either before or during the test:

- helium flow control valve 5 opening command time
- LOX tank ullage pressure during the pre-pressurization
- temperature limits of the LOX heat exchangers
- closing times of the LOX valve vent valves
- pressure and performance limits of the GN2 storage spheres
- GSE supplied gas exceeded limits at the forward compartment umbilical
- initiation of the airborne tape recorder playback exceeded the maximum time limit
- the LOX dome purge lockup pressure exceeded the maximum limit
- the engine 101 pitch actuator locks-off null position exceeded the maximum limit during pre-static checkout

The engine data recorded during the static firing was used to demonstrate that the engine thrust, specific impulse, engine manifold torus temperature, fuel pump balance cavity pressure and heat exchanger performance were within specification.

The stage data recorded during static firing was used to demonstrate that the stage thrust vector control, pressurization, electrical power, control pressure, GN2 purge, flight measurements, range safety, propellant loading, sequencing and pogo suppression systems were performing within specified limits.

Post firing checks were performed in order to check that the firing had not had an adverse effect on the stage systems. Approximately 105 test and operating procedures were used to accomplish the MTF testing.

The planned removal of the S-IC-4 stage from the test stand was delayed due to a change order to perform an engine alignment optical check. S-IC-4 was removed from the B-2 test stand and returned to MAF aboard the Pearl River, on 7 June 1967, for refurbishment and post-static checkout to verify the readiness of the stage for shipment to KSC.

At MAF it entered the Stage Test Building. During an inspection it was revealed that the welds on the engine actuator return ducts, which were part of the engine gimballing system, were marginal. The ducts were removed and returned to the supplier for rework. Additional ducts, available at Michoud, were installed in the S-IC-4 stage to minimize the schedule impact. An inspection of the LOX tank ring, during the second half of July, revealed a broken segment. The segment was removed for rework, and a subsequent study indicated a fatigue type failure. A decision was made to replace the segment with a thicker segment as used on the first three Saturn V vehicles.

On 16 August 1967 Boeing personnel at MAF completed the stage post-static checkout and placed the stage in storage. On 21 August the stage was weighed during a 16-hour operation in Test Cell # 1 of the Stage Test Building. The following day the stage was transported to the Manufacturing Building. Formal acceptance of the stage by NASA was on 28 August 1967.

On 12 September 1967 NASA asked that the S-IC-4 stage be retained in storage at MAF to permit MSFC to give work priority to activities supporting the S-IC-1 and S-IC-2 stages at KSC and the S-IC-3 stage at Michoud. On 1 November 1967 the stage was removed from storage and Boeing began installing agreed modifications. On 27 December 1967 Boeing moved the stage from post-acceptance storage in the manufacturing area at Michoud to the Stage Test Building to begin retest activities prior to shipment to KSC.

This further period of testing took place with the final test being the simulated flight test, performed on 8 March 1968. The pre-delivery review was conducted on 15 March 1968 and NASA formally accepted the stage from Boeing on 22 March 1968. However, because of changes incorporated in the stage, S-IC-4 underwent two retests following post-static checkout, the first of which started on the same day, 22 March 1968 and finished on 16 August 1968. On 23 August 1968 the stage was moved to a test cell at MAF to begin the second retest, a systems revalidation. All of this retest activity was completed on 19 September 1968 and NASA formally accepted the rework.

The stage was loaded on to the barge Orion and left port at MAF on 24 September 1968 en-route to KSC. The trip was delayed by threatening weather, before the barge together with S-IC-4 arrived at KSC dock on 30 September 1968.

The S-IC-4 stage eventually provided the first stage of AS-504 that launched Apollo 9 on 3 March 1969.

S-IC-5

Summary

This stage had an engine replaced just days before the stage was to be shipped to the Mississippi Test Facility (MTF) for static firing. Stage launched Apollo 10.

Engines

The initial engine configuration was:

Position 101: F-5035
Position 102: F-5041
Position 103: F-5040
Position 104: F-5042
Position 105: F-5036 – Removed from S-IC-5 and placed in storage.

Spare engine: F-5034 – utilized in place of F-5036

The static firing and final flown engine configuration was:

Position 101: F-5035
Position 102: F-5041
Position 103: F-5040
Position 104: F-5042
Position 105: F-5034

Stage manufacture

Fabrication and assembly of the S-IC-5 stage began in late 1965. By the end of 1965 machinists had completed three of the four Y-rings; work was in progress on both bulkheads for the fuel tank; final assembly of the thrust structure began on 8 December 1965.

67-58560 Static firing of the S-IC-5 stage at MTF 25.8.67.

67-57902 The S-IC-5 stage being installed in the B-2 test stand at MTF 29.6.67.

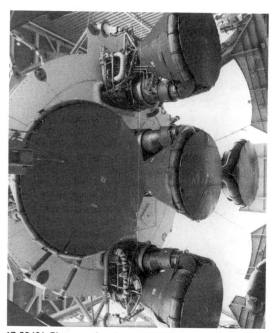

67-59481 Close up of engines as S-IC-5 stage is removed from the B-2 test stand at MTF 11.9.67.

67-58559 Static firing of the S-IC-5 stage at MTF 25.8.67.

67-59486 The S-IC-5 stage being removed from the B-2 test stand at MTF 11.9.67.

The LOX tank was mated to the inter-tank on 17 August.

Final "stack up" of the S-IC-5 stage in the vertical assembly tower at MAF was completed with the mating of the forward skirt to the top of the stage on 19 August.

Vertical assembly of the S-IC-5 stage was completed at MAF on 12 September 1966. The stage was then removed from the tower and moved to the Manufacturing Building.

The five F-1 engines were installed in the S-IC-5 stage during November 1966.

Stage testing

With assembly of the stage complete, it was moved into test cell #2 in the Checkout Building for post-manufacturing checkout on 16 December 1966. The checkout itself began on 21 December 1966. On 27 December 1966 Boeing personnel at MAF applied power to the stage for the first time to commence the checkout.

67-59485 Lowering the S-IC-5 stage from the B-2 stand at MTF onto a barge 11.9.67.

On 20 April 1967, technicians removed propulsion distributors which reliability testing revealed to be contaminated. After completion of the Engine Torque Check, the last in the series of tests, the S-IC-5 was transferred from the Michoud Stage Test Building to the Vertical Assembly Building on 8 May, for the replacement of 32 splice angle plates for the upper and lower thrust rings. Incorporation of these splice plates was delayed as a result of a NASA request to remove two engines for inspection of the LOX pump seals. Due to mechanical problems in the VAB it was decided to incorporate the splice angle changes while the stage was in the MTF Test Stand. However, change out of the F-1 engines was conducted in the Michoud Final Assembly Area before shipment.

67-59489 The S-IC-5 stage on board the Pearl River barge, leaving MTF for MAF 12.9.67.

On 31 March 1966 final assembly of the thrust structure, inter-tank and forward skirt were on schedule at MAF. However, on 29 April 1966, a test accident at MSFC severely buckled the fuel tank upper bulkhead intended for the S-IC-5 stage. Officials considered the bulkhead not repairable and ordered that it be replaced. During July hydrostatic testing, cleaning and painting of the fuel tank was completed at MAF.

On 11 August 1966 Boeing personnel in the vertical assembly tower at MAF mated the S-IC-5 inter-tank section to the fuel tank, which had been moved into the tower with the thrust structure the previous day.

During August structural testing of the S-IC-5 oxidizer tank assembly was completed successfully at MSFC.

Just prior to stage shipment the two engines were removed and inspected. One engine was replaced in the stage. However, engine 105 (F-5036) was removed and replaced with the spare (F-5034). The S-IC-5 stage was shipped by the Pearl River barge from MAF to MTF on 21 June 1967. The stage was initially stored in the MTF Booster Storage Building before being installed in the B-2 position of the B test stand on 29 June 1967. Initially pre-static firing modifications and tests were performed. These tests essentially repeated the post-manufacturing checkout (PMC) already performed at MAF prior to shipment.

On 25 July 1967 the collapse of the emergency fuel drain duct forced the termination of the first S-IC-5 tanking test. The collapse occurred when a vacuum was created in the emergency drain line by the re-circulation of fuel through the line. The fuel tank emergency drain duct, drain valve, and fuel fill and drain line were replaced.

An inspection of the LOX pre-pressurization filter revealed that an "O" ring was missing from the filter assembly. It was decided to use the filter as-is.

Test firing S-IC-5-1

S-IC-5 was the second booster flight stage to be tested at MTF. A propellant loading test was performed on 10 August 1967. During pre-static firing checks it was discovered that the actuators were not operating properly. An inspection revealed broken cam follower spiral springs in one of the actuators. The failure was due to hydrogen embrittlement, and all actuators were replaced.

The fuel for the firing was loaded on the day before the firing and the LOX on the firing day. The firing took place at 1814 CDT on 25 August 1967. The firing duration was 125.096 seconds as measured from simulated liftoff to cutoff complete (planned duration 125 +/- seconds).

The engine systems were satisfactory. The 3-2 engine cutoff sequence was successful, but the 1-2-2 start sequence was not achieved.

Thrust of engine 104 (F-5042) was not within the limits of +/- 15 kips of the predicted sea level thrust based on the Rocketdyne engine acceptance tests performed in June 1966.

The range safety system 1 did not receive the command signal.

There were no indications of stage structural deficiencies.

The following systems were satisfactory:

> the thrust vector control system
> fuel and LOX systems were satisfactory
> the onboard purge and control pressure systems
> stage environmental control systems
> electrical systems
> stage instrumentation/telemetry systems
> vibration, acoustic, and base heating environments

A number of parameters exceeded their limits either before or during the test:

> fuel loading probe voltages
> engine 104 heat exchanger effectiveness
> GSE supplied gas at the forward compartment umbilical
> low voltage indications on bus ID 11 and ID 12

The engine data recorded during the static firing was used to demonstrate that the engine thrust, specific impulse, engine manifold torus temperature, fuel pump balance cavity pressure and heat exchanger performance were within specification.

The stage data recorded during static firing was used to demonstrate that the stage thrust vector control, pressurization, electrical power, control pressure, GN2 purge, flight measurements, range safety, propellant loading, sequencing and pogo suppression systems were performing within specified limits.

Post firing checks were performed in order to check that the firing had not had an adverse effect on the stage systems. Approximately 105 test and operating procedures were used to accomplish the MTF testing.

S-IC-5 was removed from the B-2 test stand on 11 September 1967 following a three-day delay to allow for inspection of a broken turbine inlet transducer of engine 102 and location of a missing thermocouple, that was discovered in the engine throat after a two-day search. The stage was returned to MAF, departing on 12 September 1967, on the Pearl River barge, and arriving on 14 September 1967. Refurbishment and post-static checkout activities were performed at MAF in readiness for shipment of the stage to KSC. These activities included a change in the paint configuration. On 19 October the United States flag was installed on the S-IC-5 LOX tank. The flag was composed of silk screen printed decal material, and was installed in three sections.

In October the stage was placed in storage in the Manufacturing Building at Michoud. On 15 November 1967 the S-IC-5 stage was transferred to the Stage Test Building for continuation of post-static refurbishment and modification.

On 30 January 1968 S-IC-5 was removed from a further period of storage at MAF for installation of approved modifications. The modifications were undertaken in April, at the same time as post-storage checkout.

The post-static and post-storage testing at MAF was completed on 29 July 1968. A final round of tests was performed just prior to the shipment to KSC. These tests were completed on 28 October 1968. NASA took delivery of the stage from Boeing on 8 November 1968 and the stage was placed in temporary storage prior to its shipment later in the month. The stage was loaded on to the barge Orion on 21 November 1968 to start its journey to KSC. The barge left port on 22 November 1968, arriving at KSC on 27 November 1968.

The S-IC-5 stage eventually provided the first stage of AS-505 that launched Apollo 10 on 18 May 1969.

S-IC-6

Summary

This stage had an engine removed and replaced prior to the static firing. Stage launched Apollo 11.

Engines

The initial engine configuration was:

Position 101: F-6043
Position 102: F-6046
Position 103: F-TBD – removed from stage prior to static firing
Position 104: F-6047 – removed from stage prior to static firing and installed in S-IC-7, position 103
Position 105: F-6044

Spare engines: F-6054 – utilized in place of F-6047,
F-6051 – utilized in place of engine TBD.

The static firing and final flown engine configuration was:

Position 101: F-6043
Position 102: F-6046
Position 103: F-6051
Position 104: F-6054
Position 105: F-6044

Stage manufacturing

By 30 December 1966 the vertical assembly at MAF was complete and the handling ring was undergoing test in preparation for stage "laying-down" early in 1967.

On 6 January 1967 Boeing moved the S-IC-6 stage from the VAB at Michoud to the Manufacturing Plant for installation of the stage systems. During January and February 1967 the five F-1 engines were installed in the stage. The stage was returned to vertical assembly for replacement of 32 splice angle plates for the upper and lower thrust rings on 24 April.

Stage testing

On 8 May 1967 the S-IC-6 stage entered post-manufacturing checkout in the Michoud Stage Test Building. During the checkout two of the stage engines in positions 103 and 104 were removed and replaced. These were engines F-TBD (position 103) and F-6047 (position 104). F-6051 was installed on the S-IC-6 stage in position 103 on 25 May 1967, and F-6047 was replaced by F-6054 that was installed in the S-IC-6 stage on 26 May 1967.

67-61147 S-IC-6 stage being moved to Test Cell 2 of the Stage Test Building at MAF 11.67.

68-63010 The F-1 engine F-6043 being returned to the 101 position in the S-IC-6 stage at MAF after replacement of the oxidizer seal in the oxidizer pump 1.68.

68-64015 The S-IC-6 stage on the barge Pearl River, passing through the lock as it arrives at MTF from MAF 1.3.68.

68-68490 Static firing of the S-IC-6 stage in the B-2 test stand at MTF 13.8.68.

68-69295 The S-IC-6 stage returning from MTF to MAF aboard the barge Pearl River 29.8.68.

Boeing completed post-manufacturing checkout on 24 July 1967. Two days later NASA formally accepted the stage from Boeing and preparations were started for shipment of the stage from Michoud to MTF for static firing. On 14 August the stage was moved from the Stage Test Building to the factory area. Final pre-firing modifications were completed, and the stage was placed in storage in the Stage Test Building on 22 August. Instrumentation was installed on the stage at this time.

However, on 1 September 1967, NASA requested that the S-IC-6 stage, which was already prepared for shipment to MTF, be retained in storage at Michoud to allow Boeing to give priority to work supporting activities on the S-IC-1 and S-IC-2 stages at KSC and the S-IC-3 stage at Michoud.

During December 1967 pre-static test modifications were performed on the stage that was in storage in the Stage Test Building at MAF. This included temporary removal of engine 101 (F-6043) for replacement of the oxidizer seal in the oxidizer pump.

The S-IC-6 stage was subjected to further post-manufacturing checkout (PMC) tests at MAF, completing these activities on 24 January 1968. Modification work was concluded on 26 February 1968 and the stage was shipped by the barge Pearl River from MAF to MTF on 1 March 1968. Following this the stage was installed in the B-2 position of the B test stand on 4 March 1968 and pre-static firing tests were performed. At this time all three test stands at MTF were occupied by Saturn stages for the first time. S-IC-6 was in the B-2 test stand, S-II-4 was in the A-2 test stand and S-II-5 was in the A-1 test stand. The pre-static tests essentially repeated the post-manufacturing checkout (PMC) already performed at MAF prior to shipment.

Power had been turned on the stage for the first time at MAF on 4 April. AS-502 was launched the same day and experienced problems. On 22 April 1968 MSFC directed Boeing to delay the static firing of the S-IC-6 stage from the planned April date until modifications could be incorporated in the stage to alleviate the oscillations caused by the POGO longitudinal vibration that had been noticed on the AS-502 launch. Propellant tanking tests, delayed since April, were completed on 16 July 1968. Two days later NASA announced the conclusions of its investigations into the POGO problem. Tests had revealed that the natural frequencies of the vehicle structure and the propulsion system coincided, multiplying the amplitude of the oscillations. To overcome the problem small gas reservoirs would be used in the LOX pre-valves to change the frequency of the propulsion system. Minor modifications would be made to allow helium injection into the pre-valves of the F-1 engines. The solution would be incorporated into the S-IC-6 stage, initially, and verified in the upcoming static firing test on that stage.

Test firing S-IC-6-1

The pre-static firing review was conducted on 1 August 1968 which concluded that there were no reasons not to proceed towards hot fire testing. The S-IC-6 static firing countdown was originally started on 5 August 1968. However on the planned firing day, 6 August 1968, the countdown was cancelled because the configuration of the POGO suppression system could not be qualified. A subsequent configuration redesign was qualified and the countdown was rescheduled for the following week.

The countdown was restarted on 12 August 1968; fuel was loaded at 1750 gpm. On the following day the fuel was dropped to the engines and the measured ullage was 8.8%. This corresponded to 199,500 gallons of fuel on board at ignition. LOX was loaded on the firing day, 13 August 1968. Replenishment was accomplished until T-72 seconds. At ignition there were 321,388 gallons of LOX onboard, which corresponded to an ullage of 9.5%.

Firing command (T-91 seconds) was issued at 1732:43.004 CDT on 13 August 1968. Pre-pressurization was commanded on at T-89.54 seconds for the fuel tank and T-56.09 seconds for the LOX tank.

Engine ignition was successfully achieved at 1734:05.004 CDT on 13 August 1968 for the planned firing duration of 125+/- 2 seconds. Simulated lift off (T+0 seconds) occurred at 1734:14.012 CDT. Ambient temperature was 73.5F.

The gimbal program was initiated at T+0 seconds and was completed at T+111 seconds.

Engine cut off was initiated automatically by the terminal count down sequencer at T+125.005 seconds as planned. Main-stage duration was 126.504 seconds as measured from all engines running ON to all engines running OFF.

The engine data recorded during the static firing was used to demonstrate that the engine thrust, specific impulse, engine manifold torus temperature, fuel pump balance cavity pressure and heat exchanger performance were within specification.

The stage data recorded during static firing was used to demonstrate that the stage thrust vector control, pressurization, electrical power, control pressure, GN2 purge, flight measurements, range safety, propellant loading, sequencing and pogo suppression systems were performing within specified limits. All measurements indicated that the newly installed POGO suppression system had performed as planned.

Post firing checks, similar to those performed before the firing were performed in order to check that the firing had not had an adverse effect on the stage systems. Approximately 105 test and operating procedures were used to accomplish the MTF testing.

On 22 August 1968 Boeing and MSFC reviewed the failure of a LOX tank slosh baffle during tanking operations for the S-IC-6 static firing at MTF. Boeing proposed the redesign and change out of an 8 foot baffle section for S-IC-5 and subsequent stages to provide increased strength.

S-IC-6 was removed from the B-2 test stand on 28 August 1968 and returned to MAF on the following day, aboard the barge Pearl River, for refurbishment and post-static checkout to verify the readiness of the stage for shipment to KSC.

Post static checkout was completed on 9 December 1968. The final simulated flight test was run between 4 and 8 December 1968. The stage was transferred to Test Cell III at MAF on 4 January 1969 for a period of storage and modification. Finally S-IC-6 was loaded onto the barge Orion on 13 February 1969 and the stage left MAF on 16 February 1969 after being delayed two days due to bad weather. It arrived at KSC on 20 February 1969.

The S-IC-6 stage eventually provided the first stage of AS-506 that launched Apollo 11 on 16 July 1969.

S-IC-7

Summary

This stage included an engine that had previously been installed, but not fired, in the S-IC-6 stage. Launched Apollo 12.

Engines

The initial and final flown engine configuration was:

Position 101: F-6048
Position 102: F-6052
Position 103: F-6047
Position 104: F-6053
Position 105: F-6050

Stage manufacture

The S-IC-7 component assembly at Michoud continued through the months of January and February 1967, and was completed in early March.

Vertical assembly of the S-IC-7 stage began on 6 March 1967 with the movement of the thrust structure to the VAB. This thrust structure differed from previous ones installed in this position in that it contained pre-valves, the pressure volume compensating ducts for the five F-1 engines, and tubing prior to installation. With the movement of the S-IC-7 fuel tanks into the vertical assembly, the second phase of Vertical Assembly was initiated on 13 March. After alignment the tank was connected to the thrust structure with mechanical fasteners. Following the fuel tank in vertical assembly came the inter-tank, and as stack up neared completion, the LOX tank was positioned atop the inter-tank.

Vertical assembly of the major components of the stage ended on 30 March 1967 with assembly of the forward skirt. The complete vertical assembly was finished on 20 April 1967. The stage was then transferred to a horizontal attitude for component installation. The five F-1 engines were installed in the S-IC-7 stage between 4 May and 19 June 1967.

Horizontal installation of stage hardware was completed on 11 August.

Stage testing

On 14 August 1967 Boeing transferred the S-IC-7 stage from its horizontal installation position in the MAF Manufacturing Building to Test Cell 2 in the Stage Test Building for post-manufacturing checkout.

The S-IC-7 stage was subjected to post-manufacturing checkout (PMC) at MAF, completing these activities on 10 November 1967 with the simulated static firing.

The stage was then placed in storage at Michoud on 22

November 1967 due to other work priorities at KSC and Michoud.

On 1 April 1968 Boeing personnel removed the S-IC-7 stage from storage at MAF and installed it in Test Cell # 1 of the Stage Test Building for installation of modifications. On 23 August 1968 the stage was transferred to a horizontal assembly position in MAF's manufacturing building, awaiting availability of space for testing at MTF. The S-IC-7 stage was shipped by the barge Pearl River from MAF to MTF on 12 September 1968. The stage was installed in the B-2 position of the B test stand on 13 September 1968. Boeing immediately began installation of the POGO suppression system in the stage. Following this pre-static firing tests were performed. These tests essentially repeated the post-manufacturing checkout (PMC) already performed at MAF prior to shipment.

68-70773 Static firing of the S-IC-7 stage in the B-2 test stand at MTF 30.10.68.

Test firing S-IC-7-1

The S-IC-7 stage was the first one not to have a propellant loading trial prior to the loading for the static firing. This was because this operation was now considered to be mature. On 18 October 1968 NASA and Boeing officials held the pre-static firing review at MTF. They gave the go-ahead for the firing that was planned for 23 October. However, on 23 October, a malfunction in the S-IC-7 LOX depletion cutoff system during a simulated static firing test delayed the acceptance firing of the stage at MTF. The S-IC-7 stage static firing countdown was started on 28 October 1968. However, on T-1 day after the fuel tank had been loaded, a fuel leak was found and it was necessary to de-tank and repair the leak.

The countdown was restarted on 29 October 1968 and fuel was loaded at a flow rate of approximately 1970 gpm. On firing day, 30 October 1968, the fuel was dropped to the engines and the level adjusted to 1.79% ullage. There were 216,191 gallons of fuel on board at T-15 minutes.

LOX was loaded on the firing day. Replenishment was accomplished until T-72 seconds. There were 348,750 gallons of LOX on board at this time that corresponded to 2.96% ullage.

Firing command (T-91 seconds) was issued at 1514:35.004 CST on 30 October 1968. Pre-pressurization was commanded on at T-89.556 seconds for the fuel tank and T-56.332 seconds for the LOX tank.

Engine ignition was successfully achieved at 1515:57.004 CST on 30 October 1968 and simulated liftoff (T+0) occurred at 1516:06.012 CST. Barometric pressure on firing day was 14.705 psia and ambient temperature was 81.0F.

The gimbal program was initiated at T+3 seconds and was completed at T+111 seconds. Cutoff was initiated automatically by the terminal countdown sequencer at T+125.008 seconds as planned. Main-stage duration was 126.464 seconds as measured from all engines running ON to all engines running OFF.

The firing was considered successful with the planned duration of 125 +/-2 seconds being achieved.

The engine data recorded during the static firing was used to demonstrate that the engine thrust, specific impulse, engine manifold torus temperature, fuel pump balance cavity pressure and heat exchanger performance were within specification.

The stage data recorded during static firing was used to demonstrate that the stage thrust vector control, pressurization, electrical power, control pressure, GN2 purge, flight measurements, range safety, propellant loading, sequencing and pogo suppression systems were performing within specified limits.

Post firing checks were performed in order to check that the firing had not had an adverse effect on the stage systems. Approximately 105 test and operating procedures were used to accomplish the MTF testing.

Removal of the stage from the B-2 test stand could not take place until a repair was performed on the main derrick motor that had malfunctioned in September, just after the S-IC-7 stage had been installed in the test stand. On 8 November workmen completed the repair that necessitated removal and repair of an armature and the manufacture of a special winding. S-IC-7 was removed from the B-2 test stand, on 8 November, and returned to MAF by barge on 9 November 1968. Refurbishment and post-static checkout activities were performed at MAF in readiness for shipment of the stage to KSC. The testing, performed in Test Complex I at MAF, began on 9 December 1968 and concluded with the Simulated Flight Test on 20 January 1969.

The complete test and refurbishment program at MAF was completed on 21 January 1969 and the stage was temporarily placed in storage. NASA formally accepted the stage from Boeing on 17 February 1969. S-IC-7 was loaded on to the barge Orion to start its journey to KSC. The barge left port on 29 April 1969, arriving at KSC on 3 May 1969.

The S-IC-7 stage eventually provided the first stage of AS-507 that launched Apollo 12 on 14 November 1969.

S-IC-8

Summary

This stage had an engine replaced after the static firing. The replacement engine had the shortest cumulative firing total for an F-1 engine at the time of launch. Launched Apollo 13.

Engines

The initial allocation including stage static firing was:

Position 101: F-6055
Position 102: F-6058
Position 103: F-6057
Position 104: F-6059 – removed on 25 January 1969 after stage static firing, replaced by F-6078, and eventually installed in S-IC-11
Position 105: F-6056

Spare engine: F-6078

The final flown engine configuration was:

Position 101: F-6055
Position 102: F-6058
Position 103: F-6057
Position 104: F-6078
Position 105: F-6056

Stage manufacture

On 21 June 1967 the S-IC-8 fuel tank underwent a hydrostatic proof test to determine if eight cracks in the weld repair areas, discovered during inspection prior to component assembly, would propagate. The test revealed no growth in the cracks nor any leakage and the tank was returned to production for component assembly. On 14 July 1967 Boeing personnel placed the thrust structure onto the four pylons of the VAB at MAF to begin the vertical assembly of the S-IC-8 stage.

On 21 July the fuel tank was transported to the Vertical Assembly Building and mated to the thrust structure. The inter-tank was moved into the Vertical Assembly Tower over the fuel tank and mated on 25 July.

On 14 August the LOX tank was moved into the Vertical Assembly Building and mated to the inter-tank. The forward skirt was positioned on 15 August.

Vertical assembly continued until 11 September 1967 when it was completed. Five days later the S-IC-8 stage was removed from the Vertical Assembly Tower, lowered onto a transporter, and prepared for transportation to the Horizontal Assembly Position number 1 in the Michoud Manufacturing Building.

The five F-1 engines were installed in the S-IC-8 stage during September and October 1967. Boeing completed horizontal assembly of the stage on 9 May 1968 and placed the stage in storage at MAF. Verification testing of the newly assembled stage took place from early August until 22 August 1968.

Stage testing

The S-IC-8 stage was subjected to post-manufacturing checkout (PMC) at MAF, completing these activities on 29 October 1968.

The S-IC-8 stage was shipped by barge from MAF to MTF on 13 November 1968. The stage was installed in the B-2 position of the B test stand and pre-static firing tests were performed. These tests essentially repeated the post-manufacturing checkout (PMC) already performed at MAF prior to shipment.

The pre-static firing review was held on 13 December 1968. The S-IC-8 stage static firing countdown was started on 17 December 1968. Fuel was loaded on that day at a flow rate of approximately 1890 gpm. On firing day, 18 December 1968, the fuel was dropped to the engines and the level adjusted to 1.95% ullage. There were 215,953 gallons of fuel on board at T-15 minutes.

67-60478 F-1 engine installation in the S-IC-8 stage at MAF Early 10.67.

67-60490 Left to right, S-IC-8 in horizontal assembly, S-IC-4 in storage, S-IC-6 in storage; at MAF Early 10.67.

68-69297 The S-IC-8 stage being transferred to the Stage Test Building at MAF 8.68.

69-74958 Static firing of the S-IC-8 stage in the B-2 test stand at MTF 18.12.68.

67-58554 S-IC-8 vertical assembly of forward skirt 15.8.67.

Test firing S-IC-8-1

LOX was loaded on the firing day. Replenishment was accomplished until T-72 seconds. There were 348,621 gallons of LOX on board at this time that corresponded to 3.03% ullage.

Firing command (T-91 seconds) was issued at 1637:16.004 CST on 18 December 1968. Prepressurization was commanded on at T-89.544 seconds for the fuel tank and T-56.704 seconds for the LOX tank. Engine ignition was successfully achieved at 1638:40 CST on 18 December 1968 and simulated liftoff (T+0) occurred at 1638:47.020 CST. Barometric pressure on firing day was 14.698 psia and ambient temperature was 62.6F.

The gimbal program was initiated at T+3 seconds and was completed at T+111 seconds. Cutoff was initiated automatically by the terminal countdown sequencer at T+125.008 seconds as planned. Main-stage duration was 126.688 seconds as measured from all engines running ON to all engines running OFF.

The firing was considered successful with the planned duration of 125 +/-2 seconds being achieved. Test data indicated that the "POGO" suppression system performed as predicted.

The engine data recorded during the static firing was used to demonstrate that the engine thrust, specific impulse, engine manifold torus temperature, fuel pump balance cavity pressure and heat exchanger performance were within specification.

The stage data recorded during static firing was used to demonstrate that the stage thrust vector control, pressurization, electrical power, control pressure, GN2 purge, flight measurements, range safety, propellant loading, sequencing and pogo suppression systems were performing within specified limits.

Post firing checks were performed in order to check that the firing had not had an adverse effect on the stage systems. Approximately 105 test and operating procedures were used to accomplish the MTF testing.

S-IC-8 was removed from the B-2 test stand and returned to MAF by barge on 3 January 1969, arriving at MAF the following day. Refurbishment and post-static checkout activities were performed at MAF in readiness for shipment of the stage to KSC. However, on 23 January 1969, a leak was detected on the 104 engine, F-6059. Attempts to correct the problem were unsuccessful and Boeing removed the engine from the stage on 25 January 1969 and replaced it with the backup engine, F-6078, which was installed two days later.

The final post-static test was a simulated flight test performed in Test Complex I at MAF and completed on 11 March 1969. This marked what should have been the completion of the post-static tests, however following the replacement of a relay in the 115A3 distributor, it was determined that some tests had to be repeated. The retest of the S-IC-8 stage systems was completed on 24 March 1969 and preparations for shipping the stage to KSC began. On 28 March 1969 Boeing personnel completed post-static refurbishment and placed the stage in storage pending shipment. On 29 April 1969 the stage was moved to storage position #3 for further storage pending modification prior to shipment. The stage was loaded on to the barge Orion to start its journey to KSC. The ship left port on 11 June 1969, arriving at KSC on 16 June 1969.

The S-IC-8 stage eventually provided the first stage of AS-508 that launched Apollo 13 on 11 April 1970.

S-IC-9

Summary

S-IC stage that launched Apollo 14.

Engines

The original and final flown engine configuration was:

Position 101: F-6061
Position 102: F-6064
Position 103: F-6063
Position 104: F-6065
Position 105: F-6062

Stage manufacturing

The fuel tank was removed from storage in the Manufacturing Building in Michoud on 26 July 1967 and transported to the Vertical Assembly Building. The tank was then moved with a crane to Pit number 3 for a shakedown operation, prior to being moved into the Hydrostatic Test Facility.

The S-IC-9 LOX tank was removed from storage in the Manufacturing Building on 18 August and transported to the Vertical Assembly Building. A complete inspection of the LOX tank was performed, following which the LOX tank was moved into the Hydrostatic Test Facility. Following hydrostatic testing the stage was moved into the Manufacturing Building.

On 25 September the inter-tank was lifted into the assembly tower and mated with the fuel tank.

On 16 October the LOX tank was moved into the vertical assembly tower and mated to the inter-tank.

Vertical assembly of S-IC-9 was completed in the VAB at MAF on 18 October 1967 with the installation of the forward skirt.

On 22 February 1968 Boeing workmen at MAF removed the stage from the vertical assembly tower and prepared to move it to a horizontal assembly position for installation of stage systems. The five F-1 engines were installed in the S-IC-9 stage between 15 and 18 April 1968. The horizontal assembly was completed on 13 November 1968.

Stage testing

The S-IC-9 stage was subjected to post-manufacturing checkout (PMC) at MAF, completing these activities on 3 January 1969.

The S-IC-9 stage was shipped by barge from MAF to MTF on 9 January 1969. The stage was installed in the B-2 position of the B test stand on 10 January 1969. Immediate replacement of a large number of bolts found to contain flaws in the thrust posts preceded test preparations. Following this work pre-static firing tests were performed. These tests essentially repeated the post-manufacturing checkout (PMC) already performed at MAF prior to shipment.

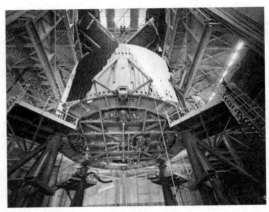

67-59510 The S-IC-9 thrust structure during vertical assembly at MAF 9.67.

67-59511 The S-IC-9 fuel tank being moved out of the hydrostatic test facility at MAF 9.67.

Test firing S-IC-9-1

The S-IC-9 stage static firing countdown was started on 18 February 1969. Fuel was loaded on that day at a flow rate of approximately 1900 gpm. On firing day, 19 February 1969, the fuel was dropped to the engines and the level adjusted to 3.086% ullage. There were 213,927 gallons of fuel on board at T-15 minutes.

LOX was loaded on the firing day. Replenishment was accomplished until T-187 seconds. There were 348,458 gallons of LOX on board at this time that corresponded to 3.039% ullage.

Firing command (T-91 seconds) was issued at 1510:53.004 CST on 19 February 1969. Pre-pressurization was commanded on at T-338.736 seconds for the fuel tank and T-71.2 seconds for the LOX tank.

Times for firing command and fuel tank pressurization were not normal with respect to T-0 because there was a 249.236 second hold at a count time indication of T-77 seconds. The count was not recycled. The reason for the hold was to clear personnel who were within the 5200 feet radius exclusion zone. It was reported that 12,000 visitors viewed the static firing.

Engine ignition was successfully achieved at 1516:27 CST on 19 February 1969 and simulated liftoff (T+0) occurred at 1516:33.260 CST. Barometric pressure on firing day was 14.828 psia and ambient temperature was 56.0F.

The gimbal program was initiated at T+3 seconds and was completed at T+111 seconds. Cutoff was initiated automatically by the terminal countdown sequencer at T+125.008 seconds as planned. Main-stage duration was 126.640 seconds as measured from all engines running ON to all engines running OFF.

The firing was considered successful with the planned duration of 125 +/-2 seconds being achieved.

The engine data recorded during the static firing was used to demonstrate that the engine thrust, specific impulse, engine manifold torus temperature, fuel pump balance cavity pressure and heat exchanger performance were within specification.

The stage data recorded during static firing was used to demonstrate that the stage thrust vector control, pressurization, electrical power, control pressure, GN2 purge, flight measurements, range safety, propellant loading, sequencing and pogo suppression systems were performing within specified limits.

Post firing checks were performed in order to check that the firing had not had an adverse effect on the stage systems. Approximately 105 test and operating procedures were used to accomplish the MTF testing.

S-IC-9 was removed from the B-2 test stand on 5 March 1969 and returned to MAF by barge on 6 March 1969, arriving at MAF the following day. On 28 March 1969 the S-IC-9 stage was transferred from Factory Storage Position 3 to Test Complex I in the Stage Test Building at MAF in readiness for refurbishment and post-static checkout activities to be performed prior to stage shipment to KSC. Testing started on 1 April 1969 with the first power-on test. The final test, a Simulated Flight test was run on 2 May 1969.

The testing at MAF was declared complete on 16 May 1969 and 4 days later the stage was moved to Factory Storage Postion 3 at Michoud. NASA formally accepted the stage from Boeing on 26 May 1969. Because of slips in the Apollo flight schedule the stage remained at MAF in storage until shipment was required some 7 months later. During this period of storage, on 17 August 1969 Hurricane Camille struck the Gulf coast, but without any damage to the S-IC stages at MAF. The stage left MAF on 6 January 1970 on board a ship, arriving at KSC on 12 January 1970.

The S-IC-9 stage eventually provided the first stage of AS-509 that launched Apollo 14 on 31 January 1971.

67-59512 The S-IC-9 fuel tank during vertical assembly onto the thrust structure 9.67.

69-74963 Static firing of the S-IC-9 stage in the B-2 test stand at MTF 19.2.69.

S-IC-10

Summary

This stage had an engine removed after the stage static firing test. A replacement engine was utilized, although this engine was never fired as part of a stage. Another engine was rejected before ever being built into the stage. Launched Apollo 15.

Engines

The original allocation was:

Position 101: F-6066
Position 102: F-6069
Position 103: F-6068
Position 104: F-6071
Position 105: F-6067 – initially allocated to S-IC-10, but replaced and never reused

The initial build and static test configuration was:

Position 101: F-6066 – removed on 17 June 1969 following checkout failure and finally used on S-IC-15 position 101
Position 102: F-6069
Position 103: F-6068
Position 104: F-6071
Position 105: F-6073

Spare engine: F-6088

The final flown engine configuration was:

Position 101: F-6088
Position 102: F-6069
Position 103: F-6068
Position 104: F-6071
Position 105: F-6073

Stage manufacturing

Work on the S-IC-10 forward skirt was in progress during August 1967. On 21 September the inter-tank was nearing structural completion, and preparations were begun to transport the inter-tank to the paint area in the Manufacturing Building. The fuel tank was lowered into the Hydrostatic Test Facility on 17 October. The tank was inspected, tested and cleaned prior to removal to the paint area.

The LOX tank was completed in December 1967, in the Vertical Assembly Building.

On 21 March 1968 workmen at MAF placed the S-IC-10 thrust structure in the Vertical Assembly Tower and prepared for fuel tank installation. By 10 May 1968 Boeing had completed the vertical assembly of the stage on schedule. The stage was removed from the Vertical Assembly Tower on 18 June 1968 and moved to the horizontal assembly position for installation of interior stage systems. The five F-1 engines were installed in S-IC-10 during August 1968. Installation of stage systems continued until the end of December 1968.

Stage testing

The S-IC-10 stage was subjected to post-manufacturing checkout (PMC) at MAF, beginning on 20 January 1969 and completing these activities on 7 March 1969. Checkout was performed in Test Complex II of the Stage Test Building at Michoud.

The S-IC-10 stage was shipped by the barge Little Lake from MAF to MTF on 11 March 1969. The stage was installed in the B-2 position of the B test stand on 12 March 1969 and pre-static firing tests were performed. These tests essentially repeated the post-manufacturing checkout (PMC) already performed at MAF prior to shipment.

68-70792 The S-IC-10, 11 and 9 stages during horizontal assembly at MAF 10.68.

Test firing S-IC-10-1

The S-IC-10 stage static firing countdown was started on 15 April 1969. Fuel was loaded on that day at a flow rate of approximately 1950 gpm. On firing day, 16 April 1969, the fuel was dropped to the engines and the level adjusted to 3.27% ullage. There were 213,230 gallons of fuel on board at T-15 minutes.

LOX was loaded on the firing day. Replenishment was not accomplished due to a leaking fill and drain umbilical line. There were 333,522 gallons of LOX on board at T-15 minutes that corresponded to 7.70% ullage.

Firing command (T-91 seconds) was issued at 1431:53.004 CST on 16 April 1969. Pre-pressurization was commanded on at T-89.528 seconds for the fuel tank and T-71.212 seconds for the LOX tank.

Engine ignition was successfully achieved at 1433:18 CST on 16 April 1969 and simulated liftoff (T+0) occurred at 1433:24.016 CST. Barometric pressure on firing day was 14.69 psia and ambient temperature was 68F.

69-01866 The S-IC-10 stage in storage position # 3 at MAF being placed inside an environmentally controlled storage shelter 8.69.

69-75265 Static firing of the S-IC-10 stage in the B-2 test stand at MTF 16.4.69.

67-61152 The S-IC-10 fuel tank being removed from the hydrostatic test facility at MAF 11.67.

The gimbal program was initiated at T+3 seconds and was completed at T+111 seconds. Cutoff was initiated automatically by the terminal countdown sequencer at T+125.004 seconds as planned. Main-stage duration was 126.372 seconds as measured from all engines running ON to all engines running OFF.

The firing was considered successful with the planned duration of 125 +/-2 seconds being achieved. The firing had been achieved about 30 minutes ahead of the scheduled time on the day. However, during the firing technicians noted that engine thrust measured lower than predicted for flight performance.

The engine data recorded during the static firing was used to demonstrate that the engine thrust, specific impulse, engine manifold torus temperature, fuel pump balance cavity pressure and heat exchanger performance were within specification.

The stage data recorded during static firing was used to demonstrate that the stage thrust vector control, pressurization, electrical power, control pressure, GN2 purge, flight measurements, range safety, propellant loading, sequencing and pogo suppression systems were performing within specified limits.

Post firing checks were performed in order to check that the firing had not had an adverse effect on the stage systems. Approximately 105 test and operating procedures were utilized to accomplish the MTF testing.

S-IC-10 was removed from the B-2 test stand on 2 May 1969 and returned to MAF by barge on the same day. On 6 May 1969 Boeing and MTF personnel initiated test analysis and investigations of the F-1 engine low thrust problem evident during the static firing of S-IC-6 and subsequent stages, but particularly evident during the S-IC-10 static firing. On 20 May 1969 the stage was moved to Test Complex I in the Stage Test Building at MAF for the commencement of refurbishment and post-static checkout activities. During this checkout, on 17 June 1969, a leak was discovered in the check valve in the gimbal filter manifold in the 101 engine, F-6066, that necessitated its removal from the stage. It was replaced by the spare engine F-6088, which was installed in the stage at MAF on 18 June 1969.

The testing at MAF was completed on 17 July 1969 and the stage was formally accepted by NASA on 8 August 1969. The day before Hurricane Camille struck the Gulf coast on 17 August 1969 the S-IC-10 stage was moved from Test Complex I to the factory for increased safety. Because of slips in the Apollo flight schedule the stage remained at MAF in storage until shipment was required nearly one year later. Finishing on 15 October 1969 Boeing used the S-IC-10 stage to checkout the recently completed Storage Position number 3 at Michoud that would be used to store the S-IC-12 stage, the only S-IC stage planned to be stored at Michoud. The S-IC-10 stage left MAF on 1 July 1970 on board a ship, arriving at KSC on 6 July 1970.

The S-IC-10 stage eventually provided the first stage of AS-510 that launched Apollo 15 on 26 July 1971.

S-IC-11

Summary

This stage had a major engine fire during the stage static firing at MTF. All the F-1 engines were replaced and the stage was re-tested one year later. Launched Apollo 16.

Engines

The original allocation in place for the initial stage static firing was:

Position 101: F-6049 – engine refurbished, became flight spare for S-IC-14
Position 102: F-6045 – engine transferred to the Smithsonian museum
Position 103: F-6072 – engine scrapped
Position 104: F-6060 – engine refurbished and used on S-IC-15
Position 105: F-6070 – engine scrapped

The final configuration for the repeat stage static firing and for flight was:

Position 101: F-6095
Position 102: F-6096
Position 103: F-6087
Position 104: F-6094
Position 105: F-6059

Stage manufacturing

Thrust structure assembly for the S-IC-11 stage was in process during July 1967. Work was continuing on the inter-tank during September.

Etching and dye-penetrant inspection of the lower LOX bulkhead welded fittings was conducted on 19 October.

On 18 December the lower LOX bulkhead and skin ring was moved over the top of tank assembly position number 1 and lowered into position on the second tank skin ring. After alignment the two tank skin rings were joined by welding.

In February 1968 Boeing workmen at MAF completed the lower LOX tank and fuel tank, placed the tanks in the vertical assembly position, and prepared for hydrostatic testing.

By April 1968 painting of the thrust structure and hydrostatic testing and inspection of the fuel tank had been completed. Assembly of the forward skirt was 95% complete at this time. In June 1968 Boeing completed component assembly and began vertical assembly of the stage.

On 27 September 1968 the stage was removed from the Vertical Assembly Tower following completion of vertical assembly operations. The stage was placed on a transporter and moved to a horizontal assembly position to be prepared for systems installation. The five F-1 engines were installed in the S-IC-11 stage around December 1968. Horizontal installations in the S-IC-11 stage were completed by Boeing at MAF on 14 March 1969.

Stage testing

The S-IC-11 stage was subjected to post-manufacturing verification (PMV) at MAF. A change letter (Program Letter No. 215) to Boeing's contract provided for consolidation of post-manufacturing checkout (PMC) with pre-static firing and post-static checkout activities at MTF. The contract change was approved on 19 February 1969 and was aimed at reducing test redundancies. This was the first stage not to be subjected to the usual post-manufacturing checkout. Verification comprised an abbreviated series of tests aimed at reducing the amount of repeated stage checks performed before and after the static firings. Financial considerations were the key drivers in this policy change.

The S-IC-11 stage was shipped by barge from MAF to MTF on 12 May 1969. The stage was installed in the B-2 position of the B test stand and pre-static firing tests were performed. These tests essentially repeated the limited post-manufacturing verification (PMV) already performed at MAF prior to shipment.

Test firing S-IC-11-1

The S-IC-11 stage static firing countdown was initially started on 24 June 1969. However, discovery of a leak in the fuel system caused a one day postponement of the static firing. In addition several problems were encountered which normally would have been detected during the full classical post manufacturing checkout at MAF. The static firing countdown was finally started on 25 June 1969. Fuel was loaded on that day at a flow rate of approximately 1960 gpm. On firing day, 26 June 1969, the fuel was dropped to the engines and the level adjusted to 3.0% ullage. There were 214,300 gallons of fuel on board at T-15 minutes.

LOX was loaded on the firing day. Replenishment was not accomplished due to a leaking fill and drain umbilical line. There were 333,400 gallons of LOX on board at T-15 minutes that corresponded to 6.6% ullage.

Firing command (T-91 seconds) was issued at 1720:48.004 CDT on 26 June 1969. Pre-pressurization was commanded on at T-89.540 seconds for the fuel tank and T-71.612 seconds for the LOX tank.

69-00104 The first static firing of the S-IC-11 stage in the B-2 test stand at MTF 26.6.69.

69-00726 The S-IC-11 stage is removed from a covered barge at MAF on its return from MTF 19.7.69.

69-00728 The engines are removed from the S-IC-11 stage at MAF following the fire 7.69.

69-00730 The fire-damaged parts from the S-IC-11 stage are displayed at MAF 8.69.

69-00732 Workers queue to view the damaged parts from the S-IC-11 stage at MAF 8.69.

69-75874 The S-IC-11 stage is installed in the B-2 test stand at MTF prior to the first static firing 13.5.69.

70-04836 New engines are installed in the S-IC-11 stage at MAF 23.1.70.

70-08319 The S-IC-11 stage is loaded aboard the barge at MAF for the return to MTF 7.5.70.

70-09124 The S-IC-11 stage is hoisted to the B-2 test stand for the second static firing 8.5.70.

Engine ignition was successfully achieved at 1722:13 CDT on 26 June 1969 and simulated liftoff (T+0) occurred at 1722:19.016 CDT. Barometric pressure on firing day was 14.70 psia and ambient temperature was 90F.

The gimbal program was initiated at T+3 seconds. However, the specified firing duration of 125 +/- 2 seconds was not completed. At T+96.8 seconds there was a premature engine cutoff resulting from a fire on engine 103 (F-6072). The test stand's water deluge system extinguished the fire in about 15 seconds after cutoff. Engine 103 was severely damaged and there was damage to the other four engines and to ground support equipment.

Initial damage assessment was completed by 30 June when officials announced that engines 101, 102 and 104 sustained minor damage and would be refurbished in place. Engines 103 and 105 received major damage and would need replacement. The heat shields near engine 103 would need replacement; the thrust structure area components in compartment 2 and the LOX interconnect duct and the GN2 control link would be replaced; and engine 103 servo actuator supply line would be replaced. The ground support electrical cables received minor damage and would need repair.

On 1 July the results of the preliminary investigation into the cause were announced. The fire resulted from a leak in the connection from the high-pressure hydraulic line to the gimbal filter manifold. The leak occurred through a GASK-O-SEAL in the line where a shipment disc dust cover had not been removed by error. Similar seals in the Apollo 11 first stage, S-IC-6 poised for launch at KSC, were checked and it was confirmed that all seals were in good condition. NASA appointed a board of inquiry to further investigate and recommend preventative measures for the remaining S-IC stages.

The S-IC-11 stage was removed from the B-2 stand at MTF on 11 July 1969 and readied for return to MAF for refurbishment. On 19 July 1969 the stage was shipped back to MAF by barge.

The refurbishment activity took one year and included removal of the engines for detailed damage assessment. It was determined that all five engines should be replaced with the two most badly damaged (F-6070 and F-6072) being scrapped. The stage with five new engines would undergo a repeat firing test. Meanwhile, workers at MAF were invited to inspect the damaged parts in a move to improve quality control.

The new engines were installed in the stage between 19 December 1969 and 23 January 1970. The new limited checkout sequence (post manufacturing verification) was completed on 23 March 1970. Finally the stage was ready to return to MTF by barge on 7 May 1970. In the meantime three other S-IC stages, S-IC-12, S-IC-13 and S-IC-14 had been test fired successfully at MTF.

The S-IC-T stage on display in the open at KSC in 1991.

70-11853 The S-II-15 stage being transferred to Station VII at Seal Beach for packaging 1.9.70. (inset top left)

65-22474 First five engine firing of the S-IC-T stage at MSFC for a duration of 6.5 seconds 16.4.65.

Saturn V 500F Located inside VAB
1.4.66 / 107-KSC-66C-2773

Erection of launch escape system for
Apollo 6 AS-502 at VAB 2.5.66

70-07728 Ceremony at Seal Beach to celebrate the S-II-14 stage completing manufacturing 17.4.70.

68-68490 Static firing of the S-IC-6 stage (Apollo 11) in the B-2 test stand at MTF 13.8.68.

67-60426 The S-II-TS-B test stage on the road from Port Huemene to SSFL 31.10.67.

68-70792 The S-IC-10, 11 and 9 stages during horizontal assembly at MAF 10.68.

68-70106 Static firing of the S-IVB-507 stage in the Beta I test stand at SACTO 16.10.68.

69-75864 The first F-1 engine being installed in the S-IC-13 stage at MAF 28.5.69.

69-71996 The S-IVB-505N stage en route from SACTO to Mather AFB for the flight to KSC 2.12.68.

Apollo Saturn V 500F enroute from VAB to Launch complex 39A 25.5.66 107-KSC-66C-4777

67-58549 S-IC-4 stage being weighed at MAF test cell 1 of the stage test building 21.8.67.

Test firing S-IC-11R-1

S-IC-11 was reinstalled in test stand B-2, a repeat test verification was performed, and the S-IC-11 (retest) static firing countdown was started on 23 June 1970. This would be the penultimate S-IC stage static firing test ever performed. As weather conditions deteriorated the actual firing was delayed from 24 June to 25 June. Fuel was loaded on 23 June with the pre-valves open and at a rate of 1850 gpm. There were 214,500 gallons of fuel on board at T-15 minutes that corresponded to a 2.92% ullage. LOX was loaded on firing day at a maximum flow rate of 4000 gpm. LOX replenishment was accomplished until T-187 seconds and automatically terminated. At T-15 minutes there were 348,294 gallons of LOX on board which corresponded to a 3.0% ullage.

Firing command (T-190 seconds) was executed at 1520:03.004 CDT on 25 June 1970. Pre-pressurization was commanded ON at T-89.6 seconds for the fuel tank and T-71.4 seconds for the LOX tank.

Engine ignition was successfully achieved at 1523:06 CDT and simulated lift off (T+0) occurred at 1523:13.020 CDT. Barometric pressure was 15.37psia and ambient temperature was 77F.

The gimbal program was initiated at T+3 seconds and was completed at T+111 seconds. However by this time the firing had been aborted. The cut off occurred after 69.344 seconds due to a redline value in the LOX ullage pressure being exceeded. The planned firing duration had been 125 +/- 2 seconds. Main-stage duration was 70.628 seconds as measured from all engines running ON to all engines running OFF.

Three special tests were successfully accomplished during the static firing countdown:

> An attempt was made to duplicate the LOX vent valve "stuck open" anomaly that had occurred on the AS-508 launch. However the anomaly did not occur and the valves remained within limits.
> LOX was replenished to evaluate the LOX probe for the second time after a similar test on the S-IC-14 static firing test. However the probe failed to record when the LOX level reaching its maximum load. The LOX overfill sensor did perform as required.
> A new venturi flow control orifice which replaced a flow control valve successfully maintained the LOX tank ullage within requirements for the test duration.

Eight anomalies were noted during the course of the countdown and firing:

> A minor hydraulic leak on the main LOX valve number 2, in engine 103 (F-6087) was discovered before the test. It was decided to proceed with the test and repair it afterwards.
> A hot gas leak was found at a connection on the systems "A" measurement tube for engine 101.
> A strut was found to be damaged at the actuator attach point. After inspection it was determined to "use-as-is".
> Time of actuation of a part in the 104 engine exceeded limits by 0.008 seconds.
> A flex flow deluge valve failed in the open position.
> Erratic operation of the Low Engine Area Purge valves.
> The Forward Environmental Conditioning Unit failed to restart on command after cutoff. Investigations revealed that the blower had separated from the motor shaft.
> Due to vibration the hand valve associated with an instrumentation transducer failed.

None of the anomalies were considered to be serious in nature although the last item had been the cause of the premature shutoff.

As all the other test requirements were met it was decided not to repeat the firing simply because of the short duration.

The engine data recorded during the static firing was used to demonstrate that the engine thrust, specific impulse, engine manifold torus temperature, fuel pump balance cavity pressure and heat exchanger performance were within specification.

The stage data recorded during static firing was used to demonstrate that the stage thrust vector control, pressurization, electrical power, control pressure, GN2 purge, flight measurements, range safety, propellant loading, sequencing and pogo suppression systems were performing within specified limits.

The stage was shipped back to MAF by barge on 14 July 1970. At MAF the stage underwent a full post static checkout (in place of the now-deleted post manufacturing checkout).

Following the post static tests and refurbishment, which were completed on 15 January 1971, the stage remained in storage at MAF until it was shipped to KSC, departing MAF on 13 September 1971 and arriving at KSC on 17 September 1971.

The S-IC-11 stage eventually provided the first stage of AS-511 that launched Apollo 16 on 16 April 1972.

S-IC-12

Summary

This was the final manned Saturn V booster stage to be launched.

Engines

The original allocation was:

Position 101: F-6079 - transferred to S-IC-13 position 101
Position 102: F-6080 - transferred to S-IC-13 position 102
Position 103: F-6082 - transferred to S-IC-13 position 103
Position 104: F-6083
Position 105: F-6081 - transferred to S-IC-13 position 105

The static firing and final flown engine configuration was:

Position 101: F-6084
Position 102: F-6076
Position 103: F-6075
Position 104: F-6083
Position 105: F-6074

Stage manufacturing

During July 1967 components for the S-IC-12 LOX and fuel cruciform baffle assemblies were being joined at Michoud.

The gore segments were structurally completed in September. An actuator rod support assembly was completed on 14 December 1967.

In October 1968 Boeing personnel at MAF placed the S-IC-12 booster in the vertical assembly area for painting and assembly operations. On 27 December 1968 the S-IC-12 vertical assembly was completed and the stage was transferred from the Vertical Assembly area at MAF to the Horizontal Assembly area and positioned onto the support tower in preparation for horizontal assembly. The five F-1 engines were installed in the S-IC-12 stage during February and March 1969. All horizontal installation activities were completed by 14 May 1969 and the stage was subjected to verification checks.

Stage testing

The S-IC-12 stage became the second stage to undergo the limited post-manufacturing verification (PMV) at MAF, completing these activities on 9 June 1969.

The S-IC-12 stage was shipped by barge from MAF to MTF on 16 July 1969, the day of the Apollo 11 launch at KSC. The stage was installed in the B-2 position of the B test stand on 18 July 1969 and pre-static firing tests were performed. These tests essentially repeated the post-manufacturing verification already performed at MAF prior to shipment. A key test was the "power up" test that was completed successfully on 28 July 1969.

69-03417 Static firing of the S-IC-12 stage in the B-2 test stand at MTF 3.11.69.

69-03488 The S-IC-12 stage entering the Stage Test Building at MAF Late 11.69.

Test firing S-IC-12-1

On 28 July 1969 it was decided that the static firing would be delayed from 2 September to 9 September in order to permit a propellant load test. When Hurricane Camille struck the Gulf coast on 17 August 1969 the S-IC-12 stage was in the B-2 test stand. Boeing performed a tie-down and padding of the stage in the test stand to prevent it being damaged. On 8 September an assessment of the downtime due to the hurricane was published and the implication for the S-IC-12 stage was that the propellant load test would be further delayed to 16 October, with the static firing on 22 October. The propellant load test was deemed necessary following the previous test at MTF (on the S-IC-11 stage) that had been aborted. The loading test was performed on the planned date. However the following discrepancies were noted:

- the flight LOX vent valve closed indication did not function properly and would be adjusted prior to the static firing
- during fuel tanking, the discrete switches at 4% and 5.5% level picked up and dropped out intermittently
- the number 3 LOX interconnect valve failed to close during the simulated wet countdown

The discrete switches were adjusted during de-tanking and performed satisfactorily; the LOX interconnect valve was replaced and sent to MAF for further analysis. As a result of this rework the static firing was further delayed to 24 October. Yet another delay was announced on 19 October because of the threat of Hurricane Laurie and because of a recurrence of the problem with the fuel loading probe circuit experienced during the propellant load test. It was deemed that the additional time would permit a more thorough check of the circuit. In the event an electronic package was replaced. This final delay pushed the firing out to 29 October.

The S-IC-12 stage static firing countdown was started on 28 October 1969. This was the first stage to be fired since the S-IC-11 abort. This contributed to the overall delay in the testing of this stage to ensure that all investigations and corrections had been performed successfully. Loading operations included for the first time a fuel and LOX electronic loading development trial. Fuel was loaded on 28 October 1969 at a flow rate of approximately 1950 gpm. On the planned firing day, 29 October 1969, the fuel was dropped to the engines. LOX was loaded on this day. The count was held at T-40 minutes for about 20 minutes because of weather conditions that were predicted to generate adverse acoustical conditions in surrounding populated areas. However, eventually, the static firing was cancelled by NASA safety for that day.

That evening the LOX was de-tanked and the fuel system was secured and fuel was drained through the engines from below the pre-valves. On 31 October 1969 a fuel re-topping operation was performed. On the actual firing day, 3 November 1969, the fuel was again dropped to the engines. There were 213,982 gallons of fuel on board at T-15 minutes that corresponded to a 3.01% ullage.

LOX was again loaded on the firing day at a maximum flow rate of 4000 gpm. LOX replenishment was accomplished until T-187 seconds at which time it was terminated automatically. There were 348,806 gallons of LOX on board at T-15 minutes that corresponded to 3.0% ullage.

Firing command (T-190 seconds) was issued at 1508:37.004 CST on 3 November 1969. Pre-pressurization was commanded on at T-89.6 seconds for the fuel tank and T-71.5 seconds for the LOX tank.

Engine ignition was successfully achieved at 1511:41 CST on 3 November 1969 and simulated liftoff (T+0) occurred at 1511:47.012 CST. Barometric pressure on firing day was 14.75 psia and ambient temperature was 61F.

The gimbal program was initiated at T+3 seconds and was completed at T+111 seconds. Cutoff was initiated automatically by the terminal countdown sequencer at T+125.100 seconds as planned. Main-stage duration was 126.328 seconds as measured from all engines running ON to all engines running OFF.

The firing was considered successful with the planned duration of 125 +/-2 seconds being achieved.

The engine data recorded during the static firing was used to demonstrate that the engine thrust, specific impulse, engine manifold torus temperature, fuel pump balance cavity pressure and heat exchanger performance were within specification.

The stage data recorded during static firing was used to demonstrate that the stage thrust vector control, pressurization, electrical power, control pressure, GN2 purge, flight measurements, range safety, propellant loading, sequencing and pogo suppression systems were performing within specified limits.

Post firing checks were performed in order to check that the firing had not had an adverse effect on the stage systems.

S-IC-12 was removed from the B-2 test stand on 17 November 1969 and returned to MAF by barge on 19 November 1969. Refurbishment and post-static checkout activities were performed at MAF in two stages. After initial activity the stage was placed in environmental storage position #3 on 30 January 1970. The stage was removed from storage on 17 March 1971 and the remainder of the tests performed.

The testing at MAF was completed on 19 August 1971. However because of slips in the Apollo flight schedule the stage remained at MAF in storage until shipment was required the following year. The stage left MAF on 6 May 1972 on board a ship, arriving at KSC on 11 May 1972.

The S-IC-12 stage eventually provided the first stage of AS-512 that launched Apollo 17 on 7 December 1972.

S-IC-13

Summary

This was the final Saturn V booster stage to be launched.

Engines

The original allocation was:

Position 101: F-6084 – transferred to S-IC-12 position 101
Position 102: F-6080
Position 103: F-6082
Position 104: F-6077
Position 105: F-6081

The final stage static firing and flown engine configuration was:

Position 101: F-6079
Position 102: F-6080
Position 103: F-6082
Position 104: F-6077
Position 105: F-6081

Stage manufacture

First component parts for the S-IC-13 stage were arriving at Michoud during the second half of 1967. On 18 October, in the NDT area, a gore base for the S-IC-13 stage was inspected for surface discontinuities. During December 1967 various parts were cleaned in the Minor Component Clean Facility at Michoud.

Boeing personnel at MAF completed the thrust assembly on 11 March 1969. The fuel tank assembly was completed two days later, to be followed a day later by the completion of the inter-tank. The LOX tank assembly followed soon after on 7 April 1969. The forward skirt was finished on 11 April 1969. Vertical assembly of the stage was completed on 13 May 1969.

The stage was transferred to the horizontal installation area in Michoud where activity continued until 4 August 1969 when the horizontal installation was concluded successfully. During this time, the five F-1 engines were installed in the S-IC-13 stage (between 28 May and 12 June 1969). When Hurricane Camille struck the Gulf coast on 17 August 1969 the S-IC-13 stage was safely stored in the Michoud factory together with 4 other booster stages. In October this stage benefited from a modification kit to the tunnel platform following stress corrosion problems on the S-IC-7 stage.

Stage testing

Between August and November the stage verification took place at MAF. The S-IC-13 stage became the third stage to undergo the limited post-manufacturing verification (PMV) at MAF, completing these activities on 13 November 1969.

Earlier, on 27 October 1969 Boeing personnel at MAF completed a "quality shakedown" of the S-IC-13 stage and began preparations for shipping the stage to MTF.

The S-IC-13 stage was shipped by barge from MAF to MTF on 20 November 1969. The stage was installed in the B-2 position of the B test stand on 24 November 1969 and pre-static firing tests were performed. These tests essentially repeated the post-manufacturing verification already performed at MAF prior to shipment. A key test was the "power up" which occurred on 8 December 1969.

69-75864 The first F-1 engine being installed in the S-IC-13 stage at MAF 28.5.69.

70-06013 Static firing of the S-IC-13 stage in the B-2 test stand at MTF 6.2.70.

Test firing S-IC-13-1

The first attempt to fire S-IC-13 was started on 12

January 1970. Fuel was loaded on that day. However, on the planned firing day, 13 January, the test was cancelled prior to the LOX loading starting because of unfavorable weather conditions. On 14 January LOX was loaded, but this attempt was aborted because LOX leakage developed through the main LOX valve number 1 on engine 102 (F-6080). At about the same time as the decision to abort was being formalized, the replenish pump on the facility barge at dock number 2 exploded.

The second attempt started on 4 February 1970 but was rescheduled to the following day because of operational delays. The S-IC-13 stage static firing countdown was started on 5 February 1970. Fuel was loaded on that day, with the pre-valves open, at a flow rate of approximately 1900 gpm. There were 214,350 gallons of fuel on board at T-15 minutes that corresponded to a 3.0% ullage.

LOX was loaded on the firing day, 6 February, at a maximum flow rate of 4000 gpm. LOX replenishment was accomplished until T-187 seconds at which time it was terminated automatically. There were 348,235 gallons of LOX on board at T-15 minutes that corresponded to 3.0% ullage.

Firing command (T-190 seconds) was issued at 1543:41.008 CST on 6 February 1970. Pre-pressurization was commanded on at T-89.6 seconds for the fuel tank and T-71.4 seconds for the LOX tank.

Engine ignition was successfully achieved at 1546:45 CST on 6 February 1970 and simulated liftoff (T+0) occurred at 1546:51.016 CST. Barometric pressure on firing day was 15.79 psia and ambient temperature was 59F.

The gimbal program was initiated at T+3 seconds and was completed at T+111 seconds. Cutoff was initiated automatically by the terminal countdown sequencer at T+125 seconds as planned. Main-stage duration was 126.432 seconds as measured from all engines running ON to all engines running OFF.

The firing was considered successful with the planned duration of 125 +/-2 seconds being achieved.

The engine data recorded during the static firing was used to demonstrate that the engine thrust, specific impulse, engine manifold torus temperature, fuel pump balance cavity pressure and heat exchanger performance were within specification.

The stage data recorded during static firing was used to demonstrate that the stage thrust vector control, pressurization, electrical power, control pressure, GN2 purge, flight measurements, range safety, propellant loading, sequencing and pogo suppression systems were performing within specified limits.

Post firing checks were performed in order to check that the firing had not had an adverse effect on the stage systems.

S-IC-13 was removed from the B-2 test stand and returned to MAF by barge on 25 February 1970. Refurbishment and post-static checkout activities were performed at MAF in two stages. After initial activity the stage was returned by barge to MTF and placed in environmental storage on 15 July 1970. The stage was removed from storage on 26 April 1971 and returned to MAF by barge on 5 May 1971 for the remainder of the tests to be performed.

The testing at MAF was completed on 10 November 1971. However because of slips in the Skylab flight schedule the stage remained at MAF in storage until shipment was required the following year. The stage was removed from storage on 13 April 1972 and left MAF on 21 July 1972 on board the barge Orion, arriving at KSC on 26 July 1972.

The S-IC-13 stage eventually provided the first stage of AS-513 that launched the Skylab space station on 14 May 1973.

S-IC-14

Summary

This was the penultimate Saturn V booster stage made and is one of only two booster stages intended for flight that are still in existence. This stage is on display at the NASA Johnson Space Center in Houston.

Engines

The original allocation was:

Position 101: F-6095 – never fitted - transferred to S-IC-11 position 101
Position 102: F-6093
Position 103: F-6087 – never fitted - transferred to S-IC-11 position 103
Position 104: F-6086
Position 105: F-6090 – never fitted – became a spare for S-IC-15

In addition engine F-6088 was formerly allocated to S-IC-14 as a spare, but was transferred to S-IC-10, position 101.

The stage static firing and final engine configuration was:

Position 101: F-6089
Position 102: F-6093

Position 103: F-6085
Position 104: F-6086
Position 105: F-6092

Spare engine: F-6049

Stage manufacture

When hurricane Camille struck the Mississippi and Louisiana Gulf Coast on 17 August 1969 the S-IC-14 booster stage was safety installed in the Michoud factory together with stages S-IC-9, S-IC-10, S-IC-11 and S-IC-13. As soon as the hurricane had passed, on 20 August 1969, Boeing began vertical assembly of the S-IC-14 stage at MAF. By 29 August 1969 the thrust structure had been completed. Inter-tank fabrication was completed on 5 September 1969.

A major milestone was reached on 19 September 1969 when Boeing completed fabrication of the LOX tank at Michoud. Three days later fabrication of the forward skirt was completed. Vertical assembly of the stage was concluded on 6 October 1969.

The stage was placed in a horizontal orientation and systems installation progressed. The five F-1 engines were installed in the S-IC-14 stage between 7 and 15 November 1969.

The stage was completed at the turn of the year and entered the verification and test phase.

Stage testing

The S-IC-14 stage became the fourth stage to undergo the limited post-manufacturing verification (PMV) at MAF, completing these activities on 9 February 1970.

The S-IC-14 stage was shipped by barge from MAF to MTF on 5 March 1970. The stage was installed in the B-2 position of the B test stand and pre-static firing tests were performed. These tests essentially repeated the post-manufacturing verification already performed at MAF prior to shipment.

Test firing S-IC-14-1

The S-IC-14 stage static firing countdown was started on 15 April 1970. Fuel was loaded on that day, with the pre-valves open, at a flow rate of approximately 1820 gpm. There were 213,750 gallons of fuel on board at T-15 minutes that corresponded to a 2.88% ullage.

LOX was loaded on the firing day, 16 April, at a maximum flow rate of 4000 gpm. LOX replenishment was accomplished until T-187 seconds at which time it was terminated automatically. There were 347,029 gallons of LOX on board at T-15 minutes that corresponded to 2.0% ullage.

The S-IC-14 stage on display at JSC in 1991

The author with S-IC-14 stage on display at JSC in 1991

70-06069 The S-IC-14 stage en route to MAF docks for shipment to MTF 5.3.70.

69-02652 The S-IC-14 stage after removal from the Vertical Assembly Position at MAF on its way to the Manufacturing Building 6.10.69.

70-06074 The S-IC-14 stage being installed in the B-2 test stand at MTF 3.70.

70-08321 The S-IC-14 and 15 stages in the horizontal assembly area at MAF 5.70.

70-09120 Removal of the S-IC-14 stage from the B-2 test stand at MTF following static firing Late 4.70.

During the loading some problems were encountered. There were malfunctions of the LOX dome purge regulators in the pneumatic console. Also there were two fuel leaks. The first one discovered on T-1 day was a minor leak at an instrumentation boss. The second was a leak on an engine that required a seal replacement. This latter leak caused approximately a four-hour hold to repair and retest. For this reason this was a rare evening static firing.

During the static firing test the LOX load was intentionally increased to cause the LOX level to be above the loading probe. The purpose was to determine if the loading probe would go "out-of-lock". However the out-of-lock signal was not attained.

Firing command (T-190 seconds) was issued at 2011:51.008 CST on 16 April 1970. Pre-pressurization was commanded on at T-89.6 seconds for the fuel tank and T-71.4 seconds for the LOX tank.

Engine ignition was successfully achieved at 2014:54 CST on 16 April 1970 and simulated liftoff (T+0) occurred at 2015:01.020 CST. Barometric pressure on firing day was 15.60 psia and ambient temperature was 72F.

The gimbal program was initiated at T+3 seconds and was completed at T+111 seconds. Cutoff was initiated automatically by the terminal countdown sequencer at T+125 seconds as planned. Main-stage duration was 126.364 seconds as measured from all engines running ON to all engines running OFF.

The firing was considered successful with the planned duration of 125 +/-2 seconds being achieved.

The engine data recorded during the static firing was used to demonstrate that the engine thrust, specific impulse, engine manifold torus temperature, fuel pump balance cavity pressure and heat exchanger performance were within specification.

The stage data recorded during static firing was used to demonstrate that the stage thrust vector control, pressurization, electrical power, control pressure, GN2 purge, flight measurements, range safety, propellant loading, sequencing and pogo suppression systems were performing within specified limits.

There were several anomalies associated with the static firing test:

- The B-1 hydraulic unit reservoir full indication failed during the static firing.
- The helium bottles emergency dump valve failed to respond to the close command after venting the bottles.
- Approximately 10 System "A" measurement excitation voltages shifted during the static test.
- Post-test inspection showed minor leakage on one flange on engine position 101.
- The LOX dome purge regulators overshot the 1200 psig maximum lock up pressure during ignition.
- Minor damage was noted to Facility Systems not critical to static firing.

Post firing checks were performed in order to check that the firing had not had an adverse effect on the stage systems.

S-IC-14 was removed from the B-2 test stand and returned to MAF by barge on 1 May 1970. Refurbishment and post-static checkout activities were performed at MAF in two stages. After initial activity the stage was returned by barge to MTF on 11 August 1970 and placed in environmental storage on 14 August 1970. The stage was removed from storage on 4 January 1971 and returned to MAF by barge on 13 January 1971 for the remainder of the tests to be performed.

The testing at MAF was completed on 21 May 1971. However because of the cancellation of Apollo 18 there was no requirement to ship this stage to KSC. The stage entered environmental storage at MAF on 24 June 1971 and was removed from this storage on 11 September 1972, in preparation as a back up for the Skylab first stage. The stage was returned to environmental storage on 15 August 1973 and on 18 October 1973 it was removed from the environmental enclosure in the Manufacturing Building at MAF and placed in the environmental enclosure in Test Cell number 3 in the Stage Test Building. The stage was transferred to CCSD on 27 November 1973, then to Mason-Rust on 19 April 1974 and to Boeing Services International on 1 January 1975.

In 1977 the stage was moved from MAF to the Johnson Space Center (JSC) in Houston, Texas. S-IC-14 left MAF on board a barge on 16 September 1977 and arrived at JSC on 19 September 1977 after traveling through the inter-coastal waterways. JSC had to dredge a channel in Clear Lake for the barge and then purchased a strip of lake waterfront for constructing the dock. The final trek across NASA Road 1 took place at 0130 on 21 September 1977. The stage was finally in place at 0930 the same day. The stage has been on display to the public at JSC ever since. Ownership of the stage was transferred to the Smithsonian museum in 1978. In 2004 plans were announced to refurbish the JSC Saturn V stages (including S-IC-14) and enclose them in a purpose built building.

S-IC-15

Summary

This was the last Saturn booster stage produced. The stage included the last F-1 engine ever made and also, interestingly, included two engines that had previously been part of Saturn booster firings on earlier stages. The stage is now visible to passing traffic along the Old Gentilly Road in New Orleans outside the NASA Michoud Assembly Facility (MAF).

Engines

The original allocation was:

Position 101: F-6066
Position 102: F-6096 – transferred to S-IC-11 position 102
Position 103: F-6097
Position 104: F-6094 – transferred to S-IC-11 position 104
Position 105: F-6098

The stage static firing and final engine configuration was:

Position 101: F-6066
Position 102: F-6091
Position 103: F-6097
Position 104: F-6060
Position 105: F-6098

Spare engine: F-6049

Stage manufacturing

Fabrication and assembly of the last Saturn V booster started in August 1968 at MAF. On 22 August 1969 NASA issued an amendment to the contract with Boeing that extended the completion and support date for the S-IC-15 stage from 30 June 1970 to 30 June 1971. The five F-1 engines were installed in the S-IC-15 stage between 13 and 29 April 1970.

Stage testing

The S-IC-15 stage became the fifth and final stage to undergo the limited post-manufacturing verification (PMV) at MAF, completing these activities on 25 July 1970.

The S-IC-15 stage was shipped by barge from MAF to MTF on 17 August 1970. The stage was installed in the B-2 position of the B test stand and pre-static firing tests were performed. These tests essentially repeated the post-manufacturing verification already performed at MAF prior to shipment.

The S-IC-15 stage on display at Michoud in 2004

The S-IC-15 stage on display at Michoud in 2004

The S-IC-15 stage on display at Michoud in 2004

70-05601 The S-IC-15 fuel tank after being mated to the thrust structure in the Vertical Assembly Tower at MAF 2.70.

70-07784 F-1 engine installation in the S-IC-15 stage at MAF Late 4.70.

Test firing S-IC-15-1

The S-IC-15 stage static firing countdown was started on 29 September 1970. Fuel was loaded on that day, with the pre-valves open, at a flow rate of approximately 2500 gpm. There were 214,514 gallons of fuel on board at T-15 minutes that corresponded to a 2.85% ullage.

LOX was loaded on the firing day, 30 September, at a maximum flow rate of 4000 gpm. LOX replenishment was accomplished until T-187 seconds at which time it was terminated automatically. There were 347,900 gallons of LOX on board at T-15 minutes that corresponded to 3.2% ullage.

T-1 day activities were accomplished with no major anomalies. A special fuel overfill test and loading probe capacitance measurement test was conducted.

One major problem occurred on firing day. 28Vdc power was lost to LOX barge number 1 during fast fill operations. The discrepancy could not be isolated and resulted in a barge-to-barge LOX transfer that extended the LOX loading operations to approximately 5.5 hours.

Firing command (T-190 seconds) was issued at 1814:05.009 CDT on 30 September 1970. Pre-pressurization was commanded on at T-89.6 seconds for the fuel tank and T-71.4 seconds for the LOX tank.

Engine ignition was successfully achieved at 1817:08 CDT on 30 September 1970 and simulated liftoff (T+0) occurred at 1817:15.020 CDT. Barometric pressure on firing day was 15.455 psia and ambient temperature was 80.3F.

The gimbal program was initiated at T+3 seconds and was completed at T+111 seconds. Cutoff was initiated automatically by the terminal countdown sequencer at T+125 seconds as planned. Main-stage duration was 126.672 seconds as measured from all engines running ON to all engines running OFF.

The firing was considered successful with the planned duration of 125 +/-2 seconds being achieved.

The engine data recorded during the static firing was used to demonstrate that the engine thrust, specific impulse, engine manifold torus temperature, fuel pump balance cavity pressure and heat exchanger performance were within specification.

The stage data recorded during static firing was used to demonstrate that the stage thrust vector control, pressurization, electrical power, control pressure, GN2 purge, flight measurements, range safety, propellant loading, sequencing and pogo suppression systems were performing within specified limits.

Three special tests were successfully conducted:

> The LOX tank vent valves were cycled every 10 minutes from 70% LOX load to T-1 hour in a controlled environment. The valve travel times were within the required limits.
> The stage LOX fill and drain valves were operated without the application of heater power and operated normally.
> LOX was loaded to the overfill sensor and the LOX loading probe capacitance was measured. The loading probes did not go "out-of-lock" and all required data was obtained.

A small F-1 engine skirt hatband fire was reported after cutoff, but was not a problem as it quickly burned itself out.

The minimum required operating pressure of 3000 psig in the helium bottles prior to ignition was not attained. The actual value was approximately 2975 psig.

Post firing checks were performed in order to check that the firing had not had an adverse effect on the stage systems.

S-IC-15 was removed from the B-2 test stand and returned to MAF by barge on 16 October 1970. Refurbishment and post-static checkout activities were completed on 19 March 1971.

As Apollo 19 had been cancelled there was no requirement to ship the stage to KSC. Instead, S-IC-15 was installed in the environmental enclosure at MAF on 5 April 1971. The environmental storage condition was certified 3 days later. The stage was removed from the environmental enclosure on 8 September 1972, in preparation for a back-up role to the Skylab first stage. The stage was placed back in environmental storage on 2 July 1973, after the Skylab launch. On 29 November 1973 it was then transferred from the Environmental Enclosure in the Manufacturing Building at MAF to the Environmental Enclosure in Test Cell number 1 in the Stage Test Building. The stage was transferred to CCSD on 10 December 1973, then to Mason-Rust on 19 April 1974 and to Boeing Services International on 1 January 1975.

On 9 December 1978 the ownership of the stage was transferred to the Smithsonian museum and the stage was moved the short distance outside the MAF buildings where it was first constructed. It is now on display in the grounds of MAF, visible to the general public through a fence.

S-II-S

Summary

S-II structural test stage destroyed in test failure in 1965.

Stage manufacture

The S-II-S, structural static test stage was built by NAA at Seal Beach, starting in 1964.

On 3 January 1964 the final meridian weld was made on the aft facing sheet for the S-II-S common bulkhead. Hydrostatic testing of the first production forward bulkhead was completed successfully during March. In April welding of the forward and aft facing sheets for the common bulkhead was completed.

On 11 April vertical assembly of the stage was initiated with the initial attempt to weld Cylinder 6 to the forward bulkhead. The weld was subsequently cut apart (because of thermal expansion problems) and re-welded successfully on 22 April.

In May the first bond was made in the Seal Beach autoclave. The core was bonded to the aft facing sheet for the S-II-S common bulkhead.

Mating of the stage's thrust structure and the aft skirt assembly was completed during June in the VAB at Seal Beach.

On 17 July 1964 the static firing condition was deleted from the S-II-S test requirements. The equivalent data would be obtained from a parallel test effort.

On 30 July bonding of the first S-II common bulkhead was completed. On 12 August Cylinder 1 was successfully welded to Cylinder 2. By September 1964 the final manufacturing operations were completed on the S-II-S common bulkhead. Also in that month Cylinder 3 was welded to Cylinder 4.

On 18 September the J-weld joining the common bulkhead to the Cylinder 1 and Cylinder 2 assembly was completed. Cylinder 5 was welded to the Cylinders 4 and 5 assembly during October.

On 28 October 1964 the first production aft bulkhead, planned for use on the S-II-S stage, ruptured during a hydrostatic proof test at an observed pressure of 50 psig while attempting to reach the required pressure of 56.5 psig (pressure acting at apex of bulkhead). Welding of a replacement bulkhead began in the first week in November. The failure originated in a manually repaired weld along the re-circulation system service plate. Design changes were initiated to make all service plates integral parts of the bulkhead gores, eliminating the service plate welds (effective on the S-II-1 and subsequent aft bulkheads and the S-II-4 and subsequent forward bulkheads and lower cylinders). A repair weld development program was initiated to develop tooling and techniques for performing out-of-station automatic weld repairs to reduce the number of instances manual repairs would be needed.

During November 1964 the Cylinder 3/4/5 assembly was welded to the Cylinder 6/forward bulkhead assembly. The S-II-S common bulkhead was successfully hydrostatically tested on 9 November 1964. Meridian welding of the replacement aft bulkhead was completed 14 days ahead of schedule on 21 November. By December 1964 the aft bulkhead was complete through dollar welding (that was described as flawless), hydrostatic test, and dye-penetrant inspection.

During January 1965 NASA and NAA finalized plans to cancel the S-II-D, dynamic test stage, and to reassign S-II-S, the structural static test stage, to the dual role of structural and dynamic testing. S-II-S/D was officially designated on 1 February 1965, and the S-II-D stage was cancelled.

NAA completed vertical assembly of S-II-S at Seal Beach on 31 January 1965, 5 days ahead of schedule, after the final weld had been performed two days earlier. On 3 February 1965 the stage was removed from the Vertical Assembly Building and placed in the Structural Static Test Fixture at Seal Beach for installation of instrumentation prior to structural testing.

Stage testing

Pretest operations began on 22 March, followed by hydrostatic testing. During April four thrust structure tests were conducted. The first ultimate load test was performed successfully on 13 April. However, on 30 April 1965, during a final test simulating a flight condition, a failure occurred in which the number 203 engine mount was torn from the structure, the skin of the structure was torn and thrust longerons broken. An investigation revealed that the structure had insufficient strength in the area of failure.

Modifications to strengthen the stage took place between 4 May and June, and repairs to the damage continued to 16 July 1965.

Center engine thrust load testing in July resulted in structural failure of rivets and attach bolts, which required repair and delayed completion of the S-II-S/D stage test program.

Further static tests on 14 August caused more structural damage including buckling of the forward skirt. Load testing continued in August and September, with local damage occurring in several areas of the stage.

On 29 September 1965, during ultimate load testing with hydrostatic pressurization, conducted at Seal Beach, the stage ruptured and was destroyed. The failure occurred at 144% of limit load on the aft skirt, simulating end of S-IC boost, thus demonstrating the optimum design and verifying the structural integrity of the stage. In spite of the failure, the conclusion of the structural testing was considered successful.

A failure evaluation team concluded that the problem probably started in the inter-stage, causing rupture of the aft LOX and common bulkheads, and released water from the LH2 tank. The sudden loss of 3.2 million pounds of water from the tanks caused the stage to collapse. Failure of the inter-stage, resulting in its loss of ability to support the stage, caused rupture of the aft LOX and common bulkheads. Failure of the LOX tankage released water in the LH2 tank, and as the water quickly receded in the lower portions of the LH2 tank a vacuum was created in the lower ullage volume causing total crushing of the LH2 tank walls and the LH2 tank forward bulkhead.

Instant release of this volume of water and its movement away from the test structure created a force of sufficient magnitude to cause collapse of the test fixture that was applying the load to the bolt ring.

Failure of the lower steel test members caused secondary fragmentation of the S-II-S/D structure. Further fragmentation of the failed stage occurred as water carried debris through the test tower. It was concluded that no explosive type failure occurred. The sudden brittle failure was due to the following:

- inability to abort 3.2 million pounds of water
- axial loading of the vehicle did not permit redistribution of the load
- loss of longitudinal stability due to local buckling of the stringers

Following the loss of the S-II-S/D stage, NASA designated the S-II-T stage, the all-system test stage, as the replacement specifically to cover dynamic testing, which the S-II-S/D stage had been intended for.

Some of the structural test objectives not attained due to the failure of this stage were subsequently achieved in February 1968 using a special S-II-3 thrust structure that was tested at MSFC in Huntsville.

S-II-TS-A

Summary

Structural test stage used at MSFC.

Stage manufacture

In March 1967 MSFC received NASA authorization (via Mod 594 CO) to fabricate three special full-scale test structures, one of which, the S-II-TS-A, was to simulate the LOX tank and lower LH2 tank assembly. The units would be used in a test program to verify the integrity of the lightweight structures of the operational S-II stages from S-II-4 to S-II-10.

On 16 February 1967 NAA workmen at Seal Beach completed welding of the S-II-8 aft facing sheet. It was then reassigned for use on the S-II-TS-A test structure.

The circumferential weld joining cylinders 1 and 2 was completed in July. Cylinder 2 used quarter panels originally intended for the S-II-8 stage. The quarter panels were modified to incorporate one S-II-5 type LH2 feed line elbow, one S-II-8 type elbow, and three blank plates. During the following month the J-weld joining the common bulkhead to cylinders 1 and 2 was accomplished. The J-weld was found to contain excessive porosity, and was subsequently cut apart and re-welded. The new J-weld was accepted on 30 August.

By the end of September mating of the simulated thrust structure and aft static-firing skirt assembly had been completed, and placed in storage. The girth weld of the LOX bulkhead was completed in Station III during the first week of October. During x-ray inspection, 124 x-rays were taken of the weld area and 53 areas were deemed to need rework.

Aft skirt and LOX tank mating was achieved during the first week of November 1967.

On 11 November 1967 NAR shipped the S-II-TS-A structural assembly including the cylinders 1 and 2, common bulkhead, LOX bulkhead, aft skirt, and the simulated thrust structure from Seal Beach to MSFC. The initial part of the journey was on board the AKD Point Barrow, leaving Seal Beach Naval Docks on 11 November and arriving at MAF on 25 November. The test stage traveled together with the S-II-4 stage that was being delivered to MTF, via MAF. For the second part of the journey the test stage was transported alone, leaving MAF on 26 November.

The S-II-TS-A test structure arrived at MSFC dock on 6 December 1967. It was unloaded and moved to the ME Laboratory to be welded to the S-IC forward bulkhead.

On 26 February 1968 Boeing personnel at MSFC completed instrumentation installation and turned over to ME Lab the S-II-TS-A stage. ME Lab would install systems hardware and complete S-IC/S-II ring frame installation.

By 25 March the ME Lab, working two daily 12-hour shifts, had completed the etch, prime and foam (insulation) operation on the stage. They also completed the foaming operation of an S-IC bulkhead for use with the structure – the first bulkhead of its size to be "foamed" at MSFC.

On 29 March the modification of the thrust structure for the S-II-TS-A had been completed. The stage was transferred to a special test tower at MSFC on 6 May 1968, following installation of all systems in the LOX and LH2 tanks. Two days later the stage was installed in the test tower where it would be subjected to flight load testing, with flight loads of LH2 and LN2.

Stage testing

On 5 July structural verification testing of the stage began. During July cryogenic testing of the stage began and proceeded through LN2 tanking. Some cracks were experienced in the bolting ring area of the test specimen.

Cryogenic cycle testing ran from 23 to 26 August at MSFC. On 5 September a limit load test was performed. However, damage to foam insulation and cracked baffles occurred. These were repaired by 17 September.

Test personnel accidentally punctured the common bulkhead forward facing sheet. However this was repaired and testing resumed on 24 October 1968.

On 1 November 1968 MSFC completed the final test

condition (ultimate pre-launch) of the S-II-TS-A test structure. The test qualified the common bulkhead for ultimate burst and ultimate collapse. There was no indication of bulkhead de-bonding.

On 6 December 1968 MSFC conducted an ambient test on the LH2 fill and drain fitting on the stage. This was followed on 11 December by a cryogenic test on the same fitting.

On 2 April 1969 low frequency oscillation tests were begun on the stage at Huntsville.

On 21 April 1969 MSFC issued Engineering Change Order 1679 requesting NAR to perform the necessary design, fabrication and test analysis support in mating the S-II-TS-A with the S-II-TS-C to determine the mechanical and fluid coupling characteristics between the aft bulkhead, the center-engine beam, and the outboard engines.

67-61129 The S-II-TS-A test stage on a modified S-IC transporter being transferred to the Seal Beach Naval docks for shipment to MSFC 9.11.67.

S-II-TS-B

Summary

Structural test stage tested at SSFL and destroyed in failure in 1968.

Stage manufacture

In March 1967 MSFC received NASA authorization (via Mod 594 CO) to fabricate three special full-scale test structures, one of which the S-II-TS-B was to simulate the upper LH2 tank and forward skirt assembly. The units would be used in a test program to verify the integrity of the lightweight structures of the operational S-II stages from S-II-4 to S-II-10.

The refurbishment of the S-II-6 forward bulkhead began on 3 June, and on 9 June the replacement of three gores began. This hardware would be used on the S-II-TS-B stage. On 19 June the installation of test instrumentation was completed and spray-on foam application was begun. Fabrication progressed through the summer with completion of meridian and dollar welding of the forward bulkhead occurring on 4 August. The assembly was joined with cylinder 6 on 12 September.

At about the same time fabrication of the forward skirt was completed at Downey, Ca. This assembly, along with the membrane structural closeout and wagon wheel, were delivered to NAA, Seal Beach on 26 September 1967. The wagon wheel was a fixture welded to the "B" structure for support purposes and the membrane was a cover, welded to the bottom of cylinder 6 making, in effect, a short LH2 tank.

Welding of the membrane to Cylinder 6 was accomplished on 9 October. Mating of the forward skirt to Cylinder 6 took place on 12 October. Mating of the wagon wheel to closeout was completed on 24 October.

On 31 October 1967 NAR transported the completed S-II-TS-B structure from Seal Beach to SSFL. Initially the structure was loaded onboard the AKD Point Barrow at the Naval Docks at Seal Beach for the short trip up the Californian coast to Port Hueneme, near Oxnard. Here the structure was transferred to a truck for the short overland journey to SSFL. The stage was installed in the Coca 4 test stand during the week of 6 December 1967 where instrument installations immediately commenced.

Stage testing

The first cycle of Phase I (pneumostatic) testing of the stage at SSFL began on 5 January 1968. Instrumentation problems delayed completion of the testing until 8 January. The four remaining pneumostatic cycles on the stage ended on 16 January. All results were satisfactory.

67-60425 The S-II-TS-B test stage being offloaded from the Point Barrow at Port Hueneme en route to SSFL 31.10.67.

67-60426 The S-II-TS-B test stage on the road from Port Hueneme to SSFL 31.10.67.

67-61137 The S-II-TS-B test stage being installed in the Coca IV test stand at SSFL 30.11.67.

68-68456 Removal of the damaged forward skirt of the S-II-TS-B stage from the Coca IV test site at SSFL 27.7.68.

69-72011 Close up of damage sustained by the S-II-TS-B stage in the Coca IV test stand at SSFL 20.12.68.

On 9 February 1968 NAR completed room temperature influence testing of the stage. On 15 March 1968 the cryogenic tanking test was accomplished using 32,000 gallons of liquid nitrogen. On the following day personnel at SSFL completed without problem the LN2 fill and drain test, using 100,000 gallons of liquid nitrogen.

On 25 March Phase V (cyclic) testing of the stage was initiated. Forty room temperature cycles were completed satisfactorily. Following a dye-penetrant inspection of the tank, thirty five cryogenic cycles were run. This was completed on 28 March and inspection revealed no defects.

NAR personnel at SSFL completed maximum dynamic pressure (Max Q) testing on the stage on 5 April 1968. This first limit load test met most test objectives. A follow up limit loading test on 12 April was performed without problem.

On 18 April Test Phase VIII (end of S-II boost) was performed on the stage at SSFL. The maximum axial load applied was one million pounds, with a tank pressure of 36 psig.

On 15 June NAR personnel at SSFL tested the stage by simulating the vehicle loads that would occur at the end of the S-IC boost during flight. This test was rerun on 24 June. Testing restarted on 3 July, following resolution of a strain gauge problem.

On 12 July 1968 the forward skirt of the S-II-TS-B failed during structural qualification testing at SSFL. The failure occurred during a simulation of the maximum AS-504 flight vehicle bending loads and maximum burst pressure on the forward skirt.

On 16 August NAR delivered the S-II-10 forward skirt to SSFL to replace the skirt that failed in the S-II-TS-B stage. However, five days later a crane operator dropped the replacement forward skirt causing some damage. Following repairs the replacement skirt was installed on the stage by NAR personnel, beginning on 4 September 1968. The rebuilt assembly was inspected on 16 October.

Ultimate burst and collapse pressures were applied successfully to the common bulkhead of the S-II-TS-B test structure on 1 November 1968. A successful test simulating "ultimate flight, end of S-II boost" (Condition XI) loads and environment was conducted on 19 November 1968. This test verified the lightweight forward bulkhead design and thus completed certification of the S-II-3 stage forward bulkhead which had formerly been assigned to the S-II-5 stage.

On 20 December 1968 (a day before the first manned launch of a Saturn V vehicle) an explosion during preparations for the final test which would assess the integrity of the forward skirt totally destroyed the S-II-TS-B stage in the Coca 4 test stand at Santa Susana. Extensive fires in the test area started immediately. It was believed that the explosion had occurred when the LH2 forward bulkhead collapsed due to negative pressure, causing a tank rupture that introduced air to the tank; the air mixed with the LH2 gas being used for the chill-down, and was ignited by some unknown source.

On 15 January 1969 NAR completed the investigation of the explosion that destroyed the stage. Results of the investigation indicated that combustion of hydrogen gas had occurred inside the tank as a result of residual oxygen not completely cleared out by the nitrogen purge. NASA recommended additional procedures and equipment for monitoring gas samples in propellant tanks and the transfer lines prior to the loading of cryogenic propellants. The changes would be implemented on the S-II-4 stage that was at KSC at the time.

Forward skirt certification was obtained by analysis of splice stringer modification test data and revision of test requirements based on the AS-504 vehicle flight plan.

S-II-TS-C

Summary

Structural test stage used at MSFC.

Stage manufacture

In March 1967 MSFC received NASA authorization (via Mod 594 CO) to fabricate three special full-scale test structures, one of which the S-II-TS-C was to simulate the aft skirt/thrust structure assembly. The units would be used in a test program to verify the integrity of the lightweight structures of the operational S-II stages from S-II-4 to S-II-10.

Assembly of the thrust core was started in early April 1967. On 20 May NAA completed the S-II-TS-C test structure and prepared it for barge shipment from California to MSFC, Huntsville. The structure left Seal Beach and arrived at MSFC on 6 June 1967. The journey was made by commercial freighter and barge.

Stage testing

Scheduled testing at MSFC was delayed in July and August due to hardware shortages in the heat control system. Problems were also encountered in cooling the aft skirt. A cooling manifold was fabricated and installed during October.

Condition I structural testing of the test structure started at MSFC on 29 November 1967.

Limit load testing to 128% completed structural qualification of the aft skirt for S-II-4 onwards on 20 December 1967.

On 20 March 1968 MSFC personnel completed temperature influence testing on the S-II-TS-C stage. Between 27 March and 2 April engineers completed a series of eight limit load tests on the stage. On 11 April ultimate load testing started at MSFC. The center engine crossbeam buckled at 110% and 115% loadings, causing the testing to be suspended. The second ultimate load test (to 130%) was run on 11 May. A crossbeam arm buckled. To rectify the problem a ¼" rivet was added in a wide space without rivets. This appeared to remedy the situation and the change was incorporated for all flight stages.

On 17 May the third ultimate load test was run. This was a repeat of the first test performed on 11 April. On 21 May the fourth ultimate load test was performed in the Load Test Annex at MSFC. This test resulted in extensive failure of the ring frame at engine 203 location, at 120% of design limit load. A failure analysis team was set up to investigate the problem.

By 17 June the thrust ring frame had been repaired. As a result of the 21 May failure, during July 1968 MSFC approved the engineering change procedure to provide for strengthening the thrust structure on flight stages S-II-4 through S-II-10 as well as the S-II-TS-C structure. The change would strengthen the thrust structure for "engine-out" capability.

With repairs completed, on 14 August personnel at MSFC conducted a room temperature test and an ultimate load test to 130% of limit load, simulating one outboard engine out without external heat applied. A slightly cracked engine stringer resulted.

The last planned qualification test, simulating one outboard engine out with external heat applied, took place at MSFC on 21 August 1968. After testing was discontinued at 122% of the limit load inspection revealed some thrust structure damage that indicated that further testing was necessary. During September NAR presented their assessment of the repair work necessary for the stage. The recommended 31 week repair plan was rejected by MSFC in favor of repairs to local areas of damage together with acceptance of minor configuration deviations.

MSFC completed repair of the S-II-TS-C stage on 4 November, 5 days ahead of schedule. On 11 November MSFC personnel completed a "two engines out" test without application of heat. The test was successfully conducted to 110% of the limit load. A successful ultimate load test on 13 November ended the scheduled testing of this stage.

On 12 December 1968 Test Lab personnel in the MSFC Load Test Annex completed the dismantling of the S-II-TS-C test article. Teardown had started on 3 December following the completion of the test program.

On 21 April 1969 MSFC issued Engineering Change Order 1679 requesting NAR to perform the necessary design, fabrication and test analysis support in mating the S-II-TS-A with the S-II-TS-C to determine the mechanical and fluid coupling characteristics between the aft bulkhead, the center-engine beam, and the outboard engines.

S-II Common Bulkhead Test Tank (CBTT)

Summary

Structural test item used for testing at SSFL before a test failure destroyed the item in 1966.

Stage manufacture

One of the special test articles in the S-II stage development and ground test program was the Common Bulkhead Test Tank (CBTT). The CBTT was a shortened version of the S-II stage, designed to help solve the special insulation and structural problems of the common bulkhead at liquid hydrogen temperatures. It consisted of a production common bulkhead, a production forward bulkhead, two production LH2 cylinders, a special forward skirt, and a simulated aft bulkhead. These components were joined to form a large tank divided by the common bulkhead into two smaller tanks.

Welding of the first two gores for the aft facing sheet of the CBTT was completed on 3 March 1964. On 5 June 1964 Rocketdyne received a telegram that gave them approval to begin construction of the CBTT test site at SSFL. This facility was completed at SSFL on 16 February 1965.

69-03462 S-II CBTT site at SSFL 24.11.69.

In July 1964 NAA completed assembly of the forward bulkhead. Assembly of both cylinders required for the tank ended in August. During September workers completed the forward facing sheet and insulated the LH2 cylinders. Insulating, bonding and testing occupied technicians in October and November. In December 1964 NAA welded together the aft and forward facing sheets to form the common bulkhead, conducted inspections and prepared to transfer the various subassemblies to the Vertical Assembly Facility at Seal Beach for circumferential welding.

The final LH2 tank weld was completed on 11 March 1965. Manufacturing work on the CBTT was completed on 4 May 1965. On 24 May 1965 NAA at Seal Beach completed assembly of the CBTT and loaded it onto a Type 1 transporter three days later. Some instrumentation installation and other work were deferred until delivery of the structure to SSFL. On 1 June the CBTT began its journey from southern Los Angeles to northern Los Angeles on the barge Orion. Initially it was shipped from Seal Beach Naval Docks to Port Huemene near Oxnard. At that point it was transferred to a truck for the short overland trip to the Santa Susana Test Facility, where it arrived on 3 June 1965.

Stage testing

NAA personnel finished installing insulation and on 8 July completed preparations for the first CBTT test. This first test was initiated on 9 July. On 11 July structural verification of the tanks was completed with a pneumatic pressurization.

Numerous insulation leaks were discovered, but attempts to repair were unsuccessful. Liquid nitrogen fill and drain tests were run successfully on 21 August and 12 September, using a mylar helium bag to prevent insulation leaks.

The final scheduled test of the CBTT program ended successfully on 6 November 1965 at SSFL, when the skirt-to-tank joint was tested to 1.3 times the design limit pressure and the common bulkhead was tested to

1.4 times the limit burst pressure. Follow on testing was planned to verify the forward skirt and tank wall joint. This supplementary testing on 24 November 1965 was completed successfully, and the CBTT was semi-mothballed.

On 2 September 1966 contract coverage was received by NAA from NASA for a biaxial test program using the CBTT. Tests would be made of structural repair methods, primarily the bolted doublers being used to reinforce certain areas of the LH2 tank. In mid-September this direction was expanded to include installation and testing of the redesigned LH2 feed line elbows.

The CBTT underwent modification to approximate the structural condition of the S-II-1, -2, and -3 LH2 tanks.

On 1 December 1966 the CBTT failed during an ultimate hydrostatic pressure test at SSFL. The LH2 tank bulkhead and wall collapsed, leaving intact the LOX tank and common bulkhead portion of the test article. The pressure test had been to certify repair doubler and feed line elbow designs. The failure occurred at 38.4 psig in the area around a re-circulation pump boss. An investigation concluded that the failure began with a prior crack at a fitting on the LH2 tank wall.

The hydrogen tank portion of the CBTT was removed from the CBTT test stand by a contractor. The portion consisted primarily of the pieces of the collapsed forward LH2 bulkhead, and the LH2 tank wall. The parts were laid out on a grid in a Rocketdyne building at Santa Susana. Some sections of the tank wall were taken to the metallurgy laboratory at Downey for examination.

Replacement of the CBTT was not envisaged as the feed line elbow testing could be continued at Huntsville with the S-II Battleship.

On 6 February 1967 testing of S-II LH2 feed line elbow designs at cryogenic temperature was begun at MSFC in parallel with NAA testing on the CBTT.

In April 1967, using the remnants of the CBTT, standard repair methods for the LH2 tank cracks were verified as structurally sound by cryogenic testing.

S-II High-Force Test Program

Summary

S-II vibration test structures, tested at Wyle, Huntsville.

Stage manufacture

The High-Force Test Program (HFTP) consisted of subjecting certain key components of the S-II stage to structural vibration loads simulating the launch condition in order to determine the dynamic response interaction characteristics. Stage components assembled by NAA for this program included:

> the forward skirt
> the aft inter-stage
> the thrust structure and a simulated aft bulkhead called the "thrust complex" (HFTC)

On 15 January 1965 the HFTP was established by Change Order 209.

In July 1965 NAA selected Wyle Laboratories in Huntsville as the test agency for this program. Assembly of the test articles began in October 1965. Assembly of the aft skirt for the High-Force Thrust Complex was begun at Seal Beach on 18 October 1965.

MSFC stopped all work on the program on 5 November except for design and fabrication of test hardware. However, on 17 November 1965, MSFC rescinded the stop order and allowed test plans to proceed, after it had been shown that there would be minimal savings if NASA assumed management responsibility.

NAA released all drawings for the hardware in December and fabrication and assembly proceeded. Firm contract negotiations were concluded on 17 December 1965.

During April 1966 the High-Force inter-stage panels were delivered by truck to Wyle Labs in Huntsville, from Tulsa. Reassembly was begun by NAA personnel on site.

On 3 April 1966 NAA at Seal Beach suspended assembly of the High Thrust Complex in order that it could be shipped on a planned voyage of the Point Barrow. The following day the High-Force Thrust Complex was shipped from the Seal Beach Naval Dock aboard the AKD Point Barrow, together with the S-II stage simulator (spacer). At MAF the High Thrust Complex was transferred to a barge for the trip to the Redstone docks, where it arrived on 25 April 1966. Two days later it was delivered to Wyle Labs in Huntsville. A period of reassembly and test preparation was begun prior to the commencement of acoustical testing.

On 20 May manufacturing operations were completed by NAA on the High-Force Thrust Complex at Wyle Labs in Huntsville. Acceptance by NASA occurred one week later.

On 7 June Wyle Labs completed construction of a test dolly assembly for the HFTP. The following day the High-Force forward skirt panels arrived at Wyle Labs by truck from Tulsa. Reassembly by NAA personnel was completed on 22 June. Also on 8 June, the HFTC and inter-stage were mated, and acoustical chamber checkout was being performed.

Stage testing

On 25 June the HFTC/inter-stage assembly was moved into the acoustic chamber for tests in the range of 25 to 300 Hz at an overall energy level of slightly less than 160 db and at 154 db from 0 to 10,000 Hz.

Acoustic testing of the S-II High Thrust Complex/inter-stage vehicle was completed on 23 July at Wyle, 13 days ahead of schedule.

On 21 October 1966 Wyle opened its High Force Vibration Testing Facility at Huntsville. The facility would be used for the S-II program and other components of the Saturn V.

High Force testing took place at Wyle in the last quarter of 1966 with completion of the thrust complex first-axis lateral testing on 30 November, the second axis lateral testing on 6 December and third axis testing on 21 December. Vibration testing of the forward skirt ended on 15 December. Vibration testing of the High-Force inter-stage was completed successfully on 30 December 1966 at Wyle Labs. No significant problems resulted from the test.

All vibration testing of the HFTC was completed successfully at Wyle Labs on 4 January 1967. Two days later the last of the HFTP test objectives were satisfied by the successful conclusion of the forward skirt acoustical tests at Wyle Labs. This completion was seven weeks ahead of schedule.

S-II Electro-Mechanical Mockup (EMM)

Summary

Mockup used for electro-mechanical checks at MSFC.

Stage manufacture

The S-II Electro-Mechanical Mockup was a stage that was built by North American Aviation at Downey in order to checkout the electrical and computer interfaces prior to the manufacture of any flight or active test stages. It consisted of two structures. One was a full-size replica of the forward skirt and LH2 tank bulkhead. The other was a full-size replica of the LOX tank, aft skirt, thrust structure and inter-stage.

On 24 January 1964 the fifth J-2 engine simulator for the EMM was received at the Downey facility of NAA. By 30 June all construction activity on the EMM facility was completed. However, activation for testing was still in process. Fabrication of the GSE for the EMM was completed in July. Buildup of the EMM at Downey was slow during July and August because of late deliveries of GSE and other equipment. On 25 November the fluid distribution system and the engine actuation system saw their first use when a J-2 engine was gimbaled on the EMM. Gimbal operations were performed on 2 and 5 December. On 17 December 1964 integration of the EMM GSE racks was begun.

In January 1965 the first J-2 engine operational simulator was received from MSFC for installation in the EMM. The last item of EMM automatic checkout equipment was delivered by the NAA Compton facility to the EMM at Downey.

By 19 February 1965 integration of the EMM computer racks had been completed. EMM systems installation and manual checkout were completed on 8 April 1965. On the following day ACE station integration was completed preparatory to initiation of EMM development testing computer programs.

On 12 April a third shift had been added to the EMM operations, with each shift working six days a week.

Stage testing

On 23 April automatic checkout of EMM systems was begun. The first run of all individual systems and integrated automatic checkout programs was initiated on the EMM on 2 June.

In July the EA 300 modification planned for the EMM was deleted. This modification would have changed the EMM to an "x" configuration close to that of the S-II-1 stage to permit validation of the S-II-1 checkout program tapes. This decision was made when it became apparent that the modification kit fabrication could not be completed in time.

By 31 July the following EMM system checkout tapes had been verified:

> electrical power system
> propellant management system
> separation system
> propellant dispersion system
> integrated system

By 5 August the development of the EMM tape baseline was completed. Electromagnetic compatibility and interface testing of EMM GSE and stage systems was conducted during August and September.

Between 9 and 14 August the S-II automatic checkout concept was successfully demonstrated to MSFC. This was a major EMM program milestone.

On 2 October EMM testing operations were terminated at the direction of NASA. However, the mockup activity would be continued. Two days later deactivation of the EMM was initiated by power shutoff. EMM computer complex and peripheral equipment responsibility was transferred from Test Operations to Engineering.

Finally, mockup activity on the EMM was brought to a finish on 29 October 1965 and during November 1965 all deactivation of the EMM was completed.

S-II Battleship

Summary

S-II Battleship used to test J-2 engines and stage systems at SSFL.

Engines

See chart for matrix of engines used.

Stage manufacture

The Battleship S-II stage was built in order to test the ability of the stage to operate as an integrated propulsion system with full tank, plumbing and engine representivity. The stage was not intended to be built to lightweight flight structural requirements.

On 18 January 1964 the LOX tank chill tests were completed successfully. On 14 February the pneumatic proof test of the Battleship LH2 tank was completed successfully. Late in February the thrust structure and aft skirt assemblies were installed. During March NAA aligned the structure and skirt with the stage's propellant tanks in the stand at Seal Beach.

On 21 May construction of the Battleship vessel was completed (Phase VIII construction).

On 26 May construction of the Battleship test facility, at the Coca I site at Santa Susana Field Laboratory (SSFL), was completed.

The first J-2 engine (J-2002) for use on the Battleship was delivered in August by Rocketdyne at Canoga Park to North American Aviation in Seal Beach, a journey of about 30 miles.

Activation of the Coca I site for Battleship testing was started by Rocketdyne personnel on 16 June 1964. The Battleship stage, with single engine installed, was transported from Seal Beach to SSFL where it was installed in the Coca I test site. Activation for the single engine testing was completed on 8 November.

Stage testing

The first single engine firing took place on 9 November 1964. Following 4 single engine firings, three of which were successful, the Battleship was configured for five engine firings. Installation of the engine cluster began on 14 January 1965.

65-27579 J-2 engines on the S-II Battleship stage at SSFL 8.65.

68-63007 Static firing of the S-II Battleship at the Coca I test site at SSFL. 31.1.68.

68-63985 The S-II-TS-B test stage in the Coca IV test stand at SSFL; the S-II Battleship stage in the Coca I test stand at SSFL. 19.2.68.

S-II Battleship firing at the Coca I Test stand at SSFL 1965 (Credit: Saturn Illustrated Chronology)

The fifth J-2 engine (J-2009) was delivered to the Battleship on 10 February. On 18 March the Battleship integrated system checkout was begun in preparation for the cluster firing test series. Integrated systems checkout was completed on 17 April and the first 5-engine firing took place on 24 April.

Milestones achieved included the first 10 second firing on 7 May, the first 25 second firing on 13 July, the first 150 second firing on 20 July and the first full duration firing on 9 August.

The Battleship and the facility were then converted for flight configuration firings. A boilermaker's strike, unusually heavy rains which resulted in soil erosion and water seepage into the equipment and checkout problems delayed completion of this conversion until 18 December. The engine configuration was also changed on 18 December 1965 with five flight configuration engines being installed. The first successful firing with the five flight configuration engines occurred on 29 December 1965.

On 29 April 1966 Change Order 425 was issued by NASA to extend the Battleship test program through to 31 July 1967.

On 1 September 1966 the next configuration change was made as the Battleship was again modified in preparation for boat tail environmental tests. All boat tail modifications were completed by 9 October 1966.

On 17 October the first in a series of boat tail environmental tests was conducted on the S-II Battleship. The tests were to verify satisfactory LOX re-circulation with existing stage equipment using a simulated inter-stage. Testing continued through November to evaluate methods of improving LOX re-circulation performance by the addition of insulation, by re-orificing and rerouting the engine compartment conditioning system, and by installing an in-flight helium injection system.

Battleship boat tail environmental testing was completed on 16 December 1966. It was determined through 159 tests that the S-II flight stages should include helium injection during S-IC boost, complete insulation of the LOX tank sump and LOX feed-lines, and special insulation on each engine.

Modifications to the S-II Battleship stage for the next series of firings began on 21 December 1966. Five up-rated J-2 engines (recalibrated for a thrust of 230,000 lbs) were integrated into the Battleship stage and static firing testing commenced again in February 1967. The test on 17 March included a 2.5 second time delay incorporated in the number 205 LOX pre-valve closing circuit. The time delay would prevent an interference problem between the LOX pre-valve relief in the number 205 feed duct and the thrust cone cruciform occurring during flight acceleration. The final Battleship firing in support of the Confidence Improvement Program was conducted on 26 May 1967, completing the overall CIP requirements.

A further series of test firings was conducted between January and September 1968, with the final Battleship firing taking place on 4 September 1968. In the middle of this series of firings, on 29 May 1968, Rocketdyne personnel installed redesigned ASI fuel and LOX lines on the S-II Battleship.

In the final firing, on 4 September 1968, the S-II Battleship was fired for full duration in order to gather data on deletion of the LOX sump baffles and on depletion cutoff. Data from this and from the 27 August test indicated that removal of the LOX sump baffles did not adversely affect LOX draw-down; engine performance degradation actually began later than it did with the LOX baffles installed. Officials decided, on the basis of the test, to delete LOX sump baffles on the S-II-3 and subsequent stages. No adverse effects on the new ASI fuel line assemblies were observed.

In total 57 firing tests were performed between 1964 and 1968. A full list of all tests is included.

Following the completion of firing tests dynamic flow verification testing of flexible fluid lines was initiated on the S-II Battleship in support of Change Order 1443 (ECP 5872). The last test of the series was completed on 31 October 1968.

A contaminated helium test was initiated on the Battleship on 5 November 1968 to evaluate the effects of using a lower grade of helium as proposed by NASA. The allowable impurity level previously was 50 ppm; in the planned tests, contamination was increased to 500 ppm in all lines.

On 31 January 1969 MSFC directed discontinuance of all S-II support operations at SSFL via CO 1600. This included mothballing of the S-II Battleship at the Coca I stand over a period of 6 months.

In June 1969 NAR accomplished termination of the S-II activity at SSFL on schedule. This activity included the transfer of material unique to Battleship testing to the upper pre-test area, and the processing of spares common to all Saturn S-II stages to the Seal Beach facility for use as required in manufacturing and test operations.

On 30 September 1969 NASA directed (Mod 1825 CO) to proceed with final deactivation of the Coca site at SSFL with a planned completion by 31 December 1969.

The S-II Battleship was likely scrapped and its engines dispersed to various museums and test stages.

S-II-F

Summary

S-II facilities and dynamic test stage used in the AS-500F first Saturn V vehicle at KSC and later in dynamic testing of the AS-500D Saturn V vehicle at MSFC. Now on display at ASRC, Huntsville.

Engines

The initial and final engine configuration was:

Position 201: Mass simulator
Position 202: Mass simulator
Position 203: Mass simulator
Position 204: Mass simulator
Position 205: Mass simulator

The engine configuration in the S-II-F/D stage displayed at the Space and Rocket Center, Huntsville is:

Position 201: J-2001
Position 202: J-2013
Position 203: J-2014
Position 204: J-2017
Position 205: J-2038

Note: although the serial numbers are correct the positions cannot be verified.

Stage manufacture

NAA started major subassembly work on the facilities checkout stage, S-II-F, late in June 1964.

The stage configuration was frozen in August 1964. NAA completed meridian welding of the common bulkhead aft facing sheet in September, except for several welds that needed repair. Problems with the welding delayed repairs until December. By the end of 1964 the thrust structure assembly had been completed and the aft skirt panels were being aligned.

On 14 December 1964 Master Program Schedule 16 was issued that reflected a change in the utilization of the S-II-F, the facilities checkout stage. The stage would not be used to verify the test facilities at MTO prior to shipment to KSC. Instead, the stage would be shipped directly from Seal Beach to KSC for verification of LC-39.

The S-II-F stage on display at ASRC in Huntsville in 2004

The S-II-F stage on display at ASRC in Huntsville in 2004

99-03176 The S-II-F/D stage arrives at MSFC aboard the Poseidon. Note the mass dummy engines 10.11.66.

Fabrication of the major assemblies for the S-II-F proceeded on schedule during early 1965. On 9 April vertical assembly was began with initiation of the Cylinder 3 to Cylinder 4 weld. By the end of June workmen had completed structural assembly of the thrust structure, forward skirt, aft LOX bulkhead, and common bulkhead. Work started on the inter-stage. Vertical buildup of the stage ended at NAA's Seal Beach plant on 16 August 1965.

Pneumatic proof tests of both propellant tanks were performed successfully, following which they were cleaned, sealed and pressurized. Insulation and stage systems were then installed. Problems with insulation closeout delayed systems installation during the last quarter of 1965. Seal Beach Station VII activation for checkout of the stage was completed on 7 December 1965.

On 7 January 1966 five J-2 mass simulators were installed on the S-II-F stage.

Stage testing

On 31 January 1966 the S-II-F stage was moved to Test and Operations at Seal Beach for system checkout, painting, final closeout and packaging for shipment.

Between 17 and 19 February 1966 workmen at Seal Beach loaded and tied down the S-II-F stage aboard the AKD Point Barrow in preparation for shipment to KSC. The Point Barrow departed Seal Beach with the S-II-F stage and its inter-stage on 20 February 1966 and arrived at Port Canaveral on 4 March 1966. Two days later the stage and its inter-stage were transferred by barge to the VAB and moved into the VAB low bay at KSC in preparation for stacking as part of the Saturn V checkout vehicle, AS-500F. The stage was officially accepted by NASA on 11 March.

Low bay checkout was finished on 24 March. On 25 March the S-II-F stage was mated with the S-IC-F stage in the VAB. Between 28 and 29 March 1966 technicians at KSC stacked the S-IVB-F atop the S-IC-F and S-II-F. The following day the S-IU-500F was also added to the rocket.

Following completion of the AS-500F, facility checkout vehicle, power was first applied on 13 May 1966. Systems tests were completed on 24 May 1966 and the complete vehicle was rolled out of the VAB towards Launch Complex 39A on the following day. The vehicle traveled on crawler-transporter No. 1 and the journey took most of the day.

On 8 June 1966 AS-500F processing and test activities at LC-39A were interrupted because of the approach of Hurricane Alma. The vehicle was rolled back into the VAB as a precaution. Two days later, with the threat of the hurricane past, the AS-500-F vehicle was again moved out to pad 39A. The journey took about eight hours.

As a result of cracks found in the flight stages at Seal Beach, a complete inspection of the S-II-F stage LH2 tank was conducted during July. Three cracks were found and repaired. The tank was closed on 3 August. Structural inspection of the LOX tank began on 26 July, and the tank was closed out on 28 July after repairs to three cracks. On 12 August 1966 full pressure tests on the S-II-F tankage were completed successfully.

On 26 August 1966 Boeing, NASA and NAA officials met at KSC to discuss the conversion of the S-II-F stage to a S-II-F/D configuration following the decision earlier not to use the S-II-T stage for dynamic testing.

On 23 September the S-II-F LOX and LH2 manual loading test was completed successfully with most major test objectives being reported as accomplished. This demonstrated the structural integrity and systems compatibility of the S-II design in the stacked condition. The S-II-F LOX and LH2 automatic loading was attempted on 3 October. The test was interrupted with 40% of the LOX loaded, due to sluggish operation of the LOX fill and drain valve.

By 12 October 1966 the AS-500F vehicle automatic LOX and LH2 loading was satisfactorily accomplished at LC-39A following an aborted attempt on 8 October. The launch vehicle drain was satisfactorily accomplished, the sequence being, LOX drain preparations, simultaneous manual S-IVB LH2 and S-II LH2 drain, simultaneous automatic S-IVB and S-II LOX drain and S-IC LOX drain. At this point, AS-500F-1 wet tests were considered complete.

Following completion of testing at the pad, AS-500F was rolled back to the VAB from LC-39A on 14 October 1966. A minor bearing overheating problem on the Crawler Transporter was encountered during the move.

Beginning on 15 October the de-erection of AS-500F took place in the VAB. On that day the CSM and IU were removed, followed by the S-IVB-F third stage and S-II-F second stage on 16 October, and finally concluding with the S-IC-F first stage on 21 October 1966.

A day after removal from the rocket NAA personnel at KSC started work on the modification of the S-II-F stage to the S-II-D configuration on 17 October in the VAB at KSC.

On 25 October technicians at KSC completed the conversion of the S-II-F stage to the S-II-D configuration and rolled the S-II-F/D stage out of the VAB. On 29 October 1966 the stage was loaded aboard the barge Poseiden at KSC and shipped to MSFC in Huntsville, where it arrived on 10 November 1966. The stage was moved to the ME Lab where dye-penetrant inspection tests were performed. Rib and stringer cracks discovered during the inspection were repaired.

The S-II-F/D stage was moved to the Saturn V Dynamic Test Facility at MSFC on 19 November, the forward skirt was attached on 20 November, and the stage was installed atop the S-IC-D stage between 21 and 22 November 1966. The S-IVB-D Aft Inter-stage stacking began on 23 November and was complete by 28 November. The S-IVB-D stage was added on 30 November 1966 and the CSM and LES were added on 3 December 1966 to complete the AS-500D vehicle assembly. Some delays were incurred due to S-II-F/D repairs necessary because damage occurred to the common bulkhead on 8 December when a worker dropped an air hose during painting of the inside of the LH2 tank. The resulting dent was found to be minor and was burnished out.

The dynamic test campaign was classified as the AS-500D Configuration I test series. Configuration I was the complete Saturn V vehicle. Testing started in early January 1967 with the roll test being completed on 7 January. However a difference in the hardware configuration compared with the flight vehicle was noticed and it was decided to repeat the test with an improved configuration.

The Configuration I test program included roll testing, completed on 16 January 1967, pitch testing, from 20 to 23 January 1967, yaw testing, completed on 15 February 1967, and longitudinal testing, completed on 26 February 1967. A LOX vent line ruptured during the final longitudinal test. On 6 March MSFC supplied a spare vent line and authorized additional Configuration I tests to verify the Flight Control System.

Testing continued until 11 March 1967 when the final test was performed to verify the flight control system.

De-stacking of the Saturn V Dynamic Vehicle was started on 28 March. The S-II-F/D was de-stacked on 29 March and prepared for LH2 tank entry for inspection by NAA personnel prior to restacking.

The S-IC-D booster was removed and shipped to MTF for storage in the Booster Storage Building.

Meanwhile the remaining stages i.e. the complete Saturn V minus the first stage were returned to the Dynamic Test Tower for the Configuration II series of dynamic tests that began on 11 May 1967.

Configuration II testing included the yaw test sequence completed on 15 May, the pitch test sequence completed on 2 June, the roll test sequence completed on 10 June and the longitudinal test sequence, started on 13 June and concluded in early July. All programmed testing in the Saturn V Configuration II Dynamic Test Series was completed on 28 July 1967. A one month extension was granted by MSFC to permit rerun of several tests. Following this the stages were separated and the vehicles were removed from the Dynamic Test Tower.

On 26 February 1968 ME Lab personnel completed the cleaning and treatment of the LOX and LH2 tanks of the S-II-F/D stage.

On 14 October 1968 MSFC performed a test in the Dynamic Test Stand on the S-II-F/D to evaluate a simplified method of removing forward skirt stringers for replacement and inspection of the skirt and bulkhead. The test was deemed successful.

Ground breaking for the Alabama Space and Rocket Center in Huntsville began in July 1968. The following year the S-II-F/D stage was moved to the outdoor exhibit area. On 26 June 1969 it was moved to a position near the Astronautics Lab at MSFC. Two days later, on Saturday 28 June at 0500 CDT, the stage was transported along Rideout Road to the museum. A number of power lines had to be disconnected, road signs taken down and some poles moved. Five J-2 engines that had been used in various ground tests were installed in the stage for display purposes. These are, J-2001, J-2013, J-2014, J-2017 and J-2038. The museum opened its doors the following year and the stage has been there ever since. On 15 July 1987 the MSFC Saturn V was designated as a National Historic Landmark. Over the following 34 years weathering took its toll and there is now a campaign to restore the complete Huntsville Saturn V and place it under cover.

Despite the fact that the stage had been at the museum since mid-1969, MSFC on 10 April 1970, requested a "quick analysis" of the risks which would be associated with converting the S-II-F/D stage to a flight stage.

S-II-D

Summary

Dynamic test stage cancelled at an early stage.

Stage manufacture

In August 1964 NAA froze the configuration design for the S-II-D, the dynamic test stage. NAA's Tulsa facility fabricated and stored ahead of schedule the non-pressurized structure for the stage. By the end of 1964 the aft skirt, thrust structure, inter-stage, center engine support, forward skirt and cruciform baffles had all been delivered.

Final assembly of S-II-D stage structures at Seal Beach was begun during January 1965, with the initiation of Cylinder 3 assembly. However, the decision was made shortly thereafter to modify the S-II-S stage to the S-II-S/D configuration and use it for dynamic testing. Accordingly, fabrication of the S-II-D terminated during January 1965.

S-II-T

Summary

S-II static firing test stage. First Saturn stage to be static fired at MTF. Destroyed in explosion in 1966.

Engines

The initial and final engine configuration was:

Position 201: J-2021
Position 202: J-2024
Position 203: J-2018
Position 204: J-2017
Position 205: J-2014

Stage manufacture

The thrust structure panels for the S-II-T, all-systems stage, were received at Seal Beach from NAA's Tulsa facility on 5 June 1964. By 19 June the S-II-T stage design was frozen. On 25 June NAA received Change Order 123 for a J-2 engine side load arresting mechanism (SLAM) for the S-II-T stage at MTF.

In August 1964 full pressure testing of the common bulkhead aft facing sheet resulted in the discovery of a minute hole, which was repaired in September. Assembly of the thrust structure started in July and ended in October with installation of the aft skirt panels. Twelve completed bulkhead gores for S-II-T were diverted for use in assembling the replacement for the S-II-S aft LOX bulkhead. On 23 December gore assembly welding was completed.

During December 1964 NASA requested MSFC to review the plans to static fire test the S-II-T stage at SSFL. Meanwhile, on 18 December 1964, installation of the computer group GSE for the S-II-T testing at SSFL was completed.

Static testing of the S-II-T thrust structure was initiated in the fixture at Seal Beach on 11 January 1965, and completed 10 days later.

On 29 January 1965 MSFC directed that all work on the S-II-T testing at SSFL be terminated. The electrical cable contractor was subsequently directed on 5 February to finish certain work before contract termination.

NAA at Seal Beach completed the S-II-T stage common bulkhead in February 1965, followed by the aft LOX bulkhead in March. Vertical assembly started on 11 February, with the welding of Cylinder 3 to Cylinder 4, and ended on 25 May 1965. After rollout from the VAB at Seal Beach, the stage underwent pneumatic proof testing, then returned to the VAB for systems installation and completion of insulation closeout.

Systems installation ended on 17 September and all manufacturing was completed by 30 September 1965.

The S-II-T stage was shipped from Seal Beach to MAF, aboard the AKD Point Barrow. It departed California on 1 October 1965, passing through the Panama Canal at 1500 on 11 October 1965 and arriving at MAF on 16 October 1965. At MAF the stage was transferred to the barge, Little Lake, for the 45 mile trip via the Intra-coastal Canal and the East Pearl river to MTF, where it arrived at 1245 on 17 October 1965. After removal of protective covering and inspection, in the S-IC Service Building, an MTF canal shuttle barge moved the stage to the new A-2 test stand on 18 October. This was the first rocket stage to be used in conjunction with the test stand.

Stage testing

The stage was hoisted directly from the barge onto the A-2 test stand on 19 October 1965. Many GSE changes were incorporated prior to the first cryogenic operation and between tests of the S-II-T vehicle as follows:

> Complete redesign of the control logic for the S7-41 to eliminate unnecessary interlocks from pressure switches and valve position indicators
> Static firing control systems modifications which incorporated auto sequence reset capabilities and safing systems for engine cutoff

On 1 January 1966 the S-II-T (the all-systems test stage) was formally re-designated the S-II-T/D (the all-systems test/dynamic test stage).

On 16 January 1966 the cold shock test of the LH2 system was accomplished. On 3 February the integrated checkout of GSE was completed. On 22 February leak detection functional checkout was performed. Stage electrical control checkout was completed on 14 March.

Two tanking tests were performed successfully on 29 March 1966 and 17 April 1966. During these tests LN2 was loaded in the LOX tank (on 29 March) and LH2 was loaded in the LH2 tank (on 17 April). The tests were designated A2-506/TA-66 and A2-507/TA-66 respectively. During the 17 April LH2 test some difficulty was experienced with the hydrogen barge pumps, but the test was accomplished without serious problem. Post-test inspections of the stage, GSE, and facility systems revealed relatively minor discrepancies except for cracks in the sidewall insulation of the stage. Repairs were hampered by heavy rainstorms.

On 5 April 1966 two cryogenic tank barges arrived at MTF carrying 196,000 gallons of LOX for the upcoming static firing of the S-II-T/D stage.

Three basic test objectives were applicable to the J-2 engine system during the S-II-T/D static firing program:

Evaluation of engine start characteristics to define engine-to-engine variations within the engine cluster

Evaluation of engine shutdown characteristics to determine the rate of thrust decay and engine variations within the engine cluster

Provision of performance data representative of the flight configuration stages

The following static firings were achieved:

Test	Test Number	Date	Total firing time	Main-stage time	Objective status
15s main-stage	A2-508/TA-66	23.4.66.	21.2 s	17.9s	Achieved. Cutoff as planned by test conductor. Minor problems with the PU computer and LH2 re-circulation pumps
150s main-stage	A2-509/TA-66	7.5.66.	-	-	Aborted prior to automatic start – PU valve failed to slew (stage malfunction)
150s main-stage	A2-509/TA-66	10.5.66.	7.4s	4.0s	Observer early cutoff due to erroneous engine helium bottle pressure indication
150s main-stage	A2-512/TA-66	11.5.66.	45.9s	42.4s	Automatic early cutoff by engine 203 gas generator erroneous temperature indication
150s main-stage	A2-513/TA-66	16.5.66.	10.7s	7.8s	Automatic early cutoff by vibration safety system (erroneous indication)
150s main-stage	A2-514/TA-66	17.5.66.	154.3s	151.3s	Achieved. Cutoff as planned by observer. First firing to use engine gimbaling program
Full duration	A2-515/TA-66	20.5.66.	357.4s	354.3s	Achieved. Cutoff as planned by LOX depletion system
Full duration	A2-516/TA-66	25.5.66.	3.1s	-	Automatic early cutoff by vibration safety system (erroneous indication)
Full duration	A2-517/TA-66	25.5.66.	198.6s	195.5s	Automatic early cutoff – LOX ASI line on engine failed; fire burned cables causing short that triggered cutoff

The J-2 engines performed satisfactorily throughout the S-II-T/D static firing program.

However, the following problems were noted during the testing:

- Stage LOX and LH2 tank pre-chill sequence anomalies resulting in difficulty in meeting the structural J-Ring temperature requirements for loading
- Difficulty in maintaining feed duct vacuums
- Massive insulation de-bond on the vehicle sidewall and forward bulkhead
- LOX vent valve flotation problem during firing
- Insufficient flow of purge gas into the sidewall causing loss of sidewall pressure during stage chill down
- Propellant Utilization (PU) valve and PU computer system malfunctions
- GSE difficulties with the Hydrogen Burn off System and Side Load Arresting Mechanism (SLAM) Arm Detach Firing System
- Failures of re-circulation pumps and converters
- Loss of LOX ullage pressure during S-IC boost simulation phase
- Numerous premature cutoffs due to the absence of sufficient redundancy in instrumentation or voting logic in automatic systems
- J-2 engine ASI line failures causing engine 205 (J-2014) compartment fire
- Massive pre-valve area leak causing engine 204 (J-2017) feed line fire
- Many countdown and auto sequence holds

The early cutoff on 25 May was due to a fire on engine 205 (J-2014). However, after the firing was terminated a serious fire broke out on engine 204 (J-2017) and persisted for some time. The stand deluge system activated and remained active through de-tanking operations, until the LH2 tank was rendered inert by induction of Helium. Leakages in the engine pre-valve had resulted in the fire. Later checks to isolate this leakage would have a catastrophic effect on the stage some three days later.

On 20 May MSFC directed NAA to cancel the scheduled modification of the S-II-T stage to the all-systems/dynamic test stage, S-II-T/D. It would revert to being the S-II-T stage.

The S-II-T stage was destroyed in an explosion at 1644 CDT on 28 May 1966 in the A-2 test stand at MTF. The explosion ruptured the empty LH2 tank during ambient work/test effort due to over-pressurization. A second-shift crew attempted to pressurize the tank with gaseous helium, not knowing that a previous work crew had disconnected the tank's pressure sensors and switches.

Six NAA workers were injured, none seriously. As well as the loss of the S-II-T stage there was an estimated $1,035,000 worth of damage done to the A-2 test stand.

NASA immediately allocated the S-II-F to the Saturn V dynamic program in place of the destroyed S-II-T. Additional test requirements were added to the Battleship, S-II-1 and S-II-2 stages to satisfy the remaining test requirements.

A board of investigation was appointed by H A Storms the day after the explosion. Chairman of the board was W F Parker, Deputy Program Manager for the S-II program. The board convened at 2045 CDT on 29 May 1966 at MTF.

Repairs to the A-2 test stand, following the explosion, took until the end of June to complete.

On 12 August Change Order 436 was issued to implement Confidence Improvement Program (CPI) objectives and hardware changes resulting from the S-II-T stage failure investigation. The findings and recommendations of the investigation board were announced on 1 September 1966.

At the time of the accident the S-II-T was being pressurized for a leak check by the Complex Coordinator, second shift, not a Certified Test Conductor. He was authorized to conduct only routine pressurizations up to 8 psig. The purpose of the test was to attempt to pressurize down to engine 204 at 8 psig so that a leak check of the engine pre-valve could be performed with a soap solution.

During the first shift the pressure sensing line had been disconnected from the LH2 tank causing the pressure indicators on the control panel to become inoperative. The configuration of the system was not determined at the start of the second shift by the Certified Test Conductor on the first shift, who set up the control panel for the Complex Coordinator to pressurize for the test.

The second shift Coordinator made five attempts to pressurize the tank, starting at 1617 CDT, but received no pressure indication on the control panel. He then authorized the closing of the facility blocking valve #2 which would normally have provided a relief capability in the event the stage over-pressurized. In fact this relief capability had worked on the first 4 pressurization cycles, protecting the tank. However he believed the tank pressure to be 0 psig (as that is what his control panel indicated). On the fifth cycle, with the control panel still reading zero, the pressure in the tank actually continued to rise as there was no relief valve in operation. Twenty six seconds after the pressure was applied the S-II-T exploded.

Data indicated that the most probable pressure at the time of the explosion in the LH2 tank was 23.4 psig, well below the design limit for the tank, which was 38 psig.

NAA study of the pressurization and structural aspects concluded that the probable origin of the structural failure was a crack located at the junction of the integral boss to support the LH2 re-circulation return line and the skin of the LH2 tank, although a failure in the weld of the fill and drain outlet plate to the tank wall was a possible cause.

The board recommended tighter control over the MTF test procedures.

The remains of the aft section of the S-II-T stage were returned to NAA for detailed inspection, via the AKD Point Barrow, that arrived at Seal Beach on 28 September 1966. The Point Barrow had departed from MAF after a barge had brought the S-II-T remains from MTF to MAF earlier in the month.

Two of the engines (J-2017 and J-2014) can still be seen today as they were later installed on the S-II-F/D stage located at the Space and Rocket museum in Huntsville.

S-II-simulator/spacer

Summary

Second stage simulator used at MTF for facility verification. Later used as a spacer in the build up of the first two Saturn V launch vehicles at KSC, due to late arrival of the flight second stages.

Stage manufacture

The Fit-up Fixture, known also as the S-II stage simulator and later the S-II spacer, and designated H7-17, was built by NAA at Seal Beach. It was representative of S-II interfaces. Its original purpose was to train technicians in the transport and handling of the S-II stage and to serve as a facility checkout item for S-II static test positions and other facilities. It was later used as a spacer in the build up of the first Saturn V-Apollo vehicle following the late delivery of the S-II-1 stage.

Its first planned use was to checkout the A-2 test stand at MTF. On 13 March 1964 fabrication of the spacer was completed at the Tulsa plant of NAA. On 11 May 1965 the stage was loaded aboard the AKD Point Barrow at Seal Beach, which departed on 13 June 1965. Also on board the ship was the S-IVB-F, third stage facilities checkout vehicle. The ship, making its first journey carrying space-related hardware, passed through the Panama Canal and arrived at MAF on 26 June 1965. The S-II simulator was transferred to a barge, the Pearl River, for the short trip to MTF. The Pearl River departed Michoud at 0610 CDT on 28 June, arriving at MTF at 1245 CDT on the same day. The simulator was taken by barge to the Booster Storage Building, where it was stored, awaiting the readiness of the A-2 test stand. On 29 August the simulator was lifted from a barge directly into the A-2 test stand to check for clearances and alignment.

Stage testing

Additional installation and removal tests during September trained operating crews and revealed some facility inadequacies that would require correction before installation of the S-II-T stage into the stand. Its mission at MTF completed, the simulator returned to Seal Beach, firstly traveling by barge from MTF to MAF in October 1965, and then traveling aboard the AKD Point Barrow from MAF, arriving at Seal Beach on 2 November 1965. It was returned to NAA for storage. On 4 April 1966 it was again shipped to MAF, aboard the AKD Point Barrow, together with the High Force thrust complex.

70-06027 The stage storage transporter being tested with the H7-17 static proof load tool the "spool" at Seal Beach 13.3.70.

On 13 August 1966 MSFC shipped the stage simulator (spacer) from MAF to KSC on board the AKD Point Barrow for temporary use in the first Saturn V vehicle, AS-501. It arrived at KSC on 16 August 1966. The spacer was unloaded and placed in the VAB high-bay transfer aisle the next day. Modifications were performed on the fixture to provide additional strength and attachments that would be necessary for stacking and supporting the required load. Spacer modifications, including installation of electrical cables were completed on 24 August. The Spacer was erected on the S-IC/S-II adapter and placed in the S-II Checkout Cell.

On 31 October 1966 the spacer was erected atop the S-IC-1 stage in the VAB at KSC. Erection of the S-IVB-501 atop the spacer took place on 1 November 1966.

The spacer was de-stacked from the AS-501 vehicle on 15 February 1967 and shipped to MTF for use in the checkout of the A-1 test stand.

Following these tests the spacer was again returned to KSC, arriving on 10 March 1967. It was to be used as a spacer in the stacking of the second Saturn V vehicle, AS-502, pending the arrival of the S-II-2 stage.

Buildup of the AS-502 vehicle on LUT number 2 took place during March with the S-II spacer stacked on the S-IC-2 stage on 22 March 1967. On 28 March the S-IVB-502 stage was erected on top of the S-II spacer.

Following testing, the vehicle was de-stacked during June, and the S-II spacer removed on 29 June 1967. Sometime during the next three years the spacer was returned to Seal Beach.

During March 1970 the S-II stage storage transporter was tested with the S-II spacer.

S-II-1

Summary

First S-II flight stage launched. Aborted stage static firing between the two successful firings. Launched Apollo 4.

Engines

The initial and final engine configuration was:

Position 201: J-2026
Position 202: J-2043
Position 203: J-2030
Position 204: J-2035
Position 205: J-2028

Stage manufacture

Fabrication of structural details for the S-II-1 first flight stage was begun at the Tulsa and Seal Beach plants of North American Aviation (NAA).

On 19 May 1964 NAA received Change Order 112 for the incorporation of recoverable cameras on stages S-II-1 and S-II-2.

By the end of August 1964 NAA was fabricating non-pressurized structural subassemblies for S-II-1. This manufacturing took place at Tulsa and Los Angeles divisions of NAA. At this time major decisions were made about the configurations of flight stages compared with the ground test versions. The LOX tank would have a new baffle design and a thicker aft bulkhead. Also the LH2 tank pressure was reduced from 39 to 36 psi. These design changed introduced a 9-week delay in the flight schedule.

Final assembly activities on the S-II-1 stage began at NAA at Seal Beach, California, on 11 February 1965. The first hardware assembled was the aft facing sheet for the common bulkhead. At this time assembly of the tank cylinders started. Hydrostatic testing of the aft facing sheet revealed a slight leak in the weld, which was repaired subsequently. By the middle of the year the thrust cone assembly and the aft skirt assembly were completed. Work continued on the common bulkhead and installation of brackets on the forward skirt was underway at NAA's plant at Tulsa.

Vertical buildup of the S-II-1 stage began on 6 July 1965 with preparations for the first circumferential weld. On 9 July NAA completed the aft skirt and thrust structure. The circumferential weld joining Cylinder 3 to Cylinder 4 was made on 19 July. A repair to this weld was required and was completed on 4 August. Completion of the forward skirt came on 10 August and the common bulkhead on 13 September.

The aft bulkhead, the first to have the 111-inch dollar weld, was hydrostatically tested successfully on 16 November. On 24 November the fifth J-2 engine was received, ahead of the Seal Beach need date. By 25 November all major subassemblies of the stage were completed with final assembly of the aft LOX bulkhead.

On 3 December MSFC directed NAA to install doubler plates over the outside of the aft bulkhead dollar weld on the S-II-1 and all subsequent stages for additional strength. Within three days drawings had been released and laboratory testing of the design was completed on 7 December. Two sets of doublers were delivered by 13 December, and installation of the first set on the S-II-1 stage was completed by 24 December, just three weeks after go-ahead. The final pass on the girth weld was completed on 14 December, clearing the way for vertical buildup to be achieved on 28 December 1965.

The S-II-1 stage was hydrostatically tested successfully in Station V on 2 January 1966. Four days later pneumostatic testing was accomplished at a remote site on the US Naval Weapons Station property, rather than in Station VII, because of the increased pressure required in the tanks during the proof test.

Between 15 and 17 January the tank structure was mated to the aft skirt/thrust structure assembly and on 27 February the forward skirt was installed on the LH2 tank.

The five J-2 engines were installed in the stage on 7 March 1966 in Station II at Seal Beach.

On 4 April bonding of 1.6 inch, improved honeycomb insulation, was completed. Systems installation on the S-II-1 stage was completed on 18 April 1966. Two days later the responsibility for the stage was transferred from manufacturing to test operations at Seal Beach, paving the way for automatic stage systems checkout initiation.

Stage testing

Power was applied to the stage systems for the first time (through Buss D-40) on 25 April.

Stage checkout operations were completed on 24 June 1966. Final painting and marking of the stage was finished on 2 July 1966 in Station VIII at Seal Beach.

The stage integrated systems checkout and acceptance proved to be troublesome and could not be completed until 22 July.

99-03169 Static firing of the S-II-1 stage in the A-2 test stand at MTF on 1, 22 or 30 December 1966 12.66.

On 14 July the LOX tank was purged, and the common bulkhead was inspected for cracks. One crack was detected in one of the fillet welds. Repairs were made by drilling out a small plate and plugging the hole with a threaded, sealed bolt. The LOX tank was closed and the necessary leak and functional checks were conducted. The stage was moved from Station VIII to Station VII on the night of 22 July. The LH2 tank was entered on 23 July and dye-penetrant inspection was undertaken to check for cracks. 24 cracks were found and repairs carried on until 29 July.

The stage was loaded on board the AKD Point Barrow at the Navy dock near Seal Beach on 30 July.

The S-II-1 stage was shipped from Seal Beach to MAF aboard the AKD Point Barrow, leaving on 31 July 1966 and arriving on 12 August 1966. Meanwhile the interstage for S-II-1 arrived at Port Canaveral (KSC) on 9 August 1966 aboard a commercial water transport barge.

On 13 August 1966 the S-II-1 stage was transferred to the Pearl River barge for the short trip to MTF. After completing receiving inspection, in the S-II service building, and replacement of an LH2 cylinder doubler, the stage was installed onto the A-2 test stand on 19 August 1966.

The major changes in the testing of this stage compared with the S-II-T stage were as follows:

Redline measurements utilized during the test were made redundant
A new static firing bay was added changing the S-II-T command-response type automatic sequence to a countdown clock time oriented automatic sequence
New controls and hardware were installed for the Hydrogen Burn-off System
Removed S7-40 compartment purge and thermal control system interlocks
A new Side Load Arresting Mechanism (SLAM) arm firing system was designed
Stage and GSE changes were incorporated to provide command through the stage switch selector to close LOX and LH2 Re-circulation valves during engine run
A LOX and LH2 vent valve checkout system was incorporated which allowed cryogenic checkout of vent valve relief and reseat pressures
A Stage Re-circulation Bottle Vent System was incorporated
An LH2 Overboard Bleed System was incorporated
Redesigned stage LH2 tank feed-line elbows to eliminate material fatigue discovered after the S-II-T explosion. (This effort also included the resolution of the LH2 tank material crack problem)
Relocation of a pre-valve solenoid due to vibration anomalies with solenoids in their previous location
Changed hardware and operational techniques as a result of the S-II-T explosion, including initiation of a comprehensive crew qualification program

Power was applied to the stage in the test stand for the first time on 26 August 1966. Bonded doublers were replaced by mechanically fastened doublers between 27 August and 3 September. Pre-static checkout activities were greatly limited and impacted by rework and modification efforts during September and October. A LN2/LH2 tanking test was performed successfully on 17 October 1966. A second tanking test was also performed successfully.

On 5 November 1966 a blast occurred in the LH2 vent line at the MTF A-2 test stand where the S-II-1 stage was being prepared for static firing. The explosion destroyed about 80 feet of the 24-inch vent line and forced foreign material into the LH2 tank of the stage.

By 17 November 1966 LH2 tank cleanup and close out was completed.

Pre-static checkouts were completed on 29 November 1966. This vehicle was planned to be static fired two times. The first firing, with a duration of 384 seconds, was performed on 1 December 1966. The firing was successful with manual cutoff as planned by the LOX depletion system at the 2% fill level. The gimbal program was accomplished on engines 201 and 203 only. The SLAM arms failed to release on engines 202 and 204 due to GSE malfunction.

The second attempt at a full duration firing occurred at 2200 CST on 22 December 1966. However the firing was aborted after 2 seconds due to engine cutoff by a faulty ignition detect probe on engine 205. The probe was found to be shorted.

The second full duration firing was performed successfully on 30 December 1966 for a duration of 363 seconds. The cutoff as planned was initiated by the LOX depletion system.

The actual cumulative firing duration of each engine in the three tests was as follows:

    Engine 201 – 739.6 seconds
    Engine 202 – 738.9 seconds
    Engine 203 – 743.2 seconds
    Engine 204 – 739.6 seconds
    Engine 205 – 739.4 seconds

The following problems were noted during the testing:

    A7-71 Heat Exchanger Level control problem
    Massive decay in stage valve actuation system
    Excessive decay in sidewall insulation purge circuits during cryogenic operations
    LH2 vent valve position indicator malfunctions
    Forward bulkhead un-insulated area membrane seal ruptures
    Major problems involved with inerting LH2 feed ducts
    Test Stand and Test Control Center AC power regulator problem due to excessive flow and usage from S7-40 engine compartment purge circuit
    A facility LH2 vent line explosion occurred near the LH2 catch tank at X+1 day post cryogenic operations

Post-static checkout of the S-II-1 stage began on 3 January 1967 at MTF. A special inspection was required to be made in the LOX tank. This inspection began on 6 January and revealed cracks on the turbo-pumps on engines 202 and 203. The pump on engine # 205 had a defective nozzle. Replacement of the pumps was completed on 9 January.

On 11 January 1967 the initial post-static checkout was terminated at MTF as program managers had made the decision to defer a majority of the inspection and testing until the stage was shipped to KSC. It was the only S-II stage to have the post-static testing completed at KSC. The reason for this is that the stage was already late for the build of the first flight Saturn V and was needed at KSC as soon as possible.

The stage was transferred from the test stand to the MTF service building on 12 January 1967 and thence to the barge Poseidon that departed MTF on 16 January 1967 for the journey to KSC, where it arrived on 21 January 1967.

S-II-1 eventually provided the second stage for the AS-501 vehicle that launched Apollo 4 on 9 November 1967.

S-II-2

Summary

Two static firings of the stage performed as planned. Launched Apollo 6.

Engines

The initial and final engine configuration was:

Position 201: J-2057
Position 202: J-2044
Position 203: J-2058
Position 204: J-2040
Position 205: J-2041

Stage manufacture

On 19 May 1964 NAA received Change Order 112 for the incorporation of recoverable cameras on stages S-II-1 and S-II-2.

During May 1965 NAA began structural assembly of the S-II-2 stage at Seal Beach by commencing the meridian welding of the common bulkhead aft facing sheet. The final weld of the bulkhead was made on 29 May, followed by x-ray and dye-penetrant Non-Destructive-Testing. By the middle of 1965 all meridian welds were complete on the forward facing sheet. During the third quarter of the year workmen fabricated and installed honeycomb insulation and prepared to fit together the aft and forward facing sheets of the common bulkhead.

Assembly of the forward bulkhead was completed and the aft bulkhead was ready for the "dollar" plate weld closure. Aft skirt panels were installed on the thrust structure, and the thrust structure assembly was completed on 15 October. The common bulkhead assembly was finished on 17 December 1965.

During February 1966 vertical buildup of the S-II-2 stage commenced following completion of the forward skirt (on 23 February) and aft LOX bulkhead (10 March) assemblies at Seal Beach. The vertical assembly was completed on 8 April.

Hydrostatic testing of the LOX and LH2 tanks was completed by 14 April.

The five J-2 engines were installed in the stage between 12 and 19 June 1966.

On 1 July 1966 workmen in Station VII at Seal Beach completed the mechanical and bracket installations for the thrust structure. The 12 cracks detected in the LH2 tank stringers were repaired while the stage was horizontal in Station VII. The stage was transferred from Station VII to Station IV on 8 July, and to Station VIII on 23 July.

Stage testing

Checkout of the stage systems started on 24 July and included control room/stage/station interface checkout during which power was applied to the stage. Checkout was completed on 24 September, ahead of schedule, as the tests had run more smoothly than those performed on the S-II-1 stage.

67-57907 The S-II-2 stage leaving MTF for KSC aboard the barge Poseidon 20.5.67.

Due to contamination, the stage LH2 feed-lines were removed and replaced on 8 and 9 October. LH2 pre-valves were also replaced as a result of contamination. The stage was then transferred from Station VIII to Station VII on 14 October.

In Station VII the stage underwent LH2 tank inspection, installation of metal doublers on the LH2 tank wall, and rework of LH2 feed-line elbows. On 26 October 1966 the stage was moved back to Station VIII, where it underwent modifications prior to painting and packaging for shipment to MTF.

On 25 January 1967 NAA completed all modifications to the S-II-2 stage following factory checkout, including replacement of all the LOX turbo-pump turbine wheels due to cracks being found in them. Following this the stage was moved into Development Station VII for final painting and packaging operations.

The stage was transported by road from the NAA plant in Seal Beach to the nearby Navy dock. The S-II-2 stage was then shipped from Seal Beach Navy dock to MAF aboard the AKD Point Barrow, leaving on 27 January 1967 and arriving on 10 February 1967. On 11 February 1967 the stage was transferred to a barge and made the 45 mile trip to MTF. The stage was initially moved to the vertical checkout position for special inspection of the LH2 tank. Approximately 25 discrepancies were found and corrected. Also the pumps for engines # 202 and # 205 required replacement. On 18 February 1967 the stage was hoisted onto the A-2 test stand. Technicians opened the LOX tank on the same day and began an inspection for excessive lubricant prior to initiating modifications and hardware instrumentation installation preceding acceptance testing. Inspection of the LOX tank was completed, and the tank was closed on 21 February.

On 4 March 1967 the S-II-2 stage pre-static and GSE checkout were initiated at MTF. A successful tanking test was performed. An x-ray inspection of the ASI chamber pressure instrumentation line for each engine was required and was completed by 27 March. The LOX vent line in the LOX tank was replaced the same day. By 30 March the pre-firing tests and checkout of the stage had been completed in preparation for the static firing scheduled for the following day.

The major changes in the testing of this stage compared with the previous one were as follows:

> Common bulkhead vacuum was programmed into the countdown sequence in an attempt to eliminate LOX ullage pressure collapse during the final minutes of the static firing countdown
> Redundancy was incorporated for major systems such as the LH2 tank fill valve control, auxiliary pressurization systems for LOX and LH2 tanks, and various others
> A feed duct purge capability was incorporated to allow the S7-41 pneumatic service console to be used remotely for pressure cycling of feed ducts
> A major stage change incorporated an onboard, in-flight Helium Injection System to replace the previously utilized ground system
> Major procedure changes were made in consideration of the LH2 dump system vent line explosion that occurred during the S-II-1 testing

Failure of the No. 1 pre-valve to close at T-8 minutes caused program officials to scrub the first attempt to static fire the S-II-2 stage on 31 March 1967. The countdown was continued to T-0 to determine if additional problems existed. This disclosed an intermittent check valve that prevented chill-down of engine 203 (J-2058).

This vehicle was planned to be static fired two times for full duration. The first firing, with a duration of 363s, was performed at 1608 CST on 6 April 1967. The firing was successful with an observer cutoff in response to an LH2 depletion indication. However, the observer action was deemed to be about 1.2 seconds premature – the desired LOX depletion cutoff signal was not transmitted and would have terminated the test as planned.

Pre-static checkouts for the second firing were

completed on 12 April 1967. The second firing, with a duration of 367s, was performed on 15 April 1967. The firing was successful with the cutoff as planned by the LOX depletion system. The gimbal program terminated prematurely by the low facility power caused by a facility transformer malfunction.

The actual cumulative firing duration of each engine in the two tests was as follows:

Engine 201 – 722.6 seconds
Engine 202 – 724.5 seconds
Engine 203 – 723.6 seconds
Engine 204 – 724.4 seconds
Engine 205 – 725.6 seconds

The following problems were noted during the testing:

LH2 pre-valve position indicator problems
LH2 barge 1 fire indications
LOX fill line drain system malfunction due to a cracked LOX fill and drain line
Experienced more LOX tank ullage pressure decay during auto sequence
Noted excessive Hydrogen content in the S-II sidewall purge gas as analyzed remotely by the leak detection system
Anomaly occurred with PU valves improperly positioning with reference to LOX tank load level during engine run
LH2 vent valve position indication malfunctions
Test stand AC power failure
A LOX barge ullage burst disc rupture due to material failure
Engine 205 (J-2041) re-circulation pump malfunction

The test on 15 April signified successful completion of that portion of the Confidence Improvement Program test requirements that had been imposed on the flight stages by the loss of the S-II-T stage.

Post-static checkout was completed on 13 May 1967 and the stage was transferred from the test stand, on 15 May 1967, to the barge Poseidon that departed MTF on 20 May 1967 for the journey to KSC, where it arrived on 25 May 1967.

S-II-2 eventually provided the second stage for the AS-502 vehicle that launched Apollo 6 on 4 April 1968.

S-II-3

Summary

Forward bulkhead damaged in an accident during manufacture and replaced. Stage returned to MTF from KSC for cryogenic proof testing to man-rate the stage. Launched Apollo 8.

Engines

The initial and final engine configuration was:

Position 201: J-2051
Position 202: J-2053
Position 203: J-2059
Position 204: J-2045
Position 205: J-2055

Stage manufacture

Assembly of the S-II-3 stage started in June 1965. By the end of September technicians had completed the dollar welds of both facing sheets of the common bulkhead. X-ray and dye-penetrant inspection as well as hydrostatic and leak tests were made on the forward facing sheet. Six meridian welds on the stage's forward bulkhead were made. Completion of the forward bulkhead was achieved in the last quarter of the year, although both common bulkhead facing sheets were being held for minor meridian weld repairs. Thrust structure components arrived before the end of 1965 from NAA's plant at Tulsa.

67-59495 Static firing of the S-II-3 stage in the A-1 test stand at MTF as viewed from the Central Contol Tower 27.9.67.

67-59499 Static firing of the S-II-3 stage in the A-1 test stand at MTF as viewed from the Canal 27.9.67.

67-59500 Static firing of the S-II-3 stage in the A-1 test stand at MTF as viewed from the Test Control Centre 27.9.67.

68-64910 The S-II-3 stage on board the barge Little Lake at MAF departing for MTF 5.5.68.

Workmen at Seal Beach completed the S-II-3 aft skirt/thrust structure assembly on 26 February 1966. The stage vertical buildup was started on 25 April 1966 following completion of the common bulkhead on 18 April. By 8 July the vertical buildup had been completed.

Inspection of the LOX tank structure during early July revealed cracks in the manufacturing purge-port closeout plate welds in the common bulkhead. The plates were removed, and threaded, sealed bolts were installed. Additional dye-penetrant inspections of the LH2 tank ribs and stringers were made. Seventeen cracks were found in the ribs, 25 in the stringers. A leak in a weld joint was discovered during hydrostatic testing of the stage, performed on 29 July. The leak developed through a crack in the weld joining the Engine 202 feed-line elbow to its mounting ring. Repair was effected by grooving and re-welding. A second hydrostatic test followed the repair.

On 31 August 1966 the S-II-3 stage was moved to Station V for LOX tank cleaning and LH2 tank helium soaking in preparation for systems installation.

During the week of 9 September, whilst thrust structure mating operations were in process, water was discovered in one set of common bulkhead purge lines. The water had entered during the hydrostatic test. Bulkhead drying operations took place during the second half of September 1966.

By 26 September 1966 three pneumostatic tests on the stage had been performed, with the final one being successful. The stage was then moved into Station VII for NDT inspections. The stage was moved to Station V on 29 September for LH2 tank cleaning.

The forward skirt was installed on the stage on 13 October. The five J-2 engines were installed in the stage on 5 and 6 November 1966. On 23 November the stage was moved from Station IV to Station VII for LH2 in-tank installations.

On 29 November 1966 the LH2 tank forward bulkhead was damaged when a ladder fell while the stage was in position for in-tank systems installation at Seal Beach. In-tank systems installation operations were completed and the personnel entrance ladder mechanism was being retracted from the LH2 tank through the access port in the forward LH2 bulkhead. During the access ladder removal operation, a weld in the retracting mechanism failed causing an estimated ten-foot section of the ladder to fall against the inside of the LH2 forward bulkhead from a height of about 15 feet. One of the 12 gores of the LH2 bulkhead was damaged by the impact of the ladder. Three cracks through the bulkhead gore were found, the longest of which was about 52 inches in length. Subsequently, the stage was moved to Station IV where the LH2 damaged bulkhead was removed.

Work continued in January on the damaged LH2 bulkhead. Cylinder 6 was removed on 2 January 1967 from the S-II-5 bulkhead for use on the S-II-3 bulkhead. Positioning of the S-II-5 bulkhead on the S-II-3 stage was completed on 8 January, and welding was completed on 12 January. X-ray inspection of the welds revealed 18 areas that required repair.

By 16 January 1967 rework of the damaged S-II-3 LH2 bulkhead had been completed. Following pneumostatic testing at a remote site on 21 January, S-II-3 would enter Station IV for completion of systems installations. On 4 February the stage was moved to Station VII for LH2 tank inspection. Following a successful inspection the tank was closed on 9 February. On the following day the stage was moved to Station VIII to undergo modifications.

Stage testing

On 27 February 1967 the stage entered systems checkout at NAA, with the remaining manufacturing effort being accomplished on a noninterference basis. Power was applied on the following day.

After a review of x-rays it was determined that it was necessary to re-enter the LOX and LH2 tanks. The LOX tank was opened on 21 March and closed again on 6 April. The fuel tank was opened on 7 April. A vent line in the LOX tank was replaced and an area of welding that lacked complete fusion in the fuel tank was repaired.

On 31 March 1967 NAA successfully completed an integrated systems evaluation run of the S-II-3 stage at Seal Beach. The automatic integrated systems checkout was completed successfully on 17 April 1967. A special inspection was started on 21 April to determine if the rivets in the LOX tank were contaminated. Following UV inspection all rivet heads and 6 isolation hangers required cleaning. The tank was again closed on 1 May. By 1 May NAA officials had completed the initial shakedown inspection. Re-inspection of the stage electrical networks, on 9 May, and the LOX tanks, on 24 May, would occur prior to a final systems retest and stage shipment to MTF.

On 25 June 1967 the stage was moved to Station VII at Seal Beach for x-ray and dye-penetrant non-destructive inspection of the LH2 tank welds, completion of the systems retest, and preparation for shipment. The stage was transferred to the Navy dock at Seal Beach on 11 July 1967 and loaded on-board the AKD Point Barrow.

The S-II-3 stage was shipped from Seal Beach to MAF aboard the AKD Point Barrow, leaving on 12 July 1967 and arriving on 26 July 1967. On 27 July 1967 the stage was transferred to a barge and made the short trip to MTF. The stage was hoisted directly from the barge onto the A-1 test stand on 28 July 1967. This was the first stage to use the newly commissioned A-1 stand. The major changes in the testing of this stage compared with the previous one were as follows:

> Procedural changes were incorporated to assure comprehensive GSE, Facility and Stage System Checkout and hydrogen systems inerting pre and post test
> Sidewall purge channel revisions were made in order to increase the efficiency of sidewall purging
> Changes were incorporated to increase the sidewall vent capability to allow increased bleed-off of sidewall pressure at lift-off
> A major LH2 tank stage weight reduction was incorporated
> Test stand and test control center AC regulators were removed to eliminate the previously encountered AC power failure

As this was the first man-rated stage processed at MTF extensive checkouts were conducted prior to cryogenic operations to ensure readiness of the new test stand to support S-II-3 testing.

A LOX/LH2 tanking test was performed successfully on 6 September 1967. During this test a piece of sidewall insulation became de-bonded in the area of cylinders No. 1 and No. 2. To permit insulation repair, test officials rescheduled the first static firing from 12 September to 19 September. The delaminated area was temporarily covered with plastic film.

By 17 September 1967 NAA had completed insulation repair and the pre-static firing checkout of the S-II-3 stage. This vehicle was planned to be static fired two times since it was the first stage in the newly activated test stand.

The first firing, with a duration of 65 seconds, was performed at 2010 CDT on 19 September 1967, and was terminated manually as planned. This firing achieved the major objectives of qualification of the A-1 Test Stand flame bucket and demonstration of the stage, stand and control room compatibility. Special objectives accomplished included evaluation of the slow chill of the LH2 tank, achievement of a 3,800 gallons per minute maximum LOX fill rate in the fast-fill mode, and a verification of the LH2 fast-fill and over-fill sensor. The test was terminated, as planned, by manual cutoff.

Pre-static checkouts for the second firing were completed on 25 September 1967. The second firing, for a full nominal duration of 358s seconds was performed at 1523 CDT on 27 September 1967 with termination being initiated automatically by LOX depletion. The firing demonstrated the functional integrity of the stage under static firing conditions and verified that the stage met specified acceptance test requirements. Special objectives achieved included evaluation of the slow chill of the LH2 tank performance in a sidewall insulation test.

The actual firing duration of each engine in the second test was as follows:

> Engine 201 – 353.4 seconds
> Engine 202 – 353.0 seconds
> Engine 203 – 353.0 seconds
> Engine 204 – 353.5 seconds
> Engine 205 – 353.6 seconds

The following problems were noted during the testing:

> Loss of AC power on the LOX barges
> Massive sidewall insulation failures at cryogenic temperatures
> Sluggish operation of LOX pre-valves
> Difficulty during test of insulation sidewall in-flight vent plugs
> LH2 vent valve position indicator cycling
> LH2 fill valve position indicator problems

Modification work was accomplished in parallel with post static firing checkout operations. These activities took place during the first two weeks of October. Stage and engine leak checks took place during the second two weeks of October.

On 12 November 1967, following post-static firing tests, special tests, and modifications, the S-II-3 stage was removed from the A-1 static test stand and placed in the horizontal bay of the S-II Service Building at MTF for insulation repair, additional modifications and further inspections.

Installation of redesigned LH2 feed-lines was completed during the week of 13 December. Horizontal post-static checkout was completed on 13 December 1967 and a final inspection was performed on 19 December 1967. The stage was transferred to a barge that departed MTF on 21 December 1967 for the journey to MAF. At MAF the stage was transferred to the AKD Point Barrow, leaving on 21 December, for KSC, where it arrived on 24 December 1967. Following the Christmas day holiday it was unloaded on 26 December 1967.

Personnel at the VAB at KSC completed post-delivery checkout of the S-II-3 stage on 21 January 1968. The stage was then placed in a horizontal position for x-ray inspection of the LH2 tank.

The S-II-3 stage was erected on top of the S-IC-3 first stage on 31 January 1968, followed the day after by the installation of the S-IVB-503N stage atop the second stage.

On 16 February 1968 limit load testing of the special S-II-3 thrust structure was completed back at MSFC. An ultimate load test was conducted on this structure on 21 February. This test demonstrated capability of the structure to withstand 130% of the design limit load without failure. The thrust structure program was completed successfully on 28 February. The test program fulfilled the thrust structure test objectives not attained at Seal Beach during the S-II-S testing in 1965.

On 27 April 1968 it was announced that the AS-503 vehicle would be manned.

Because of the need to up-rate the stage inspection criteria for the first man-use of the S-II stage it was decided to return the stage to MTF in order to perform a cryogenic proof pressure test. In addition the stage would also receive modified ASI lines while at MTF. Consequently the S-II-3 stage was de-stacked on 29 April.

S-II-3 was loaded on to the AKD Point Barrow on 30 April and left KSC on 1 May 1968, arriving at MAF on 4 May. It was transferred to the barge Little Lake and arrived at MTF on 5 May 1968.

On 11 May 1968 the S-II-3 stage was installed in the A-2 test stand at MTF and test preparations continued, including the installation of special cryogenic-proof test vent valves with a higher relief setting. The cryogenic proof pressure test was performed successfully on 29 May. For this test the LH2 fuel tank was subjected to flight pressures of 36.2 psi with LH2 aboard.

An accident on 6 June 1968 at the A-2 test stand threatened to delay the scheduled 11 June removal of the stage from the test stand. A cable on the main derrick broke during a proof load test. However, the stage was actually removed from the test stand on 10 June and installed in the Vehicle Service Building for inspection of the LH2 tank and completion of ASI line modifications on the stage. No new defects were found. S-II-3 was loaded onto the barge at MTF on 21 June. The stage once again left MTF on 22 June 1968, traveling initially to MAF. After transfer from the barge at MAF on 24 June 1968 the stage continued on to KSC, aboard the AKD Point Barrow, arriving on 26 June 1968.

S-II-3 eventually provided the second stage for the AS-503 vehicle that launched Apollo 8 on 21 December 1968.

S-II-4

Summary

Starting with this stage up rated J-2 engines were employed with a thrust of 230,000 lbs compared with 225,000 lbs on previous S-II flight stages. Also this was the first lightweight S-II stage, achieved by using thinner propellant tank walls and lighter weight structures. Stage launched Apollo 9.

Engines

The initial and final engine configuration was:

Position 201: J-2067
Position 202: J-2068
Position 203: J-2069
Position 204: J-2070
Position 205: J-2066

Stage manufacture

Assembly of the S-II-4 stage began at Seal Beach in September 1965 with fabrication of the common bulkhead. Most of the meridian welds were completed on the common bulkhead facing sheets by the end of 1965.

On 16 February 1966 cracking of 2020-T6 aluminum alloy horizontal stringers on the S-II-4 inter-stage was reported by the NAA plant at Tulsa. Rivet gun impact was believed to have been the cause. Pull-type fasteners were implemented as the solution. On 29 April a revised plan was issued, following the stringer problem, that re-designated the S-II-4 inter-stage to the S-II-T/D program. The former S-II-T/D inter-stage was transferred to the S-II-4 stage, with a completion date of 8 July.

Meridian and dollar welding of the forward bulkhead was finished early in March.
On 8 July 1966 NAA completed the assembly of the forward bulkhead at Seal Beach.

Hydrostatic testing of the S-II-4 stage aft bulkhead was completed on 3 August 1966. During hydrostatic testing of the Cylinder 6/forward bulkhead assembly on 9 August, a leak developed in Cylinder 6. Subsequent investigation revealed cracks at tank splice No 1, station 774. A repair, with mechanically attached doublers, was completed on 22 August. The assembly was then moved to Station V. Hydrostatic testing to 27 psig was completed successfully on 25 August. The assembly was moved to Station I, where the Cylinders 5 to 6 weld was completed on 28 August.

Welding of the common bulkhead to the Cylinders 1 and 2 assembly was completed on 12 August. Inspection revealed a questionable area in the J-weld and a repair was completed on 22 August. The common bulkhead/cylinders 1 and 2 assembly was moved to Station VI on 30 August for hydrostatic testing.

Hydrostatic testing of the aft bulkhead was completed on 3 August. Painting was completed on 18 August. The bulkhead was moved into Station III, positioned, leveled, and trimmed for the girth weld. The girth weld was completed on 4 September, but was faulty, with discrepancies in offset, height and porosity.

With assembly of the stage almost complete officials refused to accept the faulty girth weld of the aft bulkhead. On 28 September the material review board decided to cut it out and replace it with the equivalent bulkhead from the S-II-5 stage. The aft LOX bulkhead was cut off on 12 October. The replacement aft bulkhead was completed on 26 October 1966, and welding was completed on 5 December.

Rework of the LOX girth weld was completed on 2 January 1967. On 10 January 1967 the first stainless steel feed-line elbow was delivered to Seal Beach for installation on the S-II-4 stage. The forward bulkhead to Cylinder 6 weld was cut apart on 16 January with re-welding being completed on 18 January. X-ray inspection of the welds revealed 29 discrepancies that were reworked by 28 January.

The S-II-4 stage successfully completed pnuemostatic testing on 8 February 1967. This was the first stage of the lightweight design to be subjected to this test, and was also the first S-II stage to be tested at the higher LOX tank pressure allowing deletion of the stage hydrostatic test requirements.

67-58537 S-II-4 outside Station IX vertical checkout building at Seal Beach ready for systems installation 8.67.

67-61058 Erection of the S-II-4 stage in the A-2 test stand at MTF 27.11.67.

67-61116 The S-II-4 stage on a transporter moving to the Seal Beach VAB prior to shipment to MTF Early 11.67.

68-62056 S-II-4 firing cut off after 17 seconds during the aborted firing on the A-2 test stand at MTF 30.1.68.

68-64035 Removal of the S-II-4 stage from the A-2 stand at MTF 16.2.68.

68-64039 Second static firing of the S-II-4 stage in the A-2 test stand at MTF 10.2.68.

The stage was moved to Station VII on 9 February for dye-penetrant inspection of the LH2 tank stringer ends, bosses, and weld lands. No new cracks were found. On 13 February the stage was moved to Station V for cleaning of the LH2 tank. Cleaning was completed on 17 February 1967.

By 8 March 1967 NAA had completed structural buildup of the S-II-4 stage and moved it to Station IV for insulation closeout and systems installation. However, during March, two of the 20 LH2 cylinder quarter panels in the S-II-4 stage were determined through measurements with a Vidigauge to be too thin. Possible solutions considered were to reduce the ullage pressure or provide additional mechanical reinforcement.

The five J-2 engines were installed in the stage on 4 April 1967 in the high tool area. Rework of 164 forward skirt attachment bolt holes was completed on 25 April 1967.

During a proof pressure test at 6.8psig in July a large section of the common bulkhead insulation facing sheet separated from the honeycomb core. Repair was accomplished by the addition of rubber doublers, and a subsequent test proved the repair to be completely satisfactory.

The stage was moved to Station IX where systems installation continued. Three LH2 feed-lines were removed and returned to the vendor for rework due to weld deficiencies.

Stage testing

On 9 September 1967 the S-II-4 stage entered the Test Operations department at NAA's Seal Beach facility for systems checkout. In addition, stage modifications continued in parallel at Checkout Station IX. The LOX gas distributor and internal LOX vent line, along with the anti-vortex baffles, were replaced, and the stage was moved to Station VII for rework of frame splices, corrosion inspection and closeout of the LH2 tank.

Systems checkout was completed on 22 October 1967 by the newly named North American Rockwell (NAR), formed by the merger of North American Aviation and Rockwell-Standard Corporation on 22 September 1967.

X-ray inspection of the LH2 tank circumferential welds, during the early days of November, revealed one questionable area in the cylinder 5 to 6 union. This was later accepted as-is.

The S-II-4 stage was shipped from Seal Beach to MAF aboard the AKD Point Barrow, leaving on 11 November 1967. It passed through the Panama Canal on 21 November and arrived on 25 November 1967. The cargo onboard included the structural test stage, S-II-TS-A en-route to MSFC. On 26 November 1967 the S-II-4 stage was transferred to the barge Little Lake and made the 45 mile journey to MTF. The stage was hoisted directly from the barge onto the A-2 test stand on 27 November 1967, where pre-static tests were to begin.

During the week of 6 December LOX tank inspection was completed after some minor discrepancies were found, and cleared.

The major changes in the testing of this stage compared with the previous one were as follows:

> A two step LH2 tank vent valve system was incorporated to provide minimum LH2 tank pressure during S-IC boost phase and increased LH2 tank pressure during S-II firing
> To ensure that S-II vehicle structural integrity would meet flight requirements a new cryogenic proof pressure test of the LH2 tank was accomplished with this vehicle. Prior to conducting the cryogenic proof pressure test the LH2 tank was x-rayed and inspected for cracks. After the cryogenic proof pressure test the LH2 tank was again x-rayed and inspected for cracks to evaluate structural integrity. This method of Non-Destructive Inspection became standard for pressure vessels for many years. By performing the test at low temperatures there was a greater chance of detecting any flaws that might be present as they tended to grow in those conditions.

The LH2/LOX tanking test was performed on 16 January 1968. This eight-hour test involved loading LOX and LH2 to verify stage and facility compatibility and performing the automatic sequence test to verify stage system timing, start-tank and thrust-chamber chill-down capability, and the engine gimbal program. The engine sequence data were good, but one SLAM arm (not a flight item) failed to unlock engine 203 (the ordnance fired but the bolt wedged between the arm halves). The other three engines gimbaled properly. Severe damage to the LH2 tank sidewall insulation occurred during this operation. A second tanking test, the cryogenic proof pressure test, would be performed after the static firing.

The static firing readiness review was held successfully on 25 January 1968. Pre-static checkouts were completed on 30 January. Two attempts were made to achieve the full duration static firing. The first attempt on 30 January 1968 was aborted after 17 seconds firing time by the LOX X-Y redline observer due to a false LOX ullage pressure indication that resulted in venting of the LOX and LH2 tanks. During this firing the Tulane University Medical School had been authorized to place guinea pigs in strategic areas of sound propagation in order to study the acoustical effects of massive sound on the laboratory animals.

Following the abort, the LH2 tank was entered to determine the cause of the propellant utilization system LH2 "open" indication during the shutdown. This was traced to a broken electrical connection.

The cryogenic propellants for the successful full duration firing were loaded on the day of the firing test. The static firing was performed successfully at 1439 CST on 10 February 1968 during which all test objectives were met. The firing duration of each engine in the test was as follows:

> Engine 201 – 341.9 seconds
> Engine 202 – 341.9 seconds
> Engine 203 – 342.1 seconds
> Engine 204 – 342.0 seconds
> Engine 205 – 342.2 seconds

The cutoff was initiated manually as planned at the 2% LOX level.

The following problems were noted during the testing:

> Common bulkhead vacuum problems
> LH2 vent valve relief and reseat problems
> Hydrogen content noted in S-II sidewall insulation purge circuits
> LH2 PU probe intermittent localized to wiring inside the LH2 tank
> LH2 vent valve sense line vacuum circuit malfunction
> A7-71 Heat Exchanger fill valve malfunctioned
> Noted major LOX tank ullage pressure oscillations due to liquid entering ullage pressure sense line
> Noted common bulkhead pressure increase corresponding to LH2 tank pressurization

The stage was removed from the A-2 test stand on 16 February 1968 and placed in the VAB for LH2 tank inspection in support of the forthcoming cryogenic proof test. Dye-penetrant and x-ray inspections of the LH2 tank were performed between 19 and 25 February.

On 1 March the stage was again placed in the A-2 test stand in preparation for the cryogenic proof test that took place successfully on 22 March 1968. This was the first S-II cryogenic proof test performed. The S-II-4 successfully sustained an LH2 ullage pressure of 36.185 psig with LH2 and LN2 in the LH2 and LOX tanks respectively. There were no indications of structural anomalies.

The S-II-4 stage was finally removed from the A-2 test stand on 4 April after the completion of post-static checkout. Extensive rework and LH2 tank inspections were performed in the S-II Stage Service and Storage Building (SSSB) at MTF. In addition exposed aluminum alloy surfaces were painted to reduce their susceptibility to stress corrosion. The stage was moved, on 5 April, from the horizontal bay to the vertical bay for work in the LOX tank.

On 19 April the LOX tank was closed and the stage was moved back to the horizontal bay of the S-II service building in preparation to permit LH2 tank entry. LH2 tank entry was made on 23 April. On 29 April x-ray and dye-penetrant inspections and tank cleaning operations were finished. Upon completion of LH2 tank work, the stage was moved to the vertical bay again for insulation leak checks, installation of the flight disconnects, and sealing of the LH2 tank access-manhole-cover insulation. On 8 May the stage was transferred to the S-IC service building for painting, packaging and preparation for shipment. The stage was accepted by NASA at the turnover meeting on 9 May. The S-II-4 stage was loaded aboard the barge Orion that departed MTF at 0630 CDT on 10 May 1968 for the journey to KSC, where it arrived on 15 May 1968.

S-II-4 eventually provided the second stage for the AS-504 vehicle that launched Apollo 9 on 3 March 1969.

S-II-5

Summary

Two stage static firing aborts. Stage launched Apollo 10.

Engines

The initial and final engine configuration was:

Position 201: J-2075
Position 202: J-2077
Position 203: J-2080
Position 204: J-2081
Position 205: J-2076

Stage manufacture

During January 1966 the S-II-5 stage final assembly began at Seal Beach. Meridian welding of the common bulkhead aft facing sheet was completed at that time.

A fabrication accident at Seal Beach on 22 June 1966 damaged one of the bulkhead gores. Program officials decided to cut out and replace the damaged gore.

NAA began the vertical buildup of the S-II-5 stage with the weld joining of tank cylinders. It was reported that welding problems hindered the buildup. The dollar weld and machining operations were begun on 12 July and completed later the same day. Insulation bonding was performed during August. Final acceptance was delayed when ultrasonic inspection revealed voids that needed repairs. Also in August corrosion pitting was observed on the aft facing sheet of the common bulkhead. To eliminate schedule impact, the S-II-6 aft facing sheet was designated for use on the S-II-5 stage. The corroded aft facing sheet was burnished and prepared for hydrostatic testing. The sheet was designated for use on a later S-II stage. The dollar weld on the forward facing sheet was performed on 5 August.

Thrust structure assembly was in progress during August 1966. Assembly of the four aft skirt panels to the thrust cone was completed on 19 August. The center-engine beam was installed on 29 August. Completion of fabrication and assembly of the thrust structure and aft skirt were achieved on 1 September. Bracket and systems installations began on 30 September.

On 28 September 1966 a faulty girth weld on the aft bulkhead of the S-II-4 stage was rejected. It was decided to cut the weld apart and to replace the bulkhead with the S-II-5 aft bulkhead. This bulkhead was completed on 26 October 1966 and then transferred to the S-II-4 stage.

The S-II-5 stage LH2 tank forward bulkhead was welded to cylinder 6 in October 1966. An over-tolerance offset resulted in the welded assembly being rejected by inspectors.

On 14 December 1966 direction was received from NASA to re-designate the S-II-5 forward bulkhead to the S-II-3 stage following damage to that stage sustained on 29 November.

On 29 December 1966 the S-II-5 stage common bulkhead was cut apart from Cylinder 1 because there was excessive porosity and offset in the J-ring weld.

On 2 January 1967 the S-II-5 forward bulkhead and Cylinder 6 were cut apart, in order that the forward bulkhead could be transferred to the S-II-3 stage. It had been planned to use the S-II-6 forward bulkhead on the S-II-5 stage, however, beyond-limit corrosion pitting forced non-flight identification and storage of the S-II-6 bulkhead and utilization of the S-II-7 bulkhead on S-II-5.

On 24 January 1967 the aft LOX bulkhead was successfully hydrostatically tested following repair of 53 doublers that had leaked during tests on 13 January.

On 26 January 1967 NAA replaced cylinders No. 1 and No. 2 on the S-II-5 stage with the same ones from the S-II-6 stage. The reason was that the S-II-5 cylinders were 0.5 inches too short following the cutting apart of the J-weld during the last quarter of 1966.

Fabrication of the S-II-7 forward bulkhead was accelerated to support the S-II-5 stage. Hydrostatic tests were conducted on 8 January, and ultrasonic tests were conducted on 23 January. The forward bulkhead to Cylinder 6 weld was completed on 11 February. Final acceptance of the LH2 tank was given on 23 March 1967.

Pneumostatic testing of the S-II-5 stage was completed on 24 March 1967 at the remote site at Seal Beach. The stage was then returned to the VAB. Following vidigage inspection of the LH2 tank sidewall on 30 March the stage was moved to Station V for LOX tank and LH2 tank cleaning.

The stage was moved from Station V to the remote site where additional pneumostatic testing was accomplished on 4 April 1967. By 17 April NAA personnel had completed the LH2 in-tank systems installations and moved the stage to Assembly Station II at Seal Beach for installation of the remaining systems and insulation closeout. This activity was finished on 30 June 1967. On 16 May the stage was moved to Station III for installation of the forward skirt.

During the second week of July the LOX tank was entered and installation of the LOX vent line and gas distributor began. On 28 July the stage was moved from Station IV and positioned in Station VII for frame splice modifications of the LH2 tank. As of the first week of August, systems installation and insulation closeouts were suspended, pending completion of propellant tank inspection.

The following week the stage was temporarily moved to Station II, to complete the LH2 tank closeout. The stage was positioned in Station VI on 13 August for the remainder of systems installation work and installation of J-2 engines.

By the end of September, the LOX tank vent line had been installed and the tank was closed. Sidewall insulation proof pressure tests of cylinder 2 through cylinder 6 were performed twice in October.

Stage testing

On 23 October the newly-named North American Rockwell (NAR) transferred the S-II-5 stage to the Test Operations department at Seal Beach where preparations for systems checkout commenced in Station VIII.

By 26 December 1967 the Integrated Systems Checkout of the stage in NAR Checkout Station VIII had ended and personnel began disconnecting the stage from checkout equipment.

67-60416 The S-II-5 stage being installed on the high tool and dolly outside Station VIII at Seal Beach for systems checkout 18.10.67.

68-63979 Loading the S-II-5 stage onto the Point Barrow at Seal Beach for the journey to MAF 2.2.68.

68-64160 Erection of the S-II-5 stage in the A-1 test stand at MTF 13.3.68.

68-64045 Installation of the S-II-5 stage in the VCB at MTF 19.2.68.

The stage was moved from Station VIII to Station VII on 15 January 1968. The LH2 tank was entered and the welds were inspected by x-ray and dye-penetrant techniques. Two weld defects were found and repaired by grinding and installing a 40-inch doubler on one of the repairs.

The S-II-5 stage was shipped from Seal Beach to MAF aboard the AKD Point Barrow, leaving at 1600 PST on 2 February 1968 and arriving on 16 February 1968. The stage traveled together with the S-IC engine, F-6073. This was the first time that an S-IC engine had been transported by sea together with an S-II stage. That engine was off-loaded at MAF and the S-II-5 stage continued on to MTF on 17 February 1968 aboard a barge. At MTF it was installed in the vertical checkout bay (VCB) of the S-II service building on 19 February. Modification work took place until the stage was transferred and then hoisted directly from a barge onto the A-1 test stand on 13 March. At this time all three test positions at MTF were occupied by rocket stages for the first time (stages S-IC-6, S-II-4 and S-II-5). The major changes in the testing of this stage compared with the previous one were as follows:

Thrust structure frame reinforcing
Telemetry redundant instrumentation change
New feed duct vacuum level requirements were incorporated to improve quality of LH2 and LOX at engine start to gain additional flight margin
A new ASI line system on the J-2 engines
Insulation tests were conducted which required removal and repair of forward bulkhead and sidewall insulation circuits
Redesign of vehicle LOX sump anti-vortex baffles including a new inverted LOX dump screen
Position indicator modification for LOX and LH2 pre-valves and re-circulation valves
Changes to sidewall and forward bulkhead insulation techniques
New in-flight purge circuit for LOX tank ullage pressure sense line
Change from LAD to Parker pre-valves (installed in LH2 feed system only)
Addition of a new bay (C7-111) to work in conjunction with the digital events evaluator (C7-77)

A low frequency engine gimbal test at ambient temperature was run to investigate the phase lag and gimbal response at rates of less than 1 Hz. Entry was made into the LH2 tank to conduct an x-ray inspection of the Cylinder 5 to Cylinder 6 weld and of stringers 121 and 123. A defective area was discovered during this check and the defect was ground out successfully.

A combined tanking and cryogenic proof test was performed on 26 April 1968. The test was terminated by an abrupt loss of power to the test stand due to a failure of relays in the test stand's electrical substation. This delay created a situation wherein the test would have to be conducted after dark. This was an unacceptable condition as it would not then be possible to conduct high speed (500 frames per second) photography of the test. Prior to de-tanking, however, the cold portion of the low-frequency engine gimbal test was completed.

The cryogenic proof test was successfully completed on 30 April, following the aborted test four days earlier. Three specific tests were accomplished during the operation:

Verification of vent valve and fill and drain valve actuation system with LH2 tank pressurized was completed at 1415 CDT
Relief test of vent valves was completed at 1423 CDT
Proof pressure test of the LH2 tank was completed at 1504 CDT. A pressure of 36.22psig was achieved in the LH2 tank

On 11 July lightning struck MTF's S-II stage test complex electrical substation, causing a power failure during a critical phase of S-II-5 stage testing. It was later necessary to replace two of the five J-2 engine fuel pumps on the stage.

Pre-static checkouts were completed on 24 July 1968. Three attempts were made to achieve the full duration static firing. The first attempt on 25 July 1968 resulted in an abort during propellant loading due to LH2 vent valve failure. The second attempt on 1 August 1968 was aborted after 7.54 seconds of engine operation due to inadvertent engine cutoff by the Thrust Chamber Pressure redline observer.

The cryogenic propellants for the full duration firing were loaded on the day of the firing test. The static firing was performed successfully on 9 August 1968 during which all test objectives were met. The countdown had lasted 6 hours and 36 minutes. The firing duration of each engine in the test was as follows:

> Engine 201 – 362.7 seconds
> Engine 202 – 362.6 seconds
> Engine 203 – 362.2 seconds
> Engine 204 – 362.6 seconds
> Engine 205 – 362.4 seconds

The cutoff occurred as planned by an observer upon LOX depletion.

The following problems were noted during the testing:

> Excessive H2 leak from LH2 Engine Cutoff (ECO) seals
> Major problems in the LH2 vent valve actuation system which caused a momentary failure to open vents
> Common bulkhead pressure increase was noted corresponding with LH2 tank pressurization
> Improper operation of LH2 engine 204 LAD pre-valve
> LOX vent valve unseating during final auto sequence caused by feedback circuit from GSE LOX fill line drain system to GSE LOX vent valve checkout lines
> Pieces of a seal were found in the LH2 tank which were traced to a facility LH2 transfer system check valve failure
> Due to improperly clamped wire bundles, 207 container and associated wiring damage resulted from a short circuit
> An apparent LOX fill flex hose over-pressurization occurred which may have been due to cryogenic operations

Post-static checkout was completed on 4 September 1968 and the stage was removed from the A-1 test stand on 6 September. It was placed in the vertical checkout bay of the S-II Service Building for extensive modifications and special tests. These included reinforcement of the thrust structure, instrumentation changes, addition of the capability to read pressure in the vacuum-jacketed lines and modifications to the LH2 re-circulation system.

During September S-II stage officials concluded, following the extensive testing on the S-II-5 stage, that J-2 engine ASI fuel and oxidizer line modifications were flight-worthy. Approval was given to continue with installation of the modified lines in all engines allocated to S-II stages.

The stage was reinstalled in the A-1 test stand on 11 October for parallel post-static checkout and retest of modifications. On 9 November S-II-5 was removed from the A-1 test stand at MTF and the stage was transferred to the Vehicle Service Building for LH2 tank inspection.

On 14 November S-II-5 entered the Vertical Checkout Building at MTF for insulation modifications and replacement of LH2 pre-valves.

On 21 November the stage was formally accepted by NASA. The stage was loaded on board a barge that traveled from MTF to MAF on 6 December 1968. At MAF the stage was transferred to the AKD Point Barrow that departed MAF on 6 December 1968 for the journey to Port Canaveral, where it arrived on 10 December 1968. It was delivered to the VAB by shuttle barge the same day.

On 24 February 1969 at KSC Rocketdyne successfully completed retest of the PU valves that had failed test earlier at MTF.

S-II-5 eventually provided the second stage for the AS-505 vehicle that launched Apollo 10 on 18 May 1969.

S-II-6

Summary

Forward bulkhead scrapped during manufacture. Cork panels added to bolster the spray-on foam insulation. Launched Apollo 11.

Engines

The initial and final engine configuration was:

Position 201: J-2089
Position 202: J-2086
Position 203: J-2088
Position 204: J-2084
Position 205: J-2085

Stage manufacture

Manufacture of the S-II-6 stage started in late 1966. The main structure was made up of 6 cylinders and cylinder 6 was the first completed on 10 November 1966. Cylinder 5 was the last one completed on 23 February 1967. The start of vertical assembly was on 18 January 1967 when the first two rings were brought together. Welding of all the individual cylinders took place between 26 January 1967 and 12 April 1967. Vertical assembly of the main stage structure was completed on 19 April 1967 with the final weld between cylinder assemblies 1/2 and 3/4/5/6.

On 19 January 1967 NASA declared that the S-II-6 forward bulkhead was not flight-worthy due to excessive corrosion pitting and directed NAA to use the S-II-8 forward bulkhead for S-II-6. By the end of the month NAA had completed rework of several defects noted during the ultrasonic NDI of the S-II-6 common bulkhead earlier in the month. NAA also completed hydrostatic and helium leak testing of the aft bulkhead.

On 27 January direction was issued (ECP 3640) to apply spray foam insulation (without purge channels) to the S-II-6 forward bulkhead (formerly the S-II-8 forward bulkhead). The spray insulation was successfully applied on 23 February.

The thrust structure, forward skirt and inter-stage subassemblies were completed on 15 January, 8 February and 4 May 1967 respectively.

On 9 March 1967 NAA completed the final bonding and cure cycle on the S-II-8 forward bulkhead in preparation for its use as part of the S-II-6 stage.

The LOX tank girth weld repair was made on 28 March. Following inspection, repaired areas were accepted on 31 March.

The forward facing sheet of the common bulkhead was damaged during removal of tooling from the LH2 tank. A doubler was bonded over the damaged area.

The stage pneumostatic testing was performed on 1 May and 2 June 1967. Dye-penetrant checks, in Station VII, that followed these tests revealed 10 weld discrepancies that were reworked by 18 June. The stage was then moved into the VAB at Seal Beach, on 19 June, for LOX tank inspection.

During July NAA positioned the stage in Station VI and began preparations for starting systems installation.

On 12 July a special team was formed to evaluate electrical harness problems and issue on-the-spot engineering orders for rework and/or modifications.

During the final week of August the thrust structure was mated to the remainder of the stage. The stage was moved from Station VI to Station VII for systems installation. LH2 in-tank installations were completed, but rework and closeout of the lower organic seal was delayed due to lack of skilled personnel.

Rework of eleven cylinder frame splices was required because a substitute material was used which later proved unsatisfactory during laboratory tests.

By the end of August rework of the girth weld had been completed and accepted by NASA.

In September, while machining the foam insulation on the S-II-6 cylinder 1/bolting ring, an area of insulation, approximately 2' x 4' de-bonded. The de-bonded insulating material was removed and the area re-sprayed, after which it was hand-sanded and re-machined.

The stage was transferred to the high tool area on 28 October 1967 and the five J-2 engines were installed in the stage on 30 October 1967.

During November systems installation proceeded without interruption or significant problems. The LOX feed line was attached to the LOX pre-valve on 13 December.

68-70076 Static firing of the S-II-6 stage in the A-2 test stand at MTF 3.10.68.

On 29 December 1967 the stage was moved into Station IX. The LOX tank was opened and inspected for foreign material; the LH2 tank was inspected by x-ray. No material was found in either tank.

Stage testing

On 8 January 1968 NAR workmen completed S-II-6 systems installation and began systems checkout. This was completed on 15 February, 15 days ahead of schedule. On the following day the stage was moved from Station IX to Station VII for LH2 tank entry and inspection of the tank welds. The tank was temporarily closed on 23 February and moved back into Station IX for continuation of modification activities. While in Station IX an interference problem was resolved by modifying the fairing frame to provide adequate clearance for the LH2 re-circulation system return line bellows.

On 3 April the stage was moved to Station VII for completion of work inside the LH2 tank. Further x-ray and dye-penetrant inspections were performed at this time. On 10 April the LH2 tank was closed out and the stage was moved back into Station IX, where modifications were continued with the stage in the vertical position. An interference between the helium purge line and the radio command and telemetry coaxial cables was corrected by relocating three clamps in the forward skirt. The LOX and LH2 tanks were purged and the dew-point was measured in preparation for shipping the stage. The shipping date was adjusted from 30 April to 25 May.

The stage was moved to Station VII on 23 May where preparation for shipment was completed.

The S-II-6 stage was shipped from Seal Beach to MAF aboard the AKD Point Barrow, leaving on 25 May 1968 and arriving on 7 June 1968. The stage traveled together with the S-IC engines, F-6075 to F-6079 and seven large F-1 engine components. It was the first time that F-1 engines were shipped in quantity by water. Because of the limited space on-board, the S-II-6 inter-stage was held at Seal Beach for later transportation with the S-II-7 stage. The F-1 engines were off-loaded at MAF and the S-II-6 stage continued on to MTF on 8 June 1968 aboard the barge, Pearl River.

Initially the stage was inspected in the S-IC service building as the S-II service building was occupied by the S-II-3 stage at the time. The stage was transferred to the S-II Vertical Checkout Building on 22 June where it remained for additional modification prior to erecting in the test stand. The stage was transferred by shuttle barge to the test stand on 28 June and hoisted directly from the barge onto the A-2 test stand.

The major changes in the testing of this stage compared with the previous one were as follows:

- LH2 pre-valves were changed from LAD pre-valves to Parker pre-valves
- This was the first vehicle with a spray foam forward bulkhead area, a change from the previous purged honeycomb forward bulkhead configurations. Also included was a new forward bulkhead un-insulated debris barrier
- LOX tank anti-vortex baffles were removed
- Operational changes were incorporated to pull a common bulkhead vacuum during LH2 loading which was previously purged until auto sequence
- Post static firing leak checks were changed from detailed to the gross method

Pre-static checkouts were completed on 4 September 1968. A combined tanking and cryogenic proof test was performed on 17 September 1968. The firing was delayed by two days because of a NASA request to remove the LOX sump anti-vortex baffles prior to the static firing.

The cryogenic propellants for the firing were loaded on the day of the firing test. The static firing was performed successfully at 1522 CDT on 3 October 1968 during which all test objectives were met. The firing duration of each engine in the test was as follows:

  Engine 201 – 368.4 seconds
  Engine 202 – 368.4 seconds
  Engine 203 – 368.4 seconds
  Engine 204 – 368.5 seconds
  Engine 205 – 368.5 seconds

Cutoff was achieved as planned by the 2% LOX level signal.

The following problems were noted during the testing:

- An LH2 overfill sensor was improperly placed in the tank
- An operational problem occurred which appeared to clog the LH2 facility loading system at various screen interfaces. This problem may have been caused by insufficient purging of GN2 blanket pressures
- LH2 vent valve position indicator problems

During the firing MTF personnel successfully conducted a second test of the P&VE Lab's acoustics research program (MARL). They recorded impacts on an aft inter-stage specimen located close to the firing.

On 9 October MSFC approved an engineering change proposal submitted by NAR on 19 August requesting the application of cork to certain "hot spot" areas of the S-II stage, beginning with S-II-6. Evaluation of test data from flights of the X-15 rocket plane showed that erosion of sidewall foam insulation would occur during the Saturn V boost period because of the thermal environment and shear loads. These findings led to the conclusion that the foam insulation should be protected with cork.

Following the static firing the stage was removed from the test stand on 16 October for extensive modifications and special tests. It was placed in the Vertical Checkout Building for this work. On 23 October go-ahead was given by NASA for installation of ¼ inch cork sheeting over portions of the spray-on foam insulation on S-II-6 and subsequent stages to provide protection against aerodynamic heating and shear loads.

The stage was reinstalled in the A-2 test stand on 8 November 1968 for the resumption of post-static checkout.

The S-II-6 stage was removed from the A-2 test stand on 17 December 1968. On 27 December 1968 the S-II-6 stage entered the Vertical Checkout Building again for a second round of modifications and LH2 tank inspection prior to shipment of the stage to KSC.

Post-static checkout was completed on 15 January 1969. On 23 January failure analysis of the LH2 feed-line bellows from the S-II-6 stage was completed. Fabrication procedures were revised to preclude work-hardening prior to upsetting. Replacement of all the bellows on S-II-6 and subsequent stages prior to launch was ordered.

On 25 January 1969 a special engine propellant utilization (PU) valve test was performed on the S-II-6 stage to determine if valve problems similar to those experienced during the AS-503 countdown existed. In the S-II-6 test the valves failed to meet performance criteria on three engines. These valves were subsequently retested successfully at KSC on 24 February 1969.

The stage was transferred from the test stand to the barge Orion that departed MTF on 1 February 1969 for the journey to KSC, where it arrived on 6 February 1969 after sustaining an "extremely rough ride" enroute. Subsequent inspection of the stage did not reveal any damage caused by the rough seas.

S-II-6 eventually provided the second stage for the AS-506 vehicle that launched Apollo 11 on 16 July 1969.

S-II-7

Summary

Stage static tested at MTF. Launched Apollo 12.

Engines

The initial and final engine configuration was:

Position 201: J-2090
Position 202: J-2092
Position 203: J-2093
Position 204: J-2096
Position 205: J-2097

Stage manufacture

NAA began work on S-II-7 at Seal Beach on 28 July 1966. The first activity was to place three gores of the common bulkhead aft facing sheet on the meridian weld fixture for welding operations.

During January 1967 NAA began assembly and welding of the S-II-7 stage. At this time the S-II-7 forward bulkhead was transferred to S-II-5 as a replacement.

On 27 January recent earth tremors in the Los Angeles area resulted in the necessity for rework of the thrust structure assembly fixture at Seal Beach to regain its precision alignment. Rework was completed on 11 February limiting the impact on the S-II-7 thrust cone buildup schedule to 3 weeks.

Vertical assembly of the S-II-7 stage was started on 31 March, 28 days ahead of schedule.

Hydrostatic testing of the forward bulkhead was completed on 29 April 1967, with spray-on foam insulation being applied on 15 May.

Quarter panels from each cylinder were removed on 19 May. The panels had been machined below drawing tolerances because of malfunctioning Vidigage equipment. Quarter panels fabricated for the S-II-8 stage were used as replacement panels. Welding was completed in early June. Repair of weld defects was completed on 28 June.

In June 1967 NAA completed the S-II-7 LH2 bulkhead stud welding and placed the bulkhead in storage.

Tack welding of the S-II-7 forward bulkhead to Cylinder 6 began in station 1A at Seal Beach. Four welds had been made previously on a test cylinder to verify the new weld station. The new weld station involved a rotation of the LH2 bulkhead/cylinder 6 past the weld head as opposed to the previous configuration where the weld head itself moved and the assembly remained stationary. By the middle of July welding was complete. However, inspection revealed that repairs were necessary, and these were completed on 21 July.

The aft LOX bulkhead to common bulkhead girth weld was completed, inspected and accepted by 24 July.

Welding problems were encountered in developing weld parameters for thick to thin areas on cylinders 3-4.

LOX tank cleaning took place between the last week in August and the first week in September.

Vertical buildup of the stage was completed on 17 September 1967 with acceptance of the stage closeout weld joining cylinder No. 2 to cylinder No. 3. NAA prepared to test the weld integrity. On the following day NAA completed pneumostatic tests of the stage. Inspection revealed 11 discrepancies that required repair by re-welding. These repairs were completed by the newly named NAR on 24 October in Station VII and the stage was moved to Station IV for LH2 tank doubler installation during the final week of October. By the end of November the stage was back in Station VII for pneumostatic retest and for the start of systems installation.

67-59535 Moving S-II-6 to Station IV at Seal Beach for systems installation and S-II-7 LOX tank assembly moving to Station V 9.67.

67-59775 Vertical assembly of the S-II-7 LH2 tank at Seal Beach 9.67

67-59780 Transfer of S-II-7 to remote site at Seal Beach for pneumostatic testing 19.9.67.

68-70784 Arrival of the S-II-7 stage at MTF by the barge Little Lake 12.11.68.

The retest was completed successfully on 20 December 1967 following which the application of the spray-on foam insulation was started. Phase-1 application was completed on 2 January 1968 and Phase-2 on 7 January. The stage was then moved into Station II on 8 January for the beginning of systems installation in the LOX tank and on the thrust structure. The forward skirt was mated to the stage on 12 January and the tank was closed on 2 February.

The five J-2 engines were installed in the stage on 10 February 1968 in Station IV.

Painting of the inter-stage was completed on 16 February.

The outer one-half of the one-inch thick foam over the Cylinder 1 and bolting ring area had to be removed by hand-sanding during March. Pre-fabricated foam facing sheets with a Tedlar cover identical to those used on stages S-II-1 to S-II-6 were then bonded with scrim cloth and adhesive over the Phase-1 foam.

Sidewall insulation proof-pressure testing was completed on 2 April with only minor rework resulting. The stage was moved to Station VII on 11 April for LH2 in-tank systems installations. The stage was then moved to Station VIII on 24 April.

Stage testing

On 7 May systems checkout of the S-II-7 stage began at Seal Beach, following completion of systems installation the previous day. Numerous problems, mostly of a minor nature, were detected and resolved. Systems checkout was completed on 27 June and the stage was moved to Station VII on the following day.

On 3 July the in-tank installations and inspections of the stage were completed. Four days later the stage was released for "in-place shipment", signifying satisfactory completion of all scheduled Seal Beach activities except painting and packaging. On 8 July the stage was moved to Station VIII to advance the configuration of the stage prior to delivery to MTF for flight acceptance testing.

By 24 October NAR workers had completed a three month period of modification and retest activities on the stage.

The S-II-7 stage was shipped from Seal Beach to MAF aboard the AKD Point Barrow, leaving on 29 October 1968 and arriving on 11 November 1968. The stage traveled together with the S-II-6 inter-stage and the S-IC engines, F-6080 to F-6085. Those engines and the S-II-6 inter-stage were off-loaded at MAF and the S-II-7 stage continued on to MTF on 12 November 1968 aboard the barge Little Lake. The stage was hoisted directly from the barge onto the A-1 test stand on 13 November.

On 25 November 1968 test personnel at MTF applied power to the electrical systems of S-II-7 as they began stage systems checkout and major modifications.

The major changes in the testing of this stage compared with the previous one were as follows:

A new stage start tank emergency vent system was incorporated
A procedural revision was made to include a vacuum vent valve test in the low vent mode
A GSE change was incorporated to allow pre and post test top and bottom sampling of the LOX tank, LH2 tank, and A7-71 Heat Exchanger

A combined tanking and cryogenic proof test was performed on 15 January 1969. Pre-static checkouts were completed on 20 January 1969. The cryogenic propellants for the firing were loaded on the day of the firing test. The static firing was performed successfully on 22 January 1969 during which all test objectives were met. The firing duration of each engine in the test was as follows:

Engine 201 – 364.4 seconds
Engine 202 – 364.7 seconds
Engine 203 – 364.7 seconds
Engine 204 – 363.3 seconds
Engine 205 – 363.6 seconds

The following problems were noted during the testing:

A liquid leak developed at the dock LH2 transfer system
A pre-valve position indicator malfunction
Hydrogen leaks were noted on LH2 ECOs
Random cycling of LOX ECO sensor open indications while the LOX tank was filled. This anomaly had not been noted on previous vehicles. At ambient the ECO sensor signals performed normally
LH2 vent valve position indicator malfunctions

The special pressure and accelerometer measurements installed on the stage prior to the firing detected oscillations during the firing, but failed to provide any new data relevant to the POGO problem seen on earlier vehicles. A special engine PU valve test, performed on 26 January 1969, on the S-II-7 stage indicated that the valves performed satisfactorily on all engines.

On 17 February test stand application of spray-on foam to the S-II-7 stage insulation repair areas was halted pending resolution of a "rind" formation problem caused by inclement weather. Re-foaming was accomplished in March after transfer of the stage into the environmentally controlled S-II Service Building.

Post-static checkout was completed on 13 March 1969 and the stage was transferred on that date from the A-1 test stand to the S-II Stage and Storage Building. LH2 tank x-rays and stage inspection in the horizontal position were accomplished prior to the application of the spray foam modifications. On 11 April, after completing rework of the spray foam application, the S-II-7 stage was removed from the Vertical Checkout Building and placed on a transporter for shipment preparation. The stage departed MTF on 15 April 1969 for the journey to KSC, where it arrived on 21 April 1969.

S-II-7 eventually provided the second stage for the AS-507 vehicle that launched Apollo 12 on 14 November 1969.

S-II-8

Summary

Spray-on insulation reinforced with honeycomb/cork rails. Stage used on Apollo 13.

Engines

The initial and final engine configuration was:

Position 201: J-2082
Position 202: J-2099
Position 203: J-2102
Position 204: J-2098
Position 205: J-2100

Stage manufacture

Work began on the S-II-8 stage at Seal Beach on 19 October 1966. The aft facing sheet for the common bulkhead was placed into work at that time. On 19 January 1967 NASA decided to use the S-II-8 forward bulkhead as a replacement for a defective assembly on S-II-6.

On 16 February 1967 workmen at Seal Beach completed welding of the S-II-8 aft facing sheet. However, it was then reassigned for use on the S-II-TS-A test structure, and the S-II-9 facing sheet gores re-designated for S-II-8 use. Welding and acceptance of these replacement aft and forward facing sheets was achieved on 27 March 1967.

Work on the aft bulkhead was begun on 21 February. Dye-penetrant and x-ray inspection of the completed bulkhead were finished on 17 April. Hydrostatic testing was accomplished on 6 June.

Assembly of the thrust structure for this stage began at Seal Beach on 31 March 1967. Meridian welding of the S-II-8 forward bulkhead took place between 11 April and 24 May 1967.

67-61113 Completion of the vertical assembly of the S-II-8 stage in the Seal Beach VAB 13.11.67.

68-66843 The S-II-8 stage being installed in high tool at Seal Beach for LH2 feedline and engine installation 4.5.68.

68-66849 Installation of J-2 engines on the S-II-8 stage at Seal Beach 6.5.68.

69-74950 Static firing of the S-II-8 stage in the A-2 test stand at MTF 8.4.69.

The S-II-8 forward bulkhead underwent hydrostatic tests at Seal Beach on 23 June 1967. Foam insulation was applied to the forward bulkhead on 25 July. The forward facing sheet of the common bulkhead was welded and hydrostatically tested during July.

Foam insulation was applied to LH2 cylinder 3 on 17 July in Station VI. This was the first time that foam had been applied to a cylinder forward of the number 1 cylinder. After a cure period of 48 hours the foam was trimmed to a thickness of 0.75" and depressed areas were re-sprayed.

Vertical welding and splicing of cylinders 2, 3, 5 and 6 was completed in July, and all cylinders were covered with a heavy aluminum foil to seal out moisture until application of a suitable moisture-proof material.

On 15 September 1967 NAR began vertical assembly of the stage, which was completed on 29 November. At that time NASA accepted the final closeout weld and NAR began preparations for pneumostatic testing of the stage. This test was performed successfully on 12 December 1967.

On 20 December 1967 NAR completed the S-II-8 stage LOX tank cleaning and inspection and moved the stage to Station III for post-pneumostatic inspection of the LH2 tank.

The stage was moved to Station V on 11 January 1968. The LOX tank was cleaned and the stage was inverted on 14 January. X-ray inspection of the forward bulkhead was completed the following day, and the LH2 tank cleaning was accomplished on 17 January, on schedule. The stage was moved to Station IV on 18 January. The forward skirt was installed on 22 January following which the stage was mated to the thrust structure.

On 23 February 1968 application of spray-on foam insulation to the closeout areas on the S-II-8 stage in

Station V was completed. The stage was then moved to Station VII on 26 February, and the LH2 in-tank installations were begun. Upon completion of this effort, on 8 March, the stage was moved to Station IV where final systems installations were begun on 12 March.

LOX tank internal installations were completed on 5 April and the LOX feed lines were installed on 18 April. The stage was moved to Station VII on 29 April to inspect and map the LH2 tank welds and to install systems inside the tank. This work was completed on 4 May and the stage was moved to Station IV for installation of the LH2 feed lines, then to Station II for engine installation.

The five J-2 engines were installed in the stage on 6 May 1968.

Reinforcement of the thrust structure, that began at Seal Beach on 22 July, ended on schedule on 23 August. Also completed was replacement of the LOX vent line and installation of an inverted LOX sump screen.

The ASI fuel lines on all five engines were replaced between 27 August and 9 September.

On 24 October 1968 NAR completed systems installation and turned the stage over to test operations.

Stage testing

On 16 December the stage was moved from Station VII to Station IX for initiation of the V8-500 modification period.

Based on results of tests on honeycomb and cork insulation, MSFC directed NAR on 17 January 1969 to proceed with installation of this insulation on the S-II-8 stage in an effort to overcome insulation de-bonding problems occurring with the spray-on foam used on the S-II stages. The solution called for the foam to be reinforced with phenolic honeycomb along the ramps to give better support to the cork sheet. The foam/honeycomb/cork composite would be prefabricated into segments called "cork rails", and glued directly to the tank wall.

By early February systems test, modification, and retest of the S-II-8 stage at Seal Beach had been completed.

The S-II-8 stage was shipped from Seal Beach to MAF aboard the AKD Point Barrow, leaving on 10 February 1969 and arriving on 23 February 1969. The stage traveled together with the S-IC engines F-6086 to F-6088 and F-6090, as well as the S-II-7 inter-stage. The engines and inter-stage were off-loaded at MAF and the S-II-8 stage continued on to MTF on 24 February 1969 aboard a barge. The stage was hoisted directly from the barge onto the A-2 test stand on 25 February. The major changes in the testing of this stage compared with the previous one were as follows:

> Sidewall insulation was changed from a honeycomb purged configuration to solid spray foam making the entire vehicle, forward bulkhead and sidewall, a spray foam insulated vehicle
> Due to POGO problems encountered during the S-II flights, special modifications were made to allow the center engine to be shut down prior to all engine cutoff to test the flight characteristics of this modification
> Modifications were made to stage and GSE systems to provide an open loop PU step profile
> An LH2 ECO in-flight purge circuit was incorporated to reduce previously noted hydrogen content and leakage through the ECO seals

The measure to reduce POGO by shutting down the center engine earlier than the outer ones was initiated on 12 March and approved for use on S-II-8 on 18 March.

Cook pour foam (Gold Foam No. 402) became a candidate for field repair of spray-on foam insulation. Laboratory testing was ordered, on 13 March, to obtain all material properties. On the following day authorization was given for the temporary application of Cook pour foam on the S-II-8 stage at MTF.

A combined tanking and cryogenic proof test was performed on 28 March 1969. This was the first stage fully insulated with spray-on foam insulation. Analysis over the next 24 hours revealed 27 defects in the spray-on foam insulation (SOFI), none of which would compromise the integrity of the insulation system or constrain launch operations. Repairs would take place prior to shipment of the stage to KSC.

On 4 April environmental tolerance tests were completed for the cork rail configuration for the S-II-8 and subsequent stages. Pre-static checkouts were completed on 5 April 1969.

Firing A2-546-8A-69

The cryogenic propellants for the firing were loaded on the day of the firing test. The static firing was performed successfully at 1345:02 CST on 8 April 1969 during which all test objectives were met. The firing duration of each engine in the test was as follows:

> Engine 201 – 381.3 seconds
> Engine 202 – 381.2 seconds
> Engine 203 – 381.1 seconds
> Engine 204 – 381.0 seconds
> Engine 205 – 294.2 seconds

The following problems were noted during the testing:

> Helium Inject System regulator and relief valve failure
> LOX Engine 205 pre-valve improper operation (LAD Pre-valve)
> LH2 vent valve position indicator problems

The early center engine shutdown proved to be highly successful in eliminating POGO effects, so much so that NASA immediately issued CO 1643 to NAR authorizing modifications on stages S-II-5, S-II-6 and S-II-7 for early shut down.

Post-static checkout was completed on 10 June 1969 and modification and preparation for shipment of the stage to KSC started. On 14-15 June NAR personnel completed repair of the minor insulation damage sustained by the S-II-8 stage during the static firing test. The stage was transferred from the A-2 test stand to the horizontal position in the S-II Stage Checkout and Storage Building for completion of insulation work on 19 June 1969. The stage was shipped from MTF on 25 June 1969, arriving at KSC on 29 June 1969. The shipment had been delayed by 24 hours because of the non-availability of the NASA barge.

S-II-8 eventually provided the second stage for the AS-508 vehicle that launched Apollo 13 on 11 April 1970.

S-II-9

Summary

Stage static fired at MTF. Second stage on Apollo 14 launch.

Engines

The initial and final engine configuration was:

Position 201: J-2106
Position 202: J-2110
Position 203: J-2108
Position 204: J-2109
Position 205: J-2105

Stage manufacture

On 16 February 1967 the facing sheet gores for S-II-9 were reassigned to S-II-8 as replacements on that stage. The S-II-10 sheet was allocated as a replacement for the S-II-9 stage.

Structural assembly of the S-II-9 stage began with meridian welding of the aft facing sheet gores on 18 April 1967.

Meridian welding of the forward bulkhead began on 21 July, and was completed on 30 August. Meridian and dollar welding of the aft facing sheet was completed on 5 July, and the honeycomb core was bonded to the aft facing sheet on 30 August. Dollar welding of the forward facing sheet was accomplished initially in September, but had to be cut apart due to excessive porosity.

The aft bulkhead meridian and dollar welds were completed on 7 and 22 August respectively. Hydrostatic testing of the assembly was accomplished on 8 September.

Forward bulkhead dollar welding was completed the first week in October. Inspection revealed several defects in the bulkhead dollar weld. The defects were repaired and the unit was submitted for hydrostatic testing with acceptable results.

A defect was discovered in the components of the common bulkhead, requiring extensive machining and buildup to achieve a good fit.

Aft bulkhead stud welding and painting were completed in October.

NAR began vertical build up of the stage propellant tanks on 9 November 1967.

Vertical buildup of the overall stage commenced in January 1968. Cylinder 6 welding to the forward bulkhead in Station 1A was completed on 19 January, and this assembly was welded to the Cylinder 3/4/5 assembly. Cylinder 1 was welded to Cylinder 2, and this assembly was placed in Station 1B, where the J-weld to the common bulkhead was made in early February. Installation of the bolting ring was accomplished on 26 February and the assembly was hydrostatically tested in Station VI. The girth weld joining the aft bulkhead to this assembly was made in Station III on 6 March, with only minor repairs being needed.

During April, cork insulation was being installed on the inter-stage located in Station III. In May the inter-stage was transferred to the subassembly building where cork bonding was completed on 22 May. The effort associated with MCR 5251 (reduction of ullage motors from eight to four) was concluded on 6 June. During the remainder of June and July ablative and fungus painting of the inter-stage took place in Station VII.

Welding together of the S-II-9 LOX and LH2 tanks (stage closeout weld) occurred on 29 March. The newly assembled stage was pneumostatically tested on 2 April at Seal Beach. Following this the stage was positioned horizontally in the bonding and subsystems building (S-14) for post-pneumostatic LH2 tank inspection. Minor rework was accomplished, and a second pneumostatic test was made. Re-inspection and rework were concluded on 26 April, and the stage was moved to Station V for LOX tank post-pneumostatic inspection. Two anomalies were corrected and accepted on 30 April.

LOX tank cleaning was accomplished on 1 May, and LH2 tank cleaning on 7 May. The production forward skirt was mated in Station III on 9 May and the stage was moved to Station V for application of foam to sidewall closeout areas. Trimming of the foamed areas was concluded on 31 May 1968.

68-62995 The S-II-9 common bulkhead at Seal Beach 31.1.68.

68-66857 The S-II-9 forward skirt ready for installation on the LH2 tank at Seal Beach 9.5.68.

69-73499 The S-II-9 stage and S-II-8 inter stage en route to the Seal Beach docks for the trip to MAF. Also shown the S-II-12 stage en route to the pneumostatic test site 27.3.69.

The structure was mated to an H7-17 static firing skirt on 17 June, and to the stage tanks on the following day.

S-II-9 was transferred from Station II to Station VII at Seal Beach for LH2 tank systems installation on 11 August 1968.

At the end of August minor damage occurred to the stage during a manufacturing process at Seal Beach. An assembly station floor section struck the stage while workmen were moving it from one station to another. The damage was confined to a bracket and channel section. The accident resulted in a revision of stage movement procedures.

During September workmen at Seal Beach applied a coat of moisture barrier (Dynatherm) to the S-II-9 insulation, and moved the stage to a workstation for installation of LH2 tank systems.

The five J-2 engines were installed in the stage on 28 October 1968.

Upon completion of foam insulation installation, NAR technicians transferred the stage to the pneumostatic test site during December 1968.

Stage testing

At NAR personnel completed systems installations in the S-II-9 stage on 22 February 1969 and released the stage to Test Operations for systems checkout. This activity was completed on 18 March. Modification of the stage to complete insulation rework was accomplished on 24 March and the stage was prepared for shipment.

The S-II-9 stage was shipped from Seal Beach to MAF aboard the AKD Point Barrow, leaving on 27 March 1969 and arriving on 9 April 1969. The cargo also included the S-II-8 inter-stage. On 10 April 1969 the stage was transferred to a barge and made the short journey to MTF. The stage was hoisted directly from the barge onto the A-1 test stand on 11 April. The major changes in the testing of this stage compared with the previous one were as follows:

> Deletion of the LH2 tank slow chill down which was previously conducted for conditioning insulation system
> A new LH2 tank and feed duct purge procedure was implemented
> Parker pre-valves were installed in the LOX and LH2 feed systems
> Numerous insulation modifications were performed
> Extensive instrumentation was incorporated for POGO analysis

NAR personnel completed "power-on" tests and the telemetry automatic checkout of the stage on 23 April. A combined tanking and cryogenic proof test was performed on 23 May 1969. The stage exhibited insulation de-bond failures similar to those noted on the S-II-8 stage. NAR replaced the rails around Feedline 3 using Narmco 7343 adhesive and silane additive. The integrity of the repair would be verified during the static firing.

Pre-static checkouts were completed on 30 May 1969. The cryogenic propellants for the firing were loaded on the day of the firing test. The static firing was performed successfully on 20 June 1969 during which all test objectives were met. The firing duration of each engine in the test was as follows:

Engine 201 – 348.0 seconds
Engine 202 – 348.1 seconds
Engine 203 – 348.0 seconds
Engine 204 – 348.2 seconds
Engine 205 – 348.1 seconds

The following problems were noted during the testing:

LOX ECO cycling occurred as noted on S-II-7
Facility valve failure occurred prohibiting the pressurization of the engine start tanks during auto sequence

Visual inspection of the S-II-9 stage insulation, performed on 24 June, added assurance that the special test sections of the honeycomb insulation bonded with the improved process survived the cryogenic temperatures, vibration levels and tank pressures imposed by the static firing. Engineering support would be provided to assure effective application of the new techniques on subsequent stages.

Post-static checkout was completed on 1 August 1969 and personnel at MTF began insulation modifications. The S-II-9 stage was removed from the A-1 test stand on 7 August and transported by shuttle barge to the S-II Service Building.

On 8 August NASA requested NAR to perform an analysis of the work needed to modify the S-II-9 and S-II-10 stages to provide "terminal stage" capability for orbiting payloads (Mod 1764 CO), such as the Skylab space station.

On 17-18 August the Mississippi gulf coast was hit by Hurricane Camille, with winds of up to 190 mph and a 16 foot tide. Damage to MTF was minor and no damage was sustained by either the S-II-9 or S-II-10 stages on site at the time. S-II stage operations at MTF were suspended between 18 and 24 August to facilitate clean-up operations.

On 12 September the stage was moved from the low bay area of the S-II Service Building to the vertical checkout bay for incorporation of modifications originally planned to be performed at KSC. Also on this date, NAR completed plans for the transfer of the S-II-9 stage to Seal Beach for long term storage, although these plans were never put into force.

On 13 October 1969 NASA issued SA 1084 to NAR revising the storage schedule for S-II stages. It included plans to store S-II-9 at KSC.

On 23 October the storage environmental enclosure was removed from the S-II-9 stage in the Storage and Checkout Building at MTF and preparations began for final stage modifications prior to its shipment to KSC. Modifications were completed on 17 December 1969. On 30 December the stage was moved on a transporter from the S-II Service Building to the S-IC Service Building for shipment preparations.

During December it was determined that three Start Tank Discharge Valves (STDV) on the S-II-9 stage had been improperly reworked and would need replacement.

The stage was transferred to the ship that departed MTF on 13 January 1970 for the journey to KSC, where it arrived on 19 January 1970.

S-II-9 eventually provided the second stage for the AS-509 vehicle that launched Apollo 14 on 31 January 1971.

S-II-10

Summary

Repeat non-firing testing necessary at MTF due to effects of Hurricane Camille. Stage launched Apollo 15.

Engines

The initial and final engine configuration was:

Position 201: J-2112
Position 202: J-2113
Position 203: J-2114
Position 204: J-2115
Position 205: J-2116

Stage manufacture

On 21 August 1967 NAR completed meridian welding on the aft facing sheet gores for the S-II-10 stage common bulkhead.

By 22 October 1967 the final meridian weld on the forward facing sheet of the common bulkhead was completed and it was placed in storage awaiting the common bulkhead buildup.

All meridian welds on the forward bulkhead were completed and accepted by 1 December 1967.

On 3 January 1968 hydrostatic testing of the aft bulkhead was completed. NDI inspection was performed by the end of January, stud welding was completed on 17 February and painting was accomplished the following week. The aft bulkhead was then stored, ready for LOX tank assembly.

Post-hydrostatic test inspection of the aft facing sheet of the common bulkhead revealed defects in two meridian welds. Repairs and retest were accomplished by 26 January 1968. After another weld defect was discovered and repaired, the bulkhead was accepted on 9 February.

Hydrostatic inspection of the forward bulkhead was accomplished on 2 February and, following priming, spray-on foam insulation was applied during the first week of March. Hydrostatic testing of the forward bulkhead was performed on 6 April. On 16 April, following some rework activities and a repeat hydrostatic test, NAR officials at Seal Beach accepted the forward bulkhead. Foam application, machining, wet lay-up, and cure were completed on 20 May.

Bonding of the aft and forward facing sheets of the common bulkhead was completed on 18 March. By 22 April all purge tubes had been welded, leak-checked and accepted. Internal cleaning, dye-penetrant inspection and associated rework were completed on 17 June.

Following completion of stud welding on 21 May, the forward bulkhead was positioned on Cylinder 6 for final trimming and welding. The circumferential weld was accomplished on 28 May 1968, but due to excessive offset and weld porosity, disposition was made to cut the assembly apart.

Following negative materials analysis it was decided to cut Cylinder 3 from the Cylinder 3/4/5 assembly to remove and replace one of the quarter panels which had been fabricated from –063 material. The replacement –021 panel was received on 15 May, and after processing, was welded into the cylinder on 28 May. Meanwhile, Cylinder 2 was found to contain an LH2 feed duct elbow ring that was made of –063 material. Authorization was given to cut apart and replace the quarter panel containing the elbow.

On 26 July the common bulkhead was removed from storage and positioned for J-welding to the Cylinder 1/Cylinder 2 assembly.

The S-II-10 forward skirt was reallocated to the S-II-TS-B structure following a failure with that stage on 12 July 1968. The former S-II-10 skirt was delivered to SSFL on 16 August 1968. Hydrostatic testing of the S-II-10 common bulkhead was successfully completed on 20 August at Seal Beach.

The first pneumostatic test of the S-II-10 tankage was accomplished on schedule on 24 September. The second test was on 16 October and the third on 13 November.

On 6 December 1968 NAR personnel successfully completed manufacturing checkout of the S-II-10 tankage and began systems installation.

The five J-2 engines were installed in the stage on 8 March 1969.

NASA and NAR held the first joint status meeting for the S-II-10 stage on 23 April 1969 and determined that the stage "state of completion" was the best to date.

Stage testing

NAR personnel completed systems installation on 5 May 1969 and turned the stage over to Test Operations for post-manufacturing checkout. On 3 June the integrated systems test was performed satisfactorily.

After the S-II-8 insulation de-bonded under cryogenic temperatures at MTF it was decided to remove and re-bond the S-II-10 stage honeycomb/cork insulation rails. This task was completed on 23 June 1969 at Seal Beach.

Post-manufacturing checkout of the stage was completed on 26 June and the stage was prepared for shipment. Final painting and packing was undertaken.

The S-II-10 stage was shipped from Seal Beach to MAF aboard the AKD Point Barrow, leaving on 27 June 1969 and arriving on 10 July 1969. The stage traveled together with the S-II-9 inter-stage, and S-IC engines, F-6089 and F-6091 to F-6093. Also on board were 76 pieces of –063 plate stock and six non-flight cylinders that would be used to fabricate a space station mockup. This material was loaded onboard on 20 June and was off loaded at MAF. The F-1 engines and S-II-9 inter-stage also were off-loaded at MAF and the S-II-10 stage continued on to MTF on 10 July 1969 aboard a barge.

The stage was hoisted directly from the barge onto the A-2 test stand on 11 July. The major changes in the testing of this stage compared with the previous one were as follows:

> Water was found in the Forward Bulkhead Un-insulated (FBU) area which resulted in a modification to apply sealant to the FBU Debris Barrier and to incorporate a continuous FBU purge system
> Extensive corrosion inspections were conducted since corrosion under sidewall insulation was noted

- As a result of POGO problems experienced on S-II flights, a center engine LOX feed-line accumulator was added, supplied by a temporary GSE/facility accumulator helium charge system
- The countdown procedure was altered to conduct dual loading of the propellants
- Special tests were conducted to utilize point sensors in the A7-71 Heat Exchanger for controller level in place of the modulating control valves
- Inspections and leak checks were conducted after it was noted that the LH2 tank pressurization line located in the tunnel had cracked welds
- A special test was conducted to evaluate the feasibility of using an orifice in the thrust chamber chill circuit in place of the S7-41 regulator
- The center engine torsional bellows were replaced with redesigned units

NAR personnel at MTF completed the "power-on" test for the S-II-10 stage in Test Stand A-2 on 17 July. On 22 July NAR personnel at Downey completed the first-axis testing of a modified test feed-line to qualify the design for incorporation in the center engine LOX feed-line accumulator for S-II-10 and subsequent stages. A second qualification test would take place.

On 8 August NASA requested NAR to perform an analysis of the work needed to modify the S-II-9 and S-II-10 stages to provide "terminal stage" capability for orbiting payloads (Mod 1764 CO), such as the Skylab space station.

On 9 August MSFC issued NAR with CO 1763 which approved the installation and test of the modified LOX feed-line accumulator on the S-II-10 stage. The new feed-line was designed to reduce low frequency oscillations during stage flights.

The modified LOX feed-line accumulator was successfully installed in the S-II-10 stage at MTF by NAR personnel on 13 August 1969.

A combined tanking and cryogenic proof test was performed on 15 August 1969.

On 17-18 August the Mississippi gulf coast was hit by Hurricane Camille, with winds of up to 190 mph and a 16 foot tide. Damage to MTF was minor and no damage was sustained by either the S-II-9 or S-II-10 stages on site at the time. S-II stage operations at MTF were suspended between 18 and 24 August to facilitate clean-up operations. When work started again on 25 August it was determined that all S-II-10 stage systems would require retesting.

On 29 August third-axis vibration testing was completed successfully on the LOX feed-line accumulator qualification test specimen, relieving the constraint to S-II-10 static firing.

Pre-static checkouts were completed on 5 September 1969.

On 12 September 1969 NASA completed preliminary design of facilities that were planned to be used for the long-term storage of stages S-II-9 and S-II-10 after their static firings and return to Seal Beach. In the event this storage plan was not utilized.

On 22 September 1969 NAR personnel completed assessment and preliminary repair of insulation and corrosion damage in the S-II-10 forward skirt which resulted from accumulation of rain water during Hurricane Camille. Technicians noted that the insulation was satisfactory for the static firing but would need repair later.

The planned S-II-10 static firing was postponed on 25 September following discovery of a leak in the S-II-12 LH2 tank pressurization line. Pressure tests and leak checks of the S-II-10 stage would be needed to clear that stage. Following the delays the revised pre-static firing checkout was completed on 29 September.

The cryogenic propellants for the firing were loaded on the day of the firing test. The static firing was performed successfully on 1 October 1969 with no major problems occurring. The firing durations for each engine were:

Engine 201 – 364.5 seconds
Engine 202 – 364.4 seconds
Engine 203 – 364.3 seconds
Engine 204 – 364.6 seconds
Engine 205 – 364.5 seconds

Results of the test verified flight readiness of the stage as well as accomplishment of the following special test objectives:

- determination of individual LH2 vent valve cracking pressures
- performance verification of PU valves under cryogenic conditions
- determination of S7-41 console thrust chamber chill orifice requirements operational data
- a A7-71 heat exchanger vent valve sequencing test using point-sensor control
- determination of center engine LOX feed-line accumulator operational characteristics
- pressurization of LOX tank with ambient-temperature gaseous nitrogen
- LOX tank pressurization test using gaseous helium without the normal two minute pre-chill

On 6 October officials announced a new process and storage plan in which the S-II-10 stage would receive modified post-static checkout after firing, followed by storage in the S-II Stage Checkout and Storage Building at MTF. Finally, the stage would be placed back in the test stand just prior to shipment for the remainder of the post-static checkout.

On 13 October NASA issued SA 1084 confirming that

S-II-10 would be stored at MTF, rather than being shipped back to Seal Beach for storage. Two days later NASA issued CO 1789 stopping production of ullage motors for the S-II-10 and subsequent stages.

The stage was removed from the A-2 test stand on 4 November 1969 for insulation rework that was performed in a horizontal orientation in the S-II Stage Checkout and Storage Building at MTF.

During December 1969 personnel at MTF removed a Start Tank Discharge Valve (STDV) from the S-II-10 stage because of bellows leakage and determined that it had been reworked to an inadequate drawing. Replacement would be necessary.

The stage was returned to the test stand (this time the A-1 stand) on 16 January 1970 for the post static firing acceptance checkout to be performed. Post-static checkout was completed on 7 April 1970 and the stage was transferred from the test stand to the ship that departed MTF on 14 May 1970 for the journey to KSC, where it arrived on 18 May 1970.

S-II-10 eventually provided the second stage for the AS-510 vehicle that launched Apollo 15 on 26 July 1971.

S-II-11

Summary

Small fire in an engine during the stage static firing test. Launched Apollo 16.

Engines

The initial and final engine configuration was:

Position 201: J-2117
Position 202: J-2125
Position 203: J-2121
Position 204: J-2123
Position 205: J-2118

Stage manufacture

On 6 October 1967 the S-II-11 stage assembly activities began at Seal Beach with loading of the aft facing sheet gores on the welding fixture.

In the first week of January 1968 NAR at Seal Beach began welding together the gore segments of the S-II-11 forward bulkhead. The twelfth and final weld was accomplished on 8 February. This bulkhead was completely free of meridian weld defects. The twelfth and final meridian weld of the common bulkhead forward facing sheet was made on 19 January 1968. The dollar weld was made on 22 February and the bulkhead was stored awaiting tracing.

The dollar weld of the aft facing sheet was made on 2 February, and the hydrostatic test was accomplished on 9 February.

69-01889 The S-II-11 stage being transported to the Seal Beach docks for the trip to MAF 3.9.69.

69-75312 Installation of the J-2 engines in the S-II-11 stage at Seal Beach 2.5.69.

The twelfth and final meridian weld of the aft bulkhead was made on 21 February.

LH2 cylinder spray-foam application began on 9 January with Cylinder 4. A shortage of acceptable foam ingredients delayed completion of Cylinder 5.

Hydrostatic testing of the forward bulkhead was completed on 2 April 1968. After inspection, and a period in storage, foam insulation was applied, starting on 22 May. Foam machining was completed on 7 June.

Bonding of the common bulkhead forward facing sheet to the core was accomplished on 17 May, resulting in formation of the common bulkhead.

The aft bulkhead dollar weld was performed successfully on 24 April 1968. The bulkhead was accepted on 10 June following various rework operations. Subsequently, hydrostatic testing and the follow-up inspections were performed.

Early in April one LH2 Cylinder panel of large-grained MB0170-063 material was removed from Cylinder 3 for replacement, and Cylinder 2 quarter panels were vertically welded. On 18 April 1968 direction was given to eliminate all MB0170-063 material from the S-II-11 stage structure. All of the LH2 cylinders and bolting ring segments for this stage had been made of this material, therefore, a complete set of replacement parts of –021 material was ordered from the NAR Los Angeles Division (LAD) on 1 May. These started to arrive at Seal Beach on 23 May.

The –063 material was still used to produce a test structure, designated, S-II-NF-1. However due to lack of funds from MSFC for this unexpected assembly, no further activities on it took place after 21 May 1968.

The four forward skirt panels and the eight inter-stage panels were received from NAR Tulsa by the end of June 1968.

By the mid year point, component assembly was 80% complete.

On 6 September NAR personnel began the vertical assembly of the stage. On 27 December 1968 NAR completed vertical assembly of the stage, ahead of schedule.

Pneumostatic testing of the stage was conducted on 3 January 1969. Post-test inspection revealed two anomalies that would require re-welding prior to a second test. The repeat test was conducted successfully on 23 January.

The five J-2 engines were installed in the stage on 2 May 1969.

Stage testing

The S-II-11 stage was moved to Station IX in the Vertical Checkout Building at Seal Beach on 16 July 1969 for completion of systems modification and early start of checkout activities. Systems installation was finished on 24 July and integrated systems checkout was started on 21 August and ran for five days.

The S-II-11 stage was shipped from Seal Beach to MAF aboard the AKD Point Barrow, leaving on 3 September 1969 and arriving on 16 September 1969. The stage traveled together with S-IC engines F-6095 and F-6096 and an S-II stage horizontal storage dolly built by GE for MSFC and acceptance tested by NAR.

The engines and storage dolly were off-loaded at MAF and the S-II-11 stage continued on to MTF on 17 September 1969 aboard a barge. The stage was hoisted directly from the barge onto the A-1 test stand on the following day. The major changes in the testing of this stage compared with the previous one were as follows:

- This was the first vehicle which had all in-tank LOX and LH2 tank point sensors removed with the exception of the lower and upper Overfill Sensor (OFS) cluster and ECO sensors
- A special test was conducted to attempt stage tank loading utilizing the overfill sensors in place of the propellant utilization system
- Stage in-flight pressurization regulators for the LOX and LH2 tanks were removed and replaced by orifices

A "power-up" test on the stage was completed on 29 September in preparation for the cryogenic proof pressure test.

On 6 October it was decided that this stage, as well as two others, would receive a modified post-static checkout after static firing and would then be placed in storage. The three stages would be placed back in the test stand later, for the completion of post-static checkout.

On 13 October NASA issued SA 1084 revising the storage schedule for various S-II stages. The S-II-11 stage would be stored at MTF. On 15 October MSFC issued CO 1789 to NAR directing them to stop production of ullage motors for stages, S-II-10 through S-II-15.

Pre-static checkouts were completed on 17 October 1969. Postponement of the S-IC-12 static firing date, announced on 19 October, necessitated a change from 29 October to 31 October for the planned S-II-11 proof test. Tanking and cryogenic proof tests were performed successfully on 31 October 1969.

The cryogenic propellants for the firing were loaded on the day of the firing test. The static firing of 371.6 seconds duration was performed successfully on 14 November 1969 during which all test objectives were met.

There were, however, some problems during testing as follows:

A severe leak at the LH2 facility filter area was encountered
LOX ECO cycling occurred
LH2 re-circulation pump valves position indicator malfunctioned
AC power loss to the LH2 barges
Momentary fire indication on the 205 engine (J-2118) approximately 67 seconds into the test which was extinguished by the Firex system
LH2 pre-valve position indicator chatter

Following the fire on engine 205 MTF personnel performed leak checks on the engine between 19 November and 1 December in an effort to identify the cause of the problem. Inspection of the Gas Generator LOX injector revealed that four of the 24 holes were plugged by a LOX safe fluorocarbon resin applied to the threads of the injector during assembly (this same problem had occurred to an S-II-3 stage engine). Technicians cleaned the necessary components, reassembled the Gas Generator, and began preparations for the removal of the stage from the test stand.

During December it was determined that two Start Tank Discharge Valves (STDV) on the S-II-11 stage had been improperly reworked and would need replacement.

The stage was removed from the test stand on 15 January 1970 for necessary modifications to be performed. S-II-11 was returned to the A-1 test stand on 18 May 1970 for post-static firing acceptance checkout to be performed. Post-static checkout was completed on 15 July 1970 and the stage was removed from the test stand before being transferred to the ship that departed MTF on 22 September 1970 for the journey to KSC, where it arrived on 30 September 1970.

S-II-11 eventually provided the second stage for the AS-511 vehicle that launched Apollo 16 on 16 April 1972.

S-II-12

Summary

Five attempts needed to static fire the stage. Launched Apollo 17.

Engines

The initial and final engine configuration was:

Position 201: J-2130
Position 202: J-2126
Position 203: J-2127
Position 204: J-2129
Position 205: J-2128

Stage manufacture

The first meridian weld was made on 29 January 1968. By the end of March 8 meridian welds had been completed. Also by the end of March, Cylinder 5 was vertically welded and spliced, and was ready for spray foaming. The forward bulkhead meridian welding operations were begun on 3 April, and by the end of June all twelve welds had been completed.

By 3 June all meridian welds on both the forward facing sheet and the aft facing sheet of the common bulkhead had been completed. The aft facing sheet dollar welding was completed by 24 June, following which the hydrostatic test was performed.

The first meridian weld of the aft bulkhead was completed on 8 April. The thrust structure panels were delivered from NAR Tulsa and assembly operations began on 29 April. The four aft skirt panels were put into work on 28 May. By 30 June 1968 S-II-12 component assembly was 72% complete.

70-07737 Cork application and installation on the S-II-12 interstage 5.5.70.

Vertical assembly of the stage began on 16 December 1968.

On 28 March 1969 NAR personnel completed the S-II-12 stage pneumostatic test. Several anomalies were noted that would require rework and a second test.

On 11 April 1969 NAR completed weld repair of cracks revealed during x-ray inspection and satisfactorily completed a second pneumostatic test on the stage. On 3 May NAR personnel completed post-pneumostatic test x-ray inspection of the S-II-12 stage and began systems installation in the stage. Closeout of the foam insulation on the stage was completed on 2 July 1969.

The five J-2 engines were installed in the stage on 20 August 1969.

During September a leak in the S-II-12 LH2 tank pressurization line was discovered. This had a knock-on effect that the S-II-10 static firing, planned for 25 September, was delayed to 1 October in order to perform leak checks on the S-II-10 stage.

NASA announced, on 6 October 1969, that there would be a revised stage processing plan for the S-II-10, -11 and -12 stages. These stages would receive a modified post-static checkout after static firing and then would be placed in storage in the S-II Stage Checkout and Storage Building. Prior to shipment, the three stages would again be placed in the test stand for completion of post-static checkout.

On 13 October 1969 NASA issued SA 1084 to NAR revising the storage schedule for various stages including S-II-12. Two days later MSFC issued CO 1789 to NAR stopping production of ullage motors for second stages S-II-10 to S-II-15.

Stage testing

Systems installation on the S-II-12 stage ended at Seal Beach on 16 October. On 18 November NAR began the integrated systems checkout of the stage, that was completed on 21 November.

On 1 December 1969 the NAR Seal Beach facility suffered a momentary power dip due to an equipment failure at the Southern California Edison Company's Del Amo switch station. The dip was sufficient to cause momentary over-voltage from a 28 v dc rectifier supplying power to the S-II-12 stage; this resulted in some damage to stage systems.

On 5 December NAR completed preparations for shipping the S-II-12 stage and loaded it on to the AKD Point Barrow.

The S-II-12 stage was shipped from Seal Beach to MAF aboard the AKD Point Barrow, leaving on 6 December 1969 and arriving on 19 December 1969. The stage traveled together with the S-IC engines F-6097 and F-6098. Those engines were off-loaded at MAF and the S-II-12 stage continued on to MTF on 20 December 1969 aboard a barge.

On 15 December NASA issued CO 1878 to NAR directing deletion of the cryogenic proof testing and tanking test from the flight acceptance requirements for S-II-12 and subsequent stages.

On 17 December NASA and NAR personnel concluded analysis of low frequency oscillations observed on the S-II-7 flight. They decided to install a modified LOX feed-line accumulator on the center engine of the S-II-12 stage for evaluation during the upcoming static firing test. This was covered formally by CO 1883, issued on 30 December 1969.

During December it was determined that three Start Tank Discharge Valves (STDV) on the S-II-12 stage had been improperly reworked and would need replacement.

The stage was hoisted onto the A-2 test stand on 22 December 1969. There was no pre-firing tanking test performed with this stage. The major changes in the testing of this stage compared with the previous one were as follows:

> A center engine LOX feed-line accumulator system for POGO suppression was installed with an airborne in-flight helium charge system in place of the test temporary GSE/Facility test system used on S-II-10. In addition, an accumulator bleed valve was incorporated, redundant with the LOX Engine 205 return valve
> A major change was approved during the S-II-12 processing which eliminated the requirements for performing Cryogenic Proof Pressure Tests
> The LOX and LH2 tank pressurization orifices which were incorporated for S-II-11 were again utilized for S-II-12 with the addition of the use of the engine two coil heat exchanger
> Engine valve vacuum drying was added as a requirement

Pre-static checkouts were completed on 4 February 1970. Five attempts were made to static fire the S-II-12 stage.

The first attempt on 12 February 1970 was aborted due to a leakage in the LOX facility dump line.

The second attempt on 18 February 1970 was aborted because an additional repair to the LOX facility dump-line was needed.

A further attempt was aborted because of a facility LOX dump line failure.

The fourth attempt was aborted due to an improperly operating engine 202 main oxidizer valve. This attempt, on 26 February 1970, resulted in a Gas Generator under temperature engine cut-off and was classified as a firing as it lasted for 2 seconds.

The final attempt resulted in a successful full-duration static firing of 376.2 seconds duration on 4 March 1970 with no major problems occurring.

The main problems occurring during the test phase were as follows:

A slow Main Oxidizer Valve (MOV) movement terminated the fourth static firing test
Facility LOX dump line weld ruptures terminated three of the five tests
LOX ECO cycling occurred
LH2 re-circulation valve indicator malfunctions
Failures occurred with the facility water firex system
LH2 vent valve position indicator malfunctions
Major performance shifts on engine 203 (J-2127)

The stage was removed from the test stand on 18 March 1970 for necessary modifications to be performed. S-II-12 was installed in the A-1 test stand on 17 July 1970 for post-static firing acceptance checkout to be performed. Post-static checkout was completed on 10 September 1970 and the stage was removed from the test stand around 15 October 1970 and transferred to the ship that departed MTF on 22 October 1970 for the journey to KSC, where it arrived on 27 October 1970.

S-II-12 eventually provided the second stage for the AS-512 vehicle that launched Apollo 17 on 7 December 1972.

S-II-13

Summary

Final S-II stage to fly. Second stage on the Skylab space station launch. Remained in orbit for 606 days, the longest of any S-II stage.

Engines

The initial and final engine configuration was:

Position 201: J-2104
Position 202: J-2132
Position 203: J-2135
Position 204: J-2136
Position 205: J-2138

Stage manufacture

All forward bulkhead and forward facing sheet gores were completed by the Los Angeles Division (LAD) of NAR in the second quarter of 1968. A union strike at the Alcoa plant at Davenport, Iowa, necessitated reordering 14 pieces of the plate stock required for fabrication of LH2 cylinder quarter panels and bolting ring segments to retain schedule.

In December 1968 early phases of the assembly of the S-II-13 stage were in progress at NAR's Seal Beach facility.

NAR at Seal Beach began vertical assembly of the S-II-13 stage on 24 February 1969. This activity was completed on 11 June 1969. Pneumostatic testing of the stage was finished on 11 July. The thrust structure was completed by NAR on 30 July. Systems installation in the S-II-13 inter-stage was started on 6 August 1969 at Seal Beach.

On 13 October 1969 NASA issued SA 1084 that directed NAR to plan for storage of the S-II-13 stage at Seal Beach in the future. Two days later MSFC issued CO 1789 directing NAR to delete the production of ullage motors for a number of S-II second stages, including S-II-13. On 21 October 1969 NASA issued CO 1840 to NAR that instructed them to reconfigure the S-II-13 stage to provide the capability of injecting an S-IVB dry workshop into low Earth orbit. S-II-13 was later to perform this roll in the launch of the Skylab space station.

On 11 November 1969 NASA performed a final design review of the storage facility installations planned for this stage and others. S-II-13 would be stored in Seal Beach Station VIII.

NAR moved the S-II-13 stage to the "high-tool" area of manufacturing, on 14 November 1969, in preparation for engine installation.

The five J-2 engines were installed in the stage on 17 November 1969.

Stage testing

Seal Beach operations on the S-II-13 stage were completed on 6 March 1970. Movement of the stage to the dock was delayed until 10 March because of ammunition loading operations at the US Naval Weapons Station.

70-04755 The S-II-13 stage being lowered onto a transporter for transfer to Station VII at Seal Beach 15.1.70.

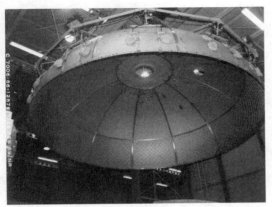

69-73077 Transfer of the S-II-13 stage aft bulkhead to inspection at Seal Beach 26.2.69.

70-07794 The S-II-13 stage being positioned in the A-2 test stand at MTF 25.3.70.

The S-II-13 stage was shipped from Seal Beach to MAF aboard the AKD Point Barrow, leaving on 11 March 1970 and arriving on 24 March 1970. On 25 March 1970 the stage was transferred to a barge and made the short trip to MTF. The stage was hoisted directly from the barge onto the A-2 test stand. The major changes in the testing of this stage compared with the previous one were as follows:

- Stage engine 205 LOX feed-line accumulator for POGO suppression was not installed on this vehicle
- Ullage motor firing circuitry Stage and GSE was deleted
- Post-static firing Skylab modifications were incorporated
- Addition of a post-firing tanking test

Pre-static checkouts were completed on 29 April 1970. The cryogenic propellants were loaded on the day of the firing test. The static firing of 364.9 seconds duration was performed successfully on 30 April 1970 with some major problems occurring as follows:

- Site-wide AC power problems
- Failure of Facility LH2 transfer system line expansion joints
- Momentary fire indication on engine 201 which appeared to be erroneous
- LH2 vent valve vent and reseat problems possibly due to high LH2 load
- Failures of Facility LOX Main Fill Valve and LH2 Emergency dump valves

Following the static firing test a special tanking test loading to about the 50% LOX and LH2 tank level was performed on 8 May 1970. This special tanking test, performed unusually after the firing test, was conducted to checkout the feed-line cork rail insulation redesign. The results were satisfactory.

The stage was removed from the test stand on 10 June 1970 to allow necessary stage modifications to be performed. The vehicle was then installed in the A-1 test stand on 16 October 1970 to allow post-static firing acceptance checkout to be accomplished. On 10 November the terminal stage modification (ECP 6621), for the Skylab mission, including associated retest effort, was successfully completed on schedule.

Post-static checkout was completed on 23 November 1970 and the stage was removed from the test stand on 18 December 1970 and transferred to the ship Poseidon that departed MTF on 30 December 1970 for the journey to KSC, via MAF, where it arrived on 6 January 1971. Shipment had been brought forward a week to preclude interference with the shipment of a Skylab dynamic test article (S-IVB-F) to Houston because of tug availability. The S-II-13 stage was the last deliverable flight stage of contract NAS 7-200.

The S-II-13 stage ultimately was launched on 14 May 1973 as the second stage of the last Saturn V rocket to be launched. This vehicle launched the Skylab space station into orbit. Because Skylab had no propulsion the second stage was required to propel Skylab to orbital altitude. Consequently the S-II-13 stage remained in orbit for some time unlike the other S-II stages that reentered immediately.

The stage received an orbital catalog number of 1973-27B, finally decaying from orbit on 11 January 1975 after an orbital lifetime of 606 days – the longest for any S-II stage.

S-II-14

Summary

Forward bulkhead damaged by accident during manufacturing. Stage on display at KSC.

Engines

The initial and final engine configuration was:

Position 201: J-2139
Position 202: J-2141
Position 203: J-2142
Position 204: J-2144
Position 205: J-2145

Stage manufacture

At Seal Beach, on 6 May 1969, NAR began vertical assembly of the S-II-14 stage. By 14 July the cylinder 5 to cylinder 6 weld had been performed. On 11 August NAR completed installation of systems and painting of the S-II-14 inter-stage.

Vertical assembly of the stage was completed at Seal Beach on 25 August 1969.

On 29 August NAR personnel completed pneumostatic testing of the stage tanks and also completed assembly and preliminary checkout of the thrust structure.

During detergent cleaning operations of the S-II-14 LH2 tank on 30 September 1969 a spray nozzle became detached from a revolving spray mast and fell 40 to 45 feet into the inverted stage causing a small crack, 1 ¼ inches in length, in a gore on the forward bulkhead. In addition five scratches were made in another gore, each measuring approximately 3/8 inch in length and 0.002 to 0.009 inch in depth.

This added a two month delay to the schedule. By 10 October MSFC and NAR had determined that the most economical means of repairing this damage was to use a doubler. Approval for this repair to proceed was given on 23 October. On 7 November the bolted-on structural doubler was successfully installed on the forward bulkhead of the S-II-14 stage completing the repair process.

Meanwhile, in order to prevent a reoccurrence of the accident, changes were put in place in the Station V spray boom assembly area, including redesign of the spray nozzles and the nozzle couplings. Drawings were issued to this effect on 24 October.

On 13 October 1969 NASA issued SA 1084 to direct NAR to plan for storage of the S-II-14 stage at Seal Beach. On 15 October NASA issued CO 1789 to direct NAR to stop production of ullage motors for a number of Saturn second stages, including S-II-14.

On 11 November NASA completed the design review of the storage facilities for S-II stages. The S-II-14 stage would be stored in Station IX at Seal Beach.

The S-II-14 stage on display in the open at KSC in 1991

70-04762 Inside the S-II-14 forward LH2 bulkhead showing the completed repair of damage caused by the spray cleaning nozzle 29.1.70.

70-07728 Ceremony at Seal Beach to celebrate the S-II-14 stage completing manufacturing 17.4.70.

70-09822 Transfer of the S-II-14 stage and the S-II-12 inter stage from the Seal Beach packaging area to the US Naval Weapons Station dock area for the trip to MAF 10.6.70.

On 12 November 1969 NAR personnel at Seal Beach completed re-pneumostatic testing of the S-II-14 stage, verifying the doubler repair performed earlier. Post-pneumostatic inspection and LOX tank cleaning on the S-II-14 stage was completed by NAR on 2 December 1969 and subsequent operations geared towards mating of the thrust structure and systems installation continued.

On 18 December 1969 NAR completed cleaning of the LH2 tank using the spray system, with deficiencies corrected, that had resulted in the nozzle accident in September 1969. NAR then resumed activity towards completion of systems installation.

The five J-2 engines were installed in the stage on 9 March 1970.

Stage testing

Systems checkout was completed on 2 June.

The S-II-14 stage, together with the S-II-11, S-II-12 and S-II-14 inter-stages, was shipped from Seal Beach to MAF aboard the AKD Point Barrow, leaving on 10 June 1970 and arriving on 23 June 1970. On 24 June 1970 the S-II-14 stage was transferred to a barge and made the 45 mile trip to MTF. The stage was hoisted directly from the barge onto the A-2 test stand. There was no pre-firing tanking test performed with this stage. The major changes in the testing of this stage compared with the previous one were as follows:

- A new post-test H2 inerting procedure was used, utilizing a newly installed GN2 purge capability for the LH2 tank
- The LOX E5 feed-line accumulator was re-installed into the vehicle
- A passive redundant stage mounted POGO center engine cut-off system was installed for evaluation (initiated by MOD 1991 CO)
- Temporary stage and GSE purge and H2 analysis circuit was installed to check cryogenically the Forward Bulkhead doubler repair

Pre-static checkouts were completed on 28 July 1970. The cryogenic propellants were loaded on the day of the firing test. The static firing of 373.2 seconds duration was performed successfully on 31 July 1970 with no major problems occurring.

However there were some minor problems during the test:

- LH2 vent valve position indicator had a problem
- A partially open GSE manual GHe supply valve caused tank pressurization problems during auto sequence
- Engine heat exchanger oscillations were noticed

Post-static checkout was completed on 25 August 1970 and the stage was removed from the test stand on 23 September 1970 and then transferred to the ship that departed MTF on 2 October 1970 for the journey to KSC, where it arrived on 8 October 1970.

The stage was placed in long term storage at KSC on 23 November 1970.

In April 1976 it was transferred outside the VAB to form a Saturn V display. It was on display in an un-mated horizontal configuration together with other Saturn V stages. In 1979 the title was turned over to the Smithsonian. The stage remained there until 1996, when on 27 April it was transferred 2 miles to its new home. It was refurbished, and located in the new indoor Saturn V center at KSC that opened its doors to the public on 5 December 1996.

S-II-15

Summary

Final S-II stage built and static tested. Stage on display at JSC.

Engines

The initial and final engine configuration was:

Position 201: J-2147
Position 202: J-2152
Position 203: J-2149
Position 204: J-2150
Position 205: J-2151

Stage manufacturing

NAR began welding the major structural subassemblies for the S-II-15 stage during February 1969.

NAR personnel formed the last gore and machined the last LH2 panels at its El Toro facility on 24 April 1969. Vertical assembly of the stage began at Seal Beach on 18 July 1969. Meridian welding of the aft bulkhead was completed on 24 July 1969. With this activity on the last S-II stage complete deactivation of the major production facilities at Seal Beach was initiated.

The forward bulkhead was completed on 18 August. The thrust structure assembly and the inter-stage assemblies were completed on 9 September 1969. The aft bulkhead finished fabrication on 15 September. The forward skirt assembly was completed on 29 September. By 1 October NAR had completed fabrication of the common bulkhead.

On 13 October 1969 NASA issued SA 1084 to NAR revising the storage schedule for the remaining S-II stages. S-II-15 would be stored at Seal Beach. Two days later NASA issued CO 1789 to NAR stopping production of S-II ullage motors for stages S-II-10 through S-II-15 as a cost saving measure. The associated mass saving per stage was 2,400 lbs permitting a payload capability increase of 243 lbs. A further change order, CO 1840 issued on 21 October, directed NAR to configure the S-II-13 and S-II-15 stages to provide the capability to inject a dry workshop (Skylab) into low Earth orbit. The S-II-13 stage was ultimately used in this role.

On 4 November 1969, during welding of the S-II-15 LOX girth joint, two holes were blown through the weld as the two machines making the weld produced unstable arcs simultaneously. Repair of this weld was completed 10 days later.

On 11 November NASA conducted a design review of the S-II stage storage facilities. S-II-15 would be stored at Station III at Seal Beach.

Vertical assembly of the stage was completed on 18 December 1969, with the final stage closeout weld (Cylinder 2 to Cylinder 3). This concluded all structural welding under the S-II stage contract between NASA and NAR.

Pneumostatic testing of the S-II-15 stage was conducted successfully on 19 December 1969. Post-test inspection later in the month did not disclose any anomalies, therefore no weld rework was deemed necessary.

The five J-2 engines were installed in the stage on 7 May 1970.

S-II-15 systems installation was completed on 14 July, two days ahead of schedule.

Stage testing

Systems checkout was completed on 14 August, nine days ahead of schedule.

The S-II-15 stage was shipped from Seal Beach to MAF aboard the AKD Point Barrow, leaving on 9 September 1970 and arriving on 23 September 1970. On 24 September 1970 the stage was transferred to a barge and made the short journey to MTF via the Intra-coastal Canal and the East Pearl River. The stage was hoisted directly from the barge onto the A-2 test stand. There was no pre-firing tanking test performed with this stage. The major changes in the testing of this stage compared with the previous one were as follows:

> Incorporation of a two step PU valve system in place of the conventional modulating valve
> A tank point sensor upgrade modification
> A new technique to more effectively load to the overfill sensors

The S-II-15 stage on display at JSC in 1991

70-04749 The S-II-15 stage in Station II at Seal Beach for etch and prime 12.1.70.

70-09927 The S-II-15 stage in Station IX at Seal Beach being viewed by Saturn S-II employees, their families and friends during open house 19.7.70.

70-11846 Arrival of the S-II-15 stage at the US Naval Weapons Station dock for loading aboard the Point Barrow for the trip to MAF 9.9.70.

70-11853 The S-II-15 stage being transferred to Station VII at Seal Beach for packaging 1.9.70.

70-11869 Farewell ceremony for the final S-II stage to leave Seal Beach, S-II-15 6.9.70.

Pre-static checkouts were completed on 19 October 1970. The cryogenic propellants were loaded on the day of the firing test. The static firing of 373.0 seconds duration was performed successfully at 1515 CST on 30 October 1970 with no major problems occurring.

Post-static checkout was completed on 20 November 1970. On 30 November MOD 2066 (SA) was received by NAR to accelerate the delivery of S-II-15 from 22 December to 17 December in order to preclude shipment of the stage to KSC during the holiday period. The stage was transferred from the test stand to the barge that traveled from MTF to MAF on 17 December 1970. At Michoud the stage was loaded on board the Point Barrow, which already had the S-IVB-512 stage on board, for the journey to KSC, where it arrived on 21 December 1970 after having left MAF on 18 December 1970.

The stage was placed in long term storage at KSC on 22 February 1971. It was removed from long term storage on 13 September 1972, in order to act as the back-up second stage for Skylab, and returned to this storage on 15 June 1973, after Skylab was launched.

In 1977 the stage was moved from KSC to the Johnson Space Center (JSC) in Houston, Texas. S-II-15 left KSC on board a barge in late 1977 and arrived at JSC a few days later after traveling through the inter-coastal waterways. The stage has been on display to the public at JSC ever since. The ownership of the stage was transferred to the Smithsonian museum in 1978. In 2004 plans were announced to refurbish the JSC Saturn V stages (including S-II-15) and enclose them in a purpose built building.

S-IVB-500FS

Summary

Flight systems test stage tested at IBM, Huntsville.

Stage testing

The S-IVB-500FS was the third stage flight systems test article.

Environmental testing of the stage with an instrument unit began at IBM Space Systems Center Simulation Laboratory in Huntsville in June 1966. Four tests were completed.

S-IVB-500ST

Summary

Third stage breadboard simulator used at SACTO and MSFC.

Stage manufacture

The S-IVB-500ST was the third-stage simulator mock-up for the Saturn V SDF (Systems Development Facility). This was a breadboard for developing integrated computer tapes for launching the Saturn V vehicle.

MSFC and DAC completed contract negotiations for the simulator on 22 November 1964. Hardware from the cancelled S-IVB All-Systems Stage was reallocated for use in the simulator.

The S-IVB-500ST was assembled at Douglas' Huntington Beach facility during 1965. Assembly continued until near the end of the year.

Final inspection and rework at Douglas' Huntington Beach facility were completed and the stage simulator was shipped by barge to Sacramento on 15 December 1965. Initially the stage traveled by barge to Courtland dock and then on to SACTO by truck, arriving on 21 December 1965. The stage was installed in the South Tower of the Vertical Checkout Laboratory on 22 December 1965.

Stage testing

Activity in the first quarter of 1966 consisted of continuity checks and completion of modifications on the S-IVB-500ST, Simulator stage, in the Vertical Checkout Tower at Sacramento. The power-on procedure was successfully completed on 7 February 1966. Post-manufacturing checkout was completed on 19 March at SACTO and the stage was removed from the stand.

The S-IVB-500ST was turned over to NASA at Courtland, California on 30 March. The stage was loaded aboard the Super Guppy and was flown to MSFC, arriving on 1 April 1966. This was the first operational flight of the Super Guppy in support of the Saturn program, following tests the previous week with the S-IVB-D stage. At MSFC the stage was installed in the Systems Development Facility.

After updating the S-IU-500ST to functional configuration, MSFC personnel positioned the IU atop the S-IVB forward section in the Saturn V SDF and conducted system tests.

S-IVB Common Bulkhead Test Article

Summary

Structural test element that failed in test.

Stage testing

The S-IVB Common Bulkhead Test Article, which was the modified All-Systems Stage, failed during reverse pressure testing at Sacramento on 13 January 1966. The Common Bulkhead was severed completely around the circumference at the aft LOX bulkhead joint. It was determined that a design deficiency existed in the joint area. The failed specimen was shipped to Santa Monica for metallurgical and fracture mechanics investigation.

S-IVB-S

Summary

Structural qualification of the third stage achieved via a number of test items.

Stage element activity

The S-IVB-S, structural test stage, was actually a series of structural elements tested and qualified separately. The S-IVB-S was under test at Huntington Beach in the Saturn IB configuration during 1964 and into 1965.

During January and February 1964 the Huntington Beach assembly tower was used for the first time for assembly of the S-IVB-S stage's LOX tank and cylindrical section. Welding of the forward dome of this assembly in February completed the basic stage structure. The aft skirt assembly was completed in May 1964.

In the second half of 1964 DAC began testing components to optimize and prove the design load-carrying capability and to establish a margin of safety beyond the maximum expected operational environment. The LOX tank passed the hydrostatic proof test, but during proofing of the LH2 tank on 14 July 1964 the cylinder and forward dome ruptured at approximately limit pressure. X-ray examination showed that failure began with a lack of fusion in one area of the weld seam on the longitudinal portion of the LH2 tank assembly. The LOX tank was returned to Santa Monica for rework in preparation for the thrust structure test. Testing of the forward skirt section continued to the end of December 1964.

The forward skirt assembly was further tested in February and March 1965. In June, DAC completed necessary redesign to correct deficiencies disclosed during tests of the thrust structure. Vibration tests of fuel vent system components, and of the LH2 instrumentation probe structure were accomplished in July. In July and August test engineers structurally qualified the aft skirt assembly, and on 26 October 1965 completed qualification of the aft inter-stage structure. Completion of the aft inter-stage/aft skirt separation joint test on 11 November left only one structural test, the aft inter-stage retrorocket installation test, to be performed in 1966. After this the stage could be converted for Saturn V configuration tests.

During the first quarter of 1966 the Saturn V S-II/S-IVB Interface Test progressed. Modifications to the S-II Forward Skirt, required to accommodate anticipated acoustic loads, were completed on 29 January 1966. The S-IVB Aft Inter-stage was positioned atop the S-II Forward Skirt in mid-March.

De-stacking of the S-II/S-IVB Interface Test unit was achieved in early July1966 after the successful completion of tests. The S-IVB/V Aft Skirt specimen was moved to a separate pad for build up for its structural test. The specimen was damaged slightly during preparations and required some rework. The bending moment parameter test was completed on 14 September 1966, with satisfactory results. The maximum ultimate load test was successfully completed on 23 September following a 3 day delay. The elevated temperature parameter test was completed on 4 October. The Main Engine Cutoff Ultimate load test was conducted on 14 October 1966. The failure test of the specimen was conducted on 19 October with satisfactory results. The design limit load was approximately 221% when the dummy aft inter-stage buckled in the vicinity of the maximum compression stringer and the aft skirt tore in the area of the maximum tension stringer.

The S-IVB/V Forward Skirt was subjected to the ultimate test at maximum conditions on 19 July 1966. Failure occurred in bending under approximately 250% design load with axial load held constant at 100% design limit.

The elevated temperature parameter test of the S-II/S-IVB Interface Joint Specimen was successfully concluded on 26 August 1966. On 12 September, during the ultimate axial load/elevated temperature test under MECO conditions, the Interface Joint Specimen failed. Data indicated that the axial load at the time of the failure was about 140.5% of the design limit while the temperature was approximately 520 F.

Testing of the first specimen of the S-II/S-IVB Interface Bolt Tension Test was completed on 11 November 1966.

Structural qualification of the elements making up the S-IVB-S stage were complete by the end of 1966.

S-IVB Battleship (MSFC)

Summary

J-2 Battleship stage that was used at MSFC over many years.

Engines

The initial engine configuration was:

Position 301: J-2013

Later configurations included the engines, J-2027, J-2048 and J-2050.

Stage manufacture

There were two S-IVB Battleship stages. The principal battleship for the stage development was tested at SACTO in California. A second battleship was installed in a dedicated test stand at MSFC where it was fired on numerous occasions. This stage was used as a backup for the DAC SACTO battleship and for development tests of the J-2 and J-2S engines at MSFC.

In January 1964 DAC shipped major elements of the battleship tank to MSFC. The battleship was erected at MSFC in February 1964. The test stand was not completed until October 1964, following which the battleship was installed in the test stand.

Stage testing

At MSFC, the Test Laboratory proceeded in June and July 1965 to activate the S-IVB test stand at Huntsville for the MSFC battleship. The battleship comprised heavy duty tankage together with a flight-type feed system and J-2 engine.

65-19302 Preparing to install the S-IVB Battleship into the West area test stand at MSFC 10.2.65.

65-22696 The S-IVB Battleship installed in the West area test stand at MSFC 4.65.

65-19304 Preparing to install the S-IVB Battleship into the West area test stand at MSFC 10.2.65.

LH2 was loaded into the S-IVB fuel tank on 24 June 1965. A successful engine thrust chamber and start bottle chill test was conducted on 29 June. On 10 July a satisfactory LOX tank loading was conducted. On 13 July an engine spin test was aborted because the power for the re-circulation pumps failed.

On 17 July liquid nitrogen was loaded in the fuel tank to check out the fuel chill down system. A countdown test was performed on 19 July, with LH2 and LOX in the battleship tanks.

An 8 second test, scheduled for 30 July 1965, was cancelled when the facility heat exchanger thrust chamber chill down coil froze. A successful thrust chamber chill test was conducted on 31 July.

The first successful ignition test for 2.1 seconds was conducted on 2 August 1965. A full duration test lasting 400 seconds followed on 15 September.

The J-2 engine used in the first series of firings (J-2013) was removed from the stage on 4 October 1965 and replaced with up-rated engine J-2027, on 11 October. Also installed at this time was engine gimballing equipment. Four battleship firings using this new engine occurred for long durations ranging from 300 to 432.4 seconds during November and December 1965. All four firings yielded normal results.

The battleship stand at MSFC was used for a total of 117 J-2 and J-2S firings up to 1971.

S-IVB Battleship SACTO

Summary

J-2 Battleship stage tested at SACTO and later at Arnold to verify the operation of stage and engine together.

Engines

The initial engine configuration for countdowns CD 614000-CD 614013 (S-IVB/IB configuration) was:

Position 301: J-2003

The engine configuration for countdowns CD 614014-CD 614032 (S-IVB/IB configuration) was:

Position 301: J-2013

The engine configuration for countdowns CD 614033-CD 614044 (S-IVB/V configuration) was:

Position 301: J-2020

Stage manufacture

There were two S-IVB Battleship stages. The principal battleship for the stage development was tested at SACTO in California. A second battleship was installed in a dedicated test stand at MSFC where it was fired on numerous occasions.

Conducted at the Douglas Aircraft Company, Sacramento, California from 18 September 1964 to 20 August 1965, the S-IVB static firing program consisted of a series of short and full-duration engine firings to prove the design parameters and to verify the integrity of the stage systems.

The battleship test vehicle comprised the flight stage systems and the battleship tank assembly - a heavy-duty stainless steel, cylindrical vessel with hemispherical heads mounted on a dummy aft inter-stage and the J-2 engine mounted on the thrust structure. The LOX and LH2 tank internal configuration resembled the S-IVB flight stage except for openings provided for special instrumentation, cameras, lighting, and emergency LOX drain provisions.

Stage testing

The battleship tank was installed on the Beta Complex test stand No. 1 at SACTO on 18 December 1963 and engineers began battleship buildup and checkout activities including test stand, Test Control Center, and facility equipment installations and checkout. Safety tests of helium spheres ended successfully at SACTO in March 1964. Battleship test operations started in April with integrated checkout of the electrical systems and pneumatic consoles. Also in April the J-2 engine was delivered from Rocketdyne.

The J-2 engine (J-2003) was installed on the battleship tank on 4 June 1964. By mid September, checkout of the battleship, GSE and support systems had been completed. Prior to static firing, a number of checks were successfully completed. These included leak checks and complete functional tests of pneumatic, propellant, aft environmental control, electrical power distribution and sequencer systems.

S-IVB/V battleship configuration testing was performed between 19 June and 20 August 1965, whilst the S-IVB/IB configuration tests were completed on 14 May 1965.

Cold flow and chill down testing consisted of a series of non-firing tests, performed in order to establish and evaluate operating procedures for propellant loading, engine purging, venting and a chill down sequence for proper engine start.

Four countdowns, CD 614000, CD 614002, CD 614003 and CD 614004 were required for these tests.

Just after main-stage signal, a failure occurred in the gas generator on CD 614005 as a result of overcooling during re-circulation chill down. The pressure rise in combustor and LH2 injector manifold destroyed the LOX poppet valve, the number 2 spark plug was blown from its threaded shell, and the LOX injector sense line was burned through and partially consumed. The poppet was blown through the LH2 turbine, where two turbine blades were subsequently destroyed. Corrective measures were taken and the gas generator operated normally during engine operation thereafter.

The first Saturn IB configuration firing occurred on 1 December 1964, and the first full duration test came before the end of that month. J-2 engine J-2003 was replaced with J-2013 on 28 January 1965. The IB firing program with a flight-type engine continued early in 1965 with satisfactory full-duration firings on 31 March, 15 April and 4 May 1965. The Saturn IB series of battleship tests at SACTO ended on 14 May 1965, with an aft inter-stage environmental conditioning test.

Conversion of the stage to the Saturn V configuration began in May and was completed early the following month. This included replacing the J-2 engine with J-2020. The modification also included the installation of 10 ambient helium bottles to the thrust structure for LOX and LH2 tank re-pressurization.

Test personnel attempted the first Saturn V development firing on 19 June, but the test ended with an automatic cutoff after 9 seconds. The second firing on 26 June consisted of a 167 second "first burn" firing followed by a 4 second restart firing which indicated good performance of all systems. During the third firing on 1 July an explosion occurred in the thrust cone area after 2 seconds of the second firing. This caused a fire that damaged wiring and instrumentation.

On 13 August fire interrupted another firing after 16 seconds, but damage was minor. The first two-burn full-duration firing of the battleship occurred on 17 August; firings were for 170 and 319 seconds with a 92 minute simulated coast period between. The final two-burn test came on 20 August for 171 and 360 seconds with a 41 minute coast period.

The S-IVB/V static firing test program consisted of seven firings - two full duration and five short duration, whilst the S-IVB/IB test program consisted of ten firings – four full duration and six short duration.

Workmen removed the battleship stage from the Beta I test stand at SACTO on 3 September 1965 in readiness for transporting the stage for further testing.

In early January 1966 the SACTO S-IVB Battleship was transferred to Tennessee. On 8 January 1966 the battleship arrived by barge at South Pittsburg, Tennessee, en route to Arnold Engineering Development Center (AEDC) in Tullahoma for a series of J-2 engine altitude firings.

The series of J-2 and J-2S development firings in the Arnold J-4 Test Cell, some utilizing the battleship stage, ran from July 1966 to October 1968.

Examples of tests run include those during January 1968, when personnel at AEDC conducted several J-2 engine tests. These included 8 low-fuel S-II net positive suction head (NPSH) hot firings and 3 blow-down tests on the S-IVB battleship. The tests also included 6 S-IVB battleship firings to investigate 8-minute restart runs of the 230,000 lb thrust engine.

At the conclusion of testing, at the end of 1968, it is assumed that the battleship stage was scrapped.

Countdown number	Date	Config/Test	Engine	Duration	Comments
CD 614000	18.9.64.	IB/non-firing	J-2003	N/A	LN2 and LH2 cryogenic loading
CD 614002	25.9.64.	IB/non-firing	J-2003	N/A	LOX and LH2 propellant loading
CD 614003	2.10.64.	IB/non-firing	J-2003	N/A	Engine chilldown test
CD 614004	24.10.64.	IB/non-firing	J-2003	N/A	Start tank blowdown test
CD 614005	7.11.64.	IB/firing attempt	J-2003	0s	3 aborted attempts at a 10s firing
CD 614006	24.11.64.	IB/non-firing	J-2003	N/A	Gas generator ignition test
CD 614007	1.12.64.	IB/firing	J-2003	10.67s	Successful shakedown firing
CD 614008	9.12.64.	IB/firing	J-2003	50.7s	Successful shakedown firing
CD 614009	15.12.64.	IB/firing	J-2003	150.4s	Successful shakedown firing
CD 614010	23.12.64.	IB/firing	J-2003	414.6s	Successful full duration firing
CD 614011	8.1.65.	IB/non-firing	J-2003	N/A	J-2 engine temp conditioning test
CD 614012	14.1.65.	IB/non-firing	J-2003	N/A	J-2 engine temp conditioning test
CD 614013	16.1.65.	IB/non-firing	J-2003	N/A	J-2 engine temp conditioning test
CD 614014	9.2.65.	IB/non-firing	J-2013	N/A	J-2 engine temp conditioning test
CD 614015	17.2.65.	IB/non-firing	J-2013	N/A	J-2 engine temp conditioning test
CD 614016	18.2.65.	IB/non-firing	J-2013	N/A	J-2 engine temp conditioning test
CD 614017	25.2.65.	IB/non-firing	J-2013	N/A	J-2 engine temp conditioning test
CD 614018	2.3.65.	IB/non-firing	J-2013	N/A	J-2 engine temp conditioning test
CD 614019	6.3.65.	IB/non-firing	J-2013	N/A	J-2 engine temp conditioning test
CD 614020	13.3.65.	IB/firing	J-2013	11.8s	Successful shakedown firing

Countdown number	Date	Config/Test	Engine	Duration	Comments
CD 614021	19.3.65.	IB/firing	J-2013	29.2s	Full duration attempt aborted due to instrumentation problem
CD 614022	25.3.65.	IB/firing attempt	J-2013	0s	High EMR, PU excursion firing – 3 aborted attempts
CD 614023	31.3.65.	IB/firing	J-2013	470s	Successful high EMR, PU excursion firing
CD 614024	7.4.65.	IB/firing	J-2013	42s	Low EMR, PU excursion firing. Aborted due to instrumentation problem
CD 614025	15.4.65.	IB/firing	J-2013	506.75s	Successful low EMR, PU excursion firing
CD 614026	22.4.65.	IB/non-firing	J-2013	N/A	Spring rate simulator verification and ambient gimbal test
CD 614028	27.4.65.	IB/firing	J-2013	374s	High EMR firing terminated due to high temperature
CD 614030	4.5.65.	IB/firing	J-2013	493.5s	Successful
CD 614031	13.5.65.	IB/non-firing	J-2013	N/A	Aft interstage environmental test
CD 614032	14.5.65.	IB/non-firing	J-2013	N/A	Aft interstage environmental test
CD 614033	19.6.65.	V/firing	J-2020	8.92s	Aborted due to facility logic problem
CD 614034	26.6.65.	V/firing	J-2020	167+3.84s	94 min coast time between burns. 2nd burn aborted due to instrumentation noise
CD 614035	1.7.65.	V/firing	J-2020	5.45+1.72s	Ist firing aborted due to control logic. 2nd firing aborted due to fire and explosion
CD 614041	12.8.65.	V/firing attempt	J-2020	0s	Aborted due to leakage in loading
CD 614042	13.8.65.	V/firing	J-2020	16s	Terminated early due to minor fire
CD 614043	17.8.65.	V/firing	J-2020	170+319s	92 min coast time between burns. 2nd burn gimbal
CD 614044	20.8.65.	V/firing	J-2020	170.9+360.2s	41 min coast time between firings. 2nd burn gimbal

S-IVB-F

Summary

Facilities checkout stage used to check test stand facilities at SACTO and launch facilities at KSC. Part of the SA-500F first Saturn V vehicle on the launch pad in 1966. Finally converted for use as the Skylab OWS dynamic test vehicle.

Engines

The stage had no engine installed.

Stage manufacture

The S-IVB-F (or S-IVB-500F), the facilities checkout stage for the Saturn V vehicle, was manufactured at Douglas' plants during 1964. Fabrication of hardware began at Santa Monica in February 1964. DAC completed all cylindrical tank panels during April. By the end of June the aft LOX dome and common bulkhead were joined, and the forward LH2 dome was ready.

During the second half of 1964 the All-Systems stage, S-IVB-T was cancelled and hardware re-allocated to the S-IVB-F vehicle.

DAC completed the thrust structure assembly and machined the attach ring in October. In late December 1964 DAC moved the stage to Assembly Tower # 2 and joined the forward skirt, aft skirt and thrust structure to the tank section. No engine was installed in the stage. The stage left the Seal Beach Naval dock on 12 February 1965, aboard the barge Orion, and arrived at SACTO on 17 February, after having been towed up the Sacramento river to Courtland dock. At SACTO, on 18 February 1965, workmen installed the stage in the Beta III Test Stand for use in qualifying the stand for S-IVB-201 acceptance testing.

Stage testing

Propellant loading in manual mode was accomplished without problems on 21 April 1965. The stage and facility checkout in Beta III ended on 1 May with a successful automatic propellant loading test. The stage was removed from the stand on 3 May and underwent post-test inspection of tanks, insulation and welds. No discrepancies were found.

The S-IVB-F stage departed SACTO on 10 June 1965 arrived at KSC on 30 June 1965. Initially, it traveled from SACTO, via Courtland dock, down the Sacramento river to Seal Beach on the Orion, arriving on 13 June. At Seal Beach it was transferred to the AKD Point Barrow for the remainder of the voyage to KSC. Joining it for this part of the journey was the S-II stage simulator. It passed through the Panama Canal on 22 June, calling in at MAF on 26 June, where the S-II stage simulator was offloaded. It was planned to use the stage for checkout of the LC-34, -37 and –39 facilities, although ultimately was only used on LC-34 and –39.

70-05519 The S-IVB-F stage (Skylab Dynamic Test Article) tank assembly being removed from tower # 1 at Huntington Beach after separation from the aft skirt 1.70.

70-14592 The S-IVB-F stage (Skylab Dynamic Test Article) entering MSC test tower in Houston 1.71.

Vehicle erection on Pad 34 began in August 1965. It had been planned to use the S-IB-D/F stage as the first stage of the facilities verification vehicle. However, that stage was damaged during dynamic testing and it was agreed to use the first flight stage, S-IB-1 for facility testing. The S-IB-1 stage was erected on the pad on 18 August. The S-IB-1 served as a spacer for the S-IVB-500F during propellant tankings to verify the LOX and LH2 loading systems. The S-IU-200F/500F was erected atop the S-IVB-500F.

Vehicle checkout of LC-34 began on 18 August and,

except for several days lost to Hurricane Betsy, progressed extremely well. Technicians completed the automatic computer controlled propellant loading of the S-IVB-500F on 23 September 1965. On completion of the tests on 29 September, KSC technicians began dismantling the vehicle from the pad. DAC personnel immediately began the process of converting the S-IVB-F stage to the Saturn V configuration.

The aft inter-stage for the Facilities Checkout stage was shipped from Seal Beach aboard the Green Mountain State ship on 19 October 1965 and arrived at KSC on 2 November 1965.

Conversion of the S-IVB-F, facilities checkout stage, to the final Saturn V configuration was accomplished at KSC in the Low Bay of the VAB. The stage was signed off and available for LC-39 checkout on 25 March 1966.

On 25 March the S-II-F stage was mated with the S-IC-F stage in the VAB. Between 28 and 29 March 1966 technicians at KSC stacked the S-IVB-F atop the S-IC-F and S-II-F. The following day the S-IU-500F was also added to the rocket.

Following completion of the AS-500F, facility checkout vehicle, power was first applied on 13 May 1966. Systems tests were completed on 24 May 1966 and the complete vehicle was rolled out of the VAB towards Launch Complex 39A on the following day. The vehicle travelled on crawler-transporter No. 1 and the journey took most of the day.

On 8 June 1966 AS-500F processing and test activities at LC-39A were interrupted because of the approach of Hurricane Alma. The vehicle was rolled back into the VAB as a precaution. Two days later, with the threat of the hurricane past, the AS-500-F vehicle was again moved out to pad 39A. The journey took about eight hours.

Power and control switching tests were performed during July. Planned tanking tests in August were delayed by the failure of the LOX supply system. Subsequent to repairs to the system, S-IVB-F manual LOX and LH2 loading tests were accomplished on 28 September. During the fast fill, loading was stopped at 52% because of a leak in the swing arm #6 umbilical connection. It was determined that it was not necessary to rerun the test as all major objectives had been achieved.

By 12 October 1966 the AS-500F vehicle automatic LOX and LH2 loading was satisfactorily accomplished at LC-39A following an aborted attempt on 8 October.

During the S-IVB-F automatic replenish, a leak in the 18 inch GH2 vehicle vent line on the Mobile Launcher developed into a fire. The fire was extinguished by closing the vehicle LH2 vents and applying a GN2 purge of the vehicle venting. The fire caused no damage to the vehicle or the facility. The leak was caused by a ruptured flex bellows in the line.

The S-IVB-F thrust chamber chill down and the terminal count were not accomplished due to LH2 leak and fire.

The launch vehicle drain was satisfactorily accomplished, the sequence being, LOX drain preparations, simultaneous manual S-IVB LH2 and S-II LH2 drain, simultaneous automatic S-IVB and S-II LOX drain and S-IC LOX drain. At this point, AS-500F–1 wet tests were considered complete.

Following completion of testing at the pad, AS-500F was rolled back to the VAB from LC-39A on 14 October 1966. A minor bearing overheating problem on the Crawler Transporter was encountered during the move.

Beginning on 15 October the de-erection of AS-500F took place in the VAB. On that day the CSM and IU were removed, followed by the S-IVB-F third stage and S-II-F second stage on 16 October, and finally concluding with the S-IC-F first stage on 21 October 1966.

The S-IVB-F was placed in storage at KSC.

In early 1967 the S-IVB-F's APS system module was subjected to inert fuel and oxidizer loadings at LC-39A on the Mobile Service Structure. These two inert loadings were followed by hot fuel loading, which was accomplished with no significant problems.

Douglas were directed to ship the stage to MSFC for use as a mockup during the AS-204 Apollo launch. A dummy J-2 engine was installed and the stage de-erected from Low Bay Cell # 2 and prepared for shipment. Prior to the scheduled ship date, the AS-204 mission was scrubbed and shipment of the S-IVB-F stage was cancelled. The stage was returned to the checkout cell at KSC and placed in storage. Early in 1969 it was, in fact, shipped to MSFC.

On 2 January 1970 MSFC shipped the S-IVB-F stage to the McDonnell Douglas plant at Huntington Beach for modification. The S-IVB stage traveled from MSFC Redstone airfield to Los Alamitos Naval Air Station aboard the Super Guppy aircraft. The stage was to be converted into a Skylab Workshop dynamic test article. On 4 December 1970, after conversion was completed, it was shipped from McDonnell Douglas at Huntington Beach to Michoud, aboard the Point Barrow, together with the S-IVB-512 stage. The S-IVB-F stage was offloaded at MAF and was transported to MSC in Houston, Texas for Skylab Workshop Dynamic testing. The journey was made, leaving MAF on 31 December 1970, onboard the Orion, and arriving at the Clear Lake dock near MSC on 5 January 1971. It was offloaded on 7 January and moved to the MSC acoustic test facility for a series of tests starting on 20 January 1971.

At MSC it underwent a series of tests to verify its bending and vibration characteristics. It was subjected to the lift-off acoustic environment for 15 seconds to qualify the OWS structural design. Acoustic testing was completed on 12 February 1971.

Following the completion of Phase I of the vibro-acoustic test program the S-IVB-F, OWS dynamic test stage, was shipped from MSC on 23 May 1971, onboard the Orion, to MSFC, where it arrived on 4 June 1971 for Skylab Workshop static testing. In June 1974 it was shipped to KSC where it was probably scrapped.

S-IVB-D

Summary

Dynamic test stage used for testing the Saturn IB and Saturn V configurations.

Engines

The engine configuration for dynamic testing of the stage was:

Position 301: J-2006

The engine configuration in the S-IVB-D stage displayed at the Space and Rocket Center, Huntsville is:

Position 301: J-204

Stage manufacture

The S-IVB-D Dynamic Test stage was built by Douglas Aircraft Company at Huntington Beach. During January 1964 DAC completed the LOX tank assembly and delivered it to the Huntington Beach assembly tower. In February the cylindrical tank section was completed. The LOX tank section and the cylindrical section were welded together during March, and the forward LH2 dome was welded in place, completing structural assembly of the tankage.

DAC also assembled the stage's thrust structure during March. In April the stage was moved into the hydrostatic test tower. During testing the LOX tank forward dome wrinkled and had to be repaired and cleaned. Successful proof testing of the stage ended in May.

During August 1964 DAC completed insulation of the S-IVB-D LH2 tank, cleaned the tank, and positioned it in Assembly Tower 2 where bonding of mounting clips in the tunnel area progressed. In August Rocketdyne delivered J-2 engine J-2006 to DAC for the S-IVB-D. The stage was placed in the Vertical Checkout Tower 5 at the end of September. This was for installation of the simulated engine and hookup of the hydraulic system. Checkout of the stage was initiated on 13 October and completed on 28 October. Buckling of the LH2 bulkhead necessitated an additional proof pressure test on 8 November. DAC then painted the stage, weighed it and attached roll rings before loading it on the States Marine ship Aloha State at Seal Beach on 8 December 1964. The Aloha State sailed on 9 December 1964 for New Orleans via the Panama Canal. On 21 December 1964 the stage was transferred to the river barge Promise and shipped up the Mississippi and Tennessee rivers to MSFC, arriving on 4 January 1965.

Stage testing

The S-IVB-D Dynamic Test stage underwent Saturn IB vehicle testing at MSFC during the first five months of 1965. The Saturn IB flight vehicle had four flight configurations that differ enough for each one to require a separate series of dynamic tests.

SA-201, SA-202, SA-204, SA-205 configuration consisted of the launch vehicle plus the Apollo spacecraft without the LM
SA-203 configuration had no spacecraft and consisted of the launch vehicle and a simple nose shroud
SA-206 configuration incorporated a LM and a boilerplate CSM
SA-207 configuration consisted of the launch vehicle and a complete Block II spacecraft

65-17815 The S-IB-D/F stage is hoisted into the Saturn I Dynamic Test Stand at MSFC to await the attachment of the S-IVB-D stage 11.1.65.

The S-IVB-D stage on display at ASRC in Huntsville in 2004

65-17991 The S-IVB-D stage is hoisted into the Saturn IB Dynamic Test Stand at MSFC atop the S-IB-D/F stage 18.1.65.

65-22399 The completely assembled Saturn IB launch vehicle sits in the MSFC Dynamic Test Stand during the second phase of dynamic testing (SA-202 configuration) Note the LM mockup at bottom. 14.4.65.

Dynamic testing was performed in the modified Saturn I Dynamic Test Stand at MSFC. The first stage of the vehicle comprised the Saturn I dynamic test stage (SA-D5) modified to the Saturn IB configuration (S-IB-D/F). The S-IVB-D stage was the upper stage of the vehicle.

Both the S-IB-D/F and the S-IVB-D stages were installed in the Dynamic Test Stand in January 1965. On 8 February the dynamic test instrument unit, S-IU-200D/500D was installed atop the S-IVB-D stage. The SLA (simulating the nosecone) was airlifted in by helicopter and placed on top of the dynamic test vehicle, although this operation was delayed due to damage sustained by the SLA on landing.

The first phase of testing (SA-203 configuration) started on 18 February 1965 and lasted until 2 March. The second phase of testing (SA-202 configuration) included Boilerplate CSM 27 and started on 15 March. On 27 March the S-IB-D/F spider-beam assembly crossbeam web cracked. This failure necessitated repair of the spider-beam and repeat of some of the SA-202 tests. Dynamic testing resumed on 2 April and continued until 19 April. The third phase of testing (covering the SA-207 configuration) ran from the end of April until 12 May. The fourth and final phase of dynamic testing (SA-206 configuration) ended on 27 May 1965, concluding the Saturn IB total vehicle test program. Vehicle disassembly began immediately in preparation for upper stage tests.

Upper stage testing in the Saturn IB configuration (S-IVB-D, IU, Payload) started on 1 August and continued until 11 September 1965. At this time the S-IVB-D stage was re-designated for Saturn V configuration testing and conversion work took place.

The aft skirt for the S-IVB-D Dynamic stage conversion to the Saturn V configuration departed Seal Beach aboard the AKD Point Barrow on 24 September 1965 and arrived at MSFC on 2 November 1965. This completed stage hardware deliveries to MSFC for this stage.

Saturn V configuration III (S-IVB, S-IU, and Payload) testing in the Saturn IB Dynamic Test Stand was successfully accomplished during the period from 15 October 1965 to 6 November 1965.

Due to the unavailability of the S-II-D stage for configuration I and II dynamic testing a decision was made to use the S-IVB-D stage for Super Guppy flight tests at the start of 1966.

Modifications were made on the S-IVB-D, Dynamic vehicle, to bring the stage to flight configuration for the Super Guppy aircraft flight test. Instrumentation was added to the stage to record aircraft environment. The S-IVB-D was loaded on the Super Guppy 20 March 1966, at MSFC. No significant problems were encountered during the test flight to or from Los Alamitos Naval Air Station, near Huntington Beach.

Data from the flight indicated acceptable conditions for transportation of S-IVB flight stages by Super Guppy air transport.

The S-IVB-D stage was prepared at MSFC for Dynamic testing. S-IVB-D Aft Inter-stage stacking began on 23 November 1966 and was completed on 28 November. The S-IVB-D stage stacking operations began on 29 November and were completed on 30 November. The stage was stacked atop the S-IC-D and S-II-F/D stages. The Lower LEM Adapter, LEM and Upper Space LEM Adapter were stacked on 1 December. The CSM and LES were attached on 3 December.

The dynamic test campaign was classified as the AS-500D Configuration I test series. Configuration I was the complete Saturn V vehicle. Testing started in early January 1967 with the roll test being completed on 7 January. However a difference in the hardware configuration compared with the flight vehicle was noticed and it was decided to repeat the test with an improved configuration.

The Configuration I test program included roll testing, completed on 16 January 1967, pitch testing, from 20 to 23 January 1967, yaw testing, completed on 15 February 1967, and longitudinal testing, completed on 26 February 1967. A LOX vent line ruptured during the final longitudinal test. On 6 March MSFC supplied a spare vent line and authorized additional Configuration I tests to verify the Flight Control System.

Testing continued until 11 March 1967 when the final test was performed to verify the flight control system. De-stacking of the AS-500D vehicle began during the second half of March. By 30 March the LES, CSM, IU, S-IVB-D, and S-II-F/D stages had been removed.

Meanwhile the remaining stages i.e. the complete Saturn V minus the first stage were returned to the Dynamic Test Tower for the Configuration II series of dynamic tests that began on 11 May 1967.

Configuration II testing included the yaw test sequence completed on 15 May, the pitch test sequence completed on 2 June, the roll test sequence completed on 10 June and the longitudinal test sequence, started on 13 June and concluded in early July. All programmed testing in the Saturn V Configuration II Dynamic Test Series was completed on 28 July 1967. A one month extension was granted by MSFC to permit rerun of several tests. Following this the stages were separated and the vehicles were removed from the Dynamic Test Tower.

Ground breaking for the Alabama Space and Rocket Center in Huntsville began in July 1968. The following year the S-IVB-D stage was moved to the outdoor exhibit area. On 26 June 1969 it was moved to a position near the Astronautics Lab at MSFC. Two days later, on Saturday 28 June at 0500 CDT, the stage was transported along Rideout Road to the museum. A number of power lines had to be disconnected, road signs taken down and some poles moved. A single R&D J-2 engine, designated J-204, was installed in the stage for display purposes. The museum opened its doors the following year and the stage has been there ever since. On 15 July 1987 the MSFC Saturn V was designated as a National Historic Landmark. Over the following 34 years weathering took its toll and there is now a campaign to restore the complete Huntsville Saturn V and place it under cover.

S-IVB-T

Summary

All-systems stage cancelled at an early phase during manufacturing.

Stage manufacture

The S-IVB-T, All-Systems Test stage was manufactured by DAC at Huntington Beach. In early 1964 DAC was halfway through fitting and bonding of the S-IVB-T common bulkhead. This procedure ended in February, and technicians were ready to join the common bulkhead and the aft dome. By the end of March the LOX tank assembly was complete. In April DAC joined the tank sections to the forward dome assembly and completed manufacture of the forward skirt panels. Hydrostatic proof testing of the tank structure occurred in May. In June the thrust structure and tankage were joined and the stage was moved to the insulation chamber for the start of insulation installation.

During the second half of 1964 the All-Systems stage, S-IVB-T was cancelled and hardware re-allocated to the S-IVB-F and S-IVB-ST vehicles.

S-IVB-501

Summary

First S-IVB-500 flight stage. Only S-IVB-500 stage to be transported from Douglas by sea. First firing attempt aborted. Launched Apollo 4. Reentered within a day.

Engine

The initial and final engine configuration was:

Position 301: J-2031

Stage manufacture

The first flight stage in the Saturn V configuration was in fabrication and assembly at Santa Monica and Huntington Beach during the latter part of 1964. LH2 tank cylindrical sections arrived at Huntington Beach, the LOX tank assembly started at Santa Monica. Design problems in the forward and aft skirt electrical installations delayed the program.

Joining of the propellant tanks, plus tank testing, occurred during the second quarter of 1965. During the third quarter workmen completed insulation installation, assembly of the thrust structure and forward skirt assembly, tank installations, cleaning, J-2 engine buildup, and painting operations.

Joining of the stage structural components started on 1 November and ended on15 November 1965. The stage was placed in the checkout tower 5 on 15 November 1965 and the engine was installed on 16 December.

Stage testing

Post-manufacturing checkout began on 22 November and proceeded slowly due to parts shortages.

On 28 January 1966 abbreviated post-manufacturing checkout of the S-IVB-501 stage ended at Douglas' Huntington Beach facility. Late component parts installation continued until time for shipment to Sacramento. The stage was weighed on 5 March and then prepared for shipment.

The S-IVB-501 stage was shipped from DAC to SACTO on 11 March 1966, arriving on 15 March 1966. The stage was initially transported the short distance to Seal Beach. From there it traveled by ship up the Pacific coast, under the Golden Gate Bridge and up the Sacramento River to Courtland docks in northern California. From there the stage was transferred the short distance to SACTO by road. Acceleration, temperature, relative humidity, stage pitch and roll, and wind velocity and direction were monitored and recorded at selected times during the shipment. It was concluded that this method of transport was acceptable, but would in the future be relegated to the back-up option in favor of direct air transport to SACTO.

Approximately 2,000 manufacturing hours were transferred to Sacramento. On 21 March the stage was installed in the Beta I Test Stand. By 30 March hookup to the facility and ground support equipment had been completed. Power was then applied to the stage. The hydrostatic proof test was performed on 8 April 1966 followed by a leak check.

On 9 May integrated systems checkout of the stage ended at SACTO's Beta I Test Stand in preparation for acceptance firing tests on the stage.

The first attempt at acceptance firing of the S-IVB-501 on 20 May 1966 (countdown 614061) resulted in an automatic cutoff after 50 seconds, as a result of a SIM interrupt. A second attempt was initiated after test stand inspection, but aborted due to a leakage condition of the cold helium crossover pneumatic valve in console B.

The final attempt at static firing the S-IVB-501 stage (countdown 614063) was successfully accomplished on 26 May 1966 at SACTO. The test consisted of a 151 second main-stage first burn, a 106 minute simulated orbital coast period, and a 301 second main-stage burn after restart. All the test objectives were completed.

Also in May the S-IVB-501 APS modules 1 and 2 were checked out and confidence-fired at SACTO in Gamma cell 3. Technicians fired module 1005-1 on 6 May and module 1005-2 on 13 May. Post-firing the modules were disassembled and inspected.

Final propulsion system leak checks were performed on 27 May 1966.

The S-IVB-501 stage was removed from the Beta I Test Stand at SACTO on 3 June following completion of the three acceptance firings. Douglas personnel moved the stage to the Vertical Checkout Laboratory for post-static checkout and modification before shipment to KSC.

The all-systems test began on 21 July, but it was necessary to repeat the test on 27 July, after replacement of the stage switch selector. The stage was removed from the checkout position of the VCL on 29 July. The stage was placed in the horizontal position and modification work and active preparations for stage shipment continued through the end of July.

Post-static checkout was completed on 9 August 1966, following which the stage turnover to NASA meeting was conducted. The stage was weighed on 11 August 1966. The weight was 28,159.18 lbs.

The stage departed SACTO on 12 August 1966, arriving at KSC on 14 August 1966. The stage was transported by Super Guppy aircraft, departing from Mather Air Force Station, but was delayed one day by bad weather on route.

The S-IVB-501 stage ultimately was launched on 9 November 1967 as the third stage for the Apollo 4 Saturn V. In free flight it was given the designation 1967-113B. It had a lifetime of 0.34 days before reentering later on 9 November 1967.

S-IVB-502

Summary

Launched Apollo 6 into Earth orbit. Had a lifetime of 22 days in orbit.

Engine

The initial and final engine configuration was:

Position 301: J-2042

Stage manufacture

DAC began fabrication of the second Saturn V flight stage early in 1965. Assembly of propellant tanks progressed satisfactorily at Santa Monica throughout the second quarter of 1965. By 1 July the LOX tank was complete and undergoing leak and proof testing at Huntington Beach. Tank assembly was completed on 6 August 1965. The stage was in the hydrostatic test tower by the end of September and proof testing was complete on 1 October. At the end of 1965 leak and dye-penetrant checks, tank insulation installation, and tank cleaning were complete.

The engine was installed in the S-IVB-502 stage on 19 February 1966.

On 20 February 1966 the S-IVB-502 assembly ended with joining of the aft skirt, forward skirt, and thrust structure. Douglas workmen then moved the stage into the checkout tower at Huntington Beach.

Stage testing

On 28 February 1966 stage checkout began at Huntington Beach. During April parts shortages hampered checkout activities.

99-03168 Static firing of the S-IVB-502 stage at the Beta I Test Stand at SACTO 28.7.66.

The S-IVB-502 stage at Mather AFB as the Super Guppy lands to pick it up 20.2.67 (Courtesy Phil Broad)

Post-manufacturing checkout of the S-IVB-502 stage was completed on 12 May 1966.

The S-IVB-502 stage was shipped from DAC to SACTO on 1 June 1966, arriving on 2 June 1966. Transport was via the Super Guppy aircraft flying from Los Alamitos Naval Air Station to Mather Air Force Base.

The stage was transported by road from Mather Air Force Base the short distance to the Sacramento Test Operations, where it was installed in the Beta I Test Stand, on 6 June 1966.

Confidence testing of APS Module 1 took place on 19 July 1966 (CD 624360) and that for APS Module 2 on 26 July 1966 (CD 624368).

Pre-static checkout was performed at SACTO, being completed on 21 July 1966 with a simulated static test at the Beta I Test Stand. All subsystem checkouts, including the integrated systems test, were completed on 13 July. The simulated static firing began on 21 July and paved the way for acceptance firing of the stage on 28 July.

At 1420:21 PDT on 28 July 1966 Douglas conducted the acceptance firings of the S-IVB-502 at SACTO. After a smooth test countdown (number 614067), initiated the day before the firing, there was a first burn of 150.7 seconds. Following a 91 minute simulated orbital coast period, the stage restarted at 1553:40 PDT on 28 July 1966, and fired for 291.2 seconds. Stage post-firing activities continued at the test stand throughout July.

On 10 August the stage was removed from the Beta I Test Stand following completion of the acceptance firing program and the stage was placed in the Vertical Checkout Laboratory. Power was applied to the stage on 17 August for post fire checkout.

Post-static checkout activity on the stage ended at SACTO on 12 September 1966, when the all-systems test was performed. The stage turnover acceptance meeting was held at Sacramento on 20 September. After acceptance of the stage it was turned back to the contractor for post-acceptance modification work and storage.

This activity was completed on 30 December 1966. However, the leak check was deferred to KSC.

Final modifications to the S-IVB-502 stage included LOX tank sump screen inspection, low pressure duct reinstallation, and installation of 12 temperature patches in the LH2 tank to provide additional data required for more accurate boil-off predictions. The contractor then deemed the stage "ready for shipment" and shipped it by Super Guppy from Mather Air Force Base to KSC.

The stage departed SACTO on 20 February 1967, arriving at KSC on 21 February 1967. Transportation was via the Super Guppy aircraft flying from Mather Air Force Base, near SACTO, to the KSC Skid Strip.

The S-IVB-502 stage ultimately was launched on 4 April 1968 as the third stage for the Apollo 6 Saturn V. In free flight it was given the designation 1968-25B. It had a lifetime of 22.01 days and reentered on 26 April 1968.

S-IVB-503

Summary

Beta III test stand top beck following the S-IVB-503 explosion 1.67 (Courtesy Phil Broad)

Stage S-IVB-503 was the third in a series of upper flight stages built for the Saturn V rocket and in hindsight would have performed a path-finding role by propelling the first manned capsule out of Earth orbit towards the moon.

In the event, it was not to be for S-IVB-503, when just days before the Apollo 1 fire the stage was completely destroyed in a massive explosion at the Sacramento Test Center just prior to the planned acceptance firing. Investigations would discover that the route cause of the failure was the rupture of a deficient Helium pressurant tank that had been weakened by an incorrect weld.

Beta I at SACTO with the S-IVB-503N stage installed. View looking North. 2.67 (Courtesy Phil Broad)

Two test stands were badly damaged in the explosion. The follow on stage, S-IVB-504, was rapidly brought forward and renamed S-IVB-503N (N for new), and this is the stage which ultimately boosted Apollo 8 to the moon.

Engine

The initial and final engine configuration was:

Position 301: J-2061

Stage manufacturing

Fabrication and assembly of the third production stage began at Santa Monica in the third quarter of 1965. Fabrication of propellant tanks was completed early in 1966.

The stage was assembled at the Douglas Space Systems Center at Huntington Beach, California, from piece-parts manufactured at various subcontractors

Beta III test stand engine deck and J-2 engine following the S-IVB-503 explosion. 1.67 (Courtesy Phil Broad)

The hydrostatic proof test first attempt on 24 February 1966 was aborted when the LOX tank, S/No 1007, aft dome failed the test. Cracks appeared at a pressure of 50psi and the tank was subsequently reworked for later use. A replacement tank assembly, S/No 1007A was tested successfully on 29 and 30 March 1966. The stage tank assembly was then moved to the insulation chamber on 7 April 1966. In May 1966 officials directed that the S-IVB-503 tank assembly be used as the hydrostatic test program unit and that the S-IVB-504 tankage be assigned to S-IVB-503.

Stage testing

Following the completion of the manufacturing the assembled stage underwent a comprehensive checkout or acceptance testing at the Douglas Huntington Beach facility, consuming 44 two-shift working days.

The stage was installed in tower 6 of the Space Systems Center (SSC) Vertical Checkout Laboratory (VCL) on 15 July 1966 and system checkouts were initiated on 21 July 1966.

The all systems test was the final, and all-encompassing, test of the stage prior to shipment to Sacramento for the planned hot fire test. All stage systems were operated under conditions simulating pre-launch, liftoff, powered flight, hydraulic gimballing, restart and coast. The first attempt on 8 September 1966 was terminated due to a facility power failure. The repeat run on 9 September 1966 was completed successfully.

All tests were reviewed and accepted finally on 27 September 1966, clearing the way for the flight to Sacramento. Stage painting was completed on 3 October 1966.

The final inspection on 4 October 1966 revealed 95 mechanical and 120 electrical defects. All were resolved except for three that required failure reports to be written. These covered damage to coax cables, the aft skirt and the exterior of the thrust chamber.

On 4 and 5 October 1966 the stage was prepared for horizontal weighing which was accomplished on 6 October 1966 in the VCL. The horizontal center of gravity was also determined. The measured weight in air was 28,011.1 pounds. Between 5 and 10 October 1966 the stage was purged and dried.

The stage was loaded on to an air carry pallet/transporter and on 11 October 1966 was transported by road the few miles from the Douglas SSC plant to the nearest airport at the Los Alamitos Naval Air Station. Here, the stage and pallet were loaded on to the Supper Guppy aircraft with the aid of a cargo lift trailer.

The Super Guppy flew from Los Alamitos in southern Los Angeles to Mather Air Force Base, just outside Sacramento on 11 October 1966. The stage was offloaded and transported by road the short distance to the Douglas Sacramento Test Center.

Prior to scheduled hot fire testing the stage was installed in the VCL on 12 October 1966 and then transferred to the Beta Complex test stand III on 14 October 1966 where it was subjected to a number of handling and checkout tests.

The countdown activities included sixty separate tasks designed to ensure stage preparation, propellant loading, static firing, residual propellant offloading, and stage securing. The countdown, number 614078, was initiated on 19 January 1967.

Performed on 20 January 1967, the integrated systems test verified the functional readiness of the stage and the facility systems to proceed with countdown operations and acceptance firing.

Propellant loading activities were performed between 0718 PST and 1124:50 PST on 20 January 1967. These comprised,

> LH2 pretest purge,
> LOX loading,
> LH2 loading,
> cold and ambient helium fill,
> LOX and LH2 tank overfill sensor checks,
> flow checks of all cold gas circuits,
> LH2 and LOX umbilicals purge

Task 43, "Static Firing Preparations" was started at 1144:30 PST on 20 January 1967. With a wind reported at 12 to 15 mph from the southeast, an inspection crew proceeded to the test stand for stage and stand inspection. At 1245 PST, all personnel were cleared from the test stand. Static firing preparations program was initiated at 1257 PST, and at 1340 PST Task 44, "Terminal Countdown and Firing", was started.

The computer gave a cutoff indication at 1424:30 PST (T-150 seconds of run 1A, T being the time of simulated Saturn lift off). The cutoff was established as a computer summation error that resulted in a termination of tape readings. To determine the source of the problem, the tape was read on the Beta I tape unit. After successful verification the tape was transferred back to the Beta III computer.

Task 43 was started for a second time at 1531 PST. This run was designated as 1B. Terminal countdown was started at 1555:50 PST.

Following is the sequence of events from T-20 minutes:

1602:10 PST Resume at T-20 minutes 30 seconds
1616:20 PST Begin thrust chamber chill down
1617:20 PST Re-pressurize engine start bottle
1619:50 PST LOX and LH2 tank pre-pressurization
1620:50 PST LOX flow
1622:20 PST Place stage on internal power
1622:40 PST Explosion resulting in destruction of stage S-IVB-503. The explosion occurred when the count had reached the point of simulated lift off minus 11 seconds. T-11 seconds was 522 seconds prior to actual engine start command.

The explosion completely destroyed the Saturn third stage, blasting fragments of the stage in all directions. The Beta III test stand was badly damaged by debris as were surrounding support buildings. The two adjacent test stands also suffered. Test stand Beta II was damaged by flying material and required significant repair work. Beta I test stand was a little further away, but had a flight stage, S-IVB-208, in the stand at the time of the explosion. The stage was carefully inspected for damage but none was found. The conclusion reached was that this stage had not been hit by any debris.

Calculations were made of the pressure wave which would have propagated the 1,860 feet from the explosion to the Beta I stand containing the S-IVB-208 stage. The calculated peak pressure to reach the forward dome of the S-IVB-208 stage (the most critical part) was 0.16 psid. Compared with the design limit of 0.45 psid, a positive safety margin of safety existed. As a double check, an extra leak check was performed and this confirmed the lack of damage.

As for the failure itself, a Douglas investigating team under Jack Bromberg started operations the next morning. NASA convened a failure investigation board three days later at the Douglas Sacramento facility. Board members included Dr Kurt Debus, Chairman; Karl Heimberg, MSFC Test Laboratory Director; and T J Gordon, Douglas Aircraft representative.

The board of inquiry reported on 10 February 1967 that the cause of the failure was the rupture of one of the eight ambient temperature helium storage tanks located on the engine thrust structure. The exploding sphere ruptured the propellant fill lines, allowing LH2 and LOX propellants to mix and ignite, producing the destructive explosion. A weld on the helium tank had been weakened by the use of the wrong weld filler material (pure Titanium instead of a Titanium alloy). This had induced embrittlement in the weld, degrading the capability of the weldment to withstand repeated pressure cycles. As part of the corrective action NASA demanded that the welding records of all existing tanks be inspected for similar process defects. In addition, all future Helium tanks were manufactured by Douglas themselves instead of the previously-used supplier.

Following the loss of the S-IVB-503 stage NASA officials amended identification numbers of subsequent S-IVB stages to fill the vacancy created. The S-IVB-504 became the S-IVB-503N, S-IVB-505 became S-IVB-504N, S-IVB-506 became S-IVB-505N, a replacement stage using an old S-IVB-507 tankage became S-IVB-506N, and S-IVB-507 and subsequent stages retained the old identification.

S-IVB-503N

Summary

Converted from the former S-IVB-504 stage. Super Guppy flight from SACTO to KSC was aborted once. Stage still in orbit.

Engine

The initial and final engine configuration was:

Position 301: J-2071

Stage manufacture

The fourth Saturn V flight stage entered fabrication and assembly operations near the end of 1965. By the end of 1965 the aft common bulkhead was awaiting fit-up of honeycomb insulation and the forward common bulkhead was being prepared for the dollar weld.

The LOX tank assembly was completed at Santa Monica and the tank was shipped to Huntington Beach on 3 March 1966. The LH2 tank forward dome was shipped to Huntington Beach the day before. Because of manufacturing problems the Clevis Probe mount fitting in the forward dome had to be replaced because of a crack in the parent metal around the fitting.

During May 1966 S-IVB project officials directed that the S-IVB-503 tank assembly be used as the hydrostatic test program unit and that the S-IVB-504 tankage be assigned to S-IVB-503 (none of this hardware would become part of the S-IVB-503N stage).

All tankage for the stage had been assembled and hydrostatically tested by mid-July. Leak and dye-penetrant checks were performed and the LH2 tank was moved to the insulation chamber for insulation installation. Installations on the forward and aft skirts began on 25 July. LH2 tank insulation work was completed and the stage was removed from the Insulation Chamber on 29 July. All major sections (aft skirt, forward skirt, thrust structure and tankage) were located in Tower 2 and stage joining operations began in late August. Some delays were experienced due to hardware shortages.

The stage was erected in Tower 6 on 22 September for Megger checks and final installations, prior to J-2 engine installation.

The J-2 engine was installed in the S-IVB-504 stage on 30 September 1966.

Stage testing

Post-manufacturing checkout of the S-IVB-504 stage was completed on 9 December 1966, having started on 4 October.

The stage was moved to Tower 6 for a post checkout modification period prior to the high pressure leak checks.

On 29 December 1966 DAC technicians at Huntington Beach completed high pressure leak checks of the S-IVB-504 stage and moved it to another assembly tower for final installations. Early in January 1967 it was in Checkout Tower 7 undergoing final installations in preparation for shipment to SACTO.

On 18 January 1967 DAC completed preparations for airlifting the S-IVB-504 stage from the Space Systems Center to SACTO. On the following day it was transported to the Los Alamitos Naval Air Station.

The S-IVB-504 stage was shipped from DAC to SACTO on 24 January 1967, arriving on 25 January 1967. Transport was by Super Guppy from Los Alamitos Naval Air Station to Mather Air Force Base. Transportation had been delayed by one week (from 19 January) because of engine problems with the Super Guppy, which delayed its arrival from its base at Santa Barbara. In the meantime the S-IVB-504 stage was temporarily returned to DAC at Huntington Beach.

Just as the stage arrived at SACTO it was renamed the S-IVB-503N stage following the loss of the original S-IVB-503 stage in an explosion on 20 January.

On 27 January the stage was installed in the Beta I Test Stand at SACTO where modification work began in preparation for its acceptance firing.

Pre-static checkout was performed at SACTO, being completed on 20 April 1967. Stage power-on was accomplished on 24 February. The 96 hour Common Bulkhead purge with Argon and the Oxygen-Hydrogen Burner gaseous nitrogen altitude simulated system checkout were conducted on 23 March. Subsystem checkout was completed by the end of March. Simulated static firing was initiated on 19 April 1967.

The first attempt to static fire the S-IVB-503N stage was scrubbed on 26 April because of erratic operations of the propellant utilization (PU) system. Replacement of the PU system began immediately.

The single static firing of the S-IVB-503N stage in the Beta I Test Stand at SACTO took place on 3 May 1967 with a burn duration of 446.9 seconds, with cutoff initiated by the PU processor due to imminent LOX depletion. The Oxygen-Hydrogen burner performance verification test was aborted because of an instrumentation malfunction. However, an Oxygen-Hydrogen burner firing of 230 seconds duration was accomplished on 8 May. A special liquid Hydrogen re-circulation pump fairing purge test was also conducted.

On 12 May the stage was transferred from the Beta I Test Stand to the VCL for post-acceptance firing modifications and checkout. On 14 July the first run of the All Systems Test (AST) was completed in the VCL at SACTO. The stage was placed on a roll dolly in the South Tower for modification work on 27 July. Anti-flutter kit modifications began on 31 July. The stage was placed in the horizontal storage position in the North Bay early in August, where flutter-kit modifications continued. The stage was transferred to the South Tower in September.

Post-storage checkout of the stage began on 25 September and was completed on 21 November 1967 with a successful All Systems Test.

On 29 November the stage was removed from the South Tower of the VCL at SACTO and placed on a roll dolly for modifications prior to shipment to KSC.

On 27 December modification work on the stage at SACTO was stopped and the stage was prepared for shipment to KSC where remaining modifications would take place. A total of 596 hours installation hours were to be transferred to KSC.

The stage departed SACTO on 27 December 1967. Transport was by Super Guppy aircraft from Mather Air Force Base outside SACTO. Shortly after takeoff the plane crew reported loud popping and rushing air noises and returned the plane with its cargo to Mather Air Force Base. Investigations revealed a structural problem with one of the latches on the plane's nose section. Following repairs the aircraft departed Mather for a second time on 29 December, arriving at KSC on 30 December 1967.

The S-IVB-503N stage ultimately was launched on 21 December 1968 as the third stage for the Apollo 8 Saturn V. In free flight it was given the designation 1968-118B. It is still in free flight with an indefinite lifetime in a heliocentric orbit.

S-IVB-504N

Summary

Manufactured from the former S-IVB-505 stage. First stage firing aborted. Still in orbit.

Engine

The initial and final engine configuration was:

Position 301: J-2094

Stage manufacture

During August 1966 workmen at Santa Monica completed the common bulkhead for the S-IVB-505 stage. Welding of the LH2 cylinder to the LOX tank began in Tower 1 in mid-September. Tank assembly and proof testing were completed during September. X-ray and dye-penetrant inspection of the joining tank welds were completed early in October. The tankage was placed in the Insulation Chamber and a special bulkhead drying test was performed.

Between 13 and 15 October workmen at Huntington Beach performed hydrostatic testing of the S-IVB-505 tankage. The 10psi leak check was performed on 20 and 21 October, and the stage was moved to the Insulation Chamber for processing.

67-59473 S-IVB-504N stage being transported from Beta I test stand at SACTO to the VCL for storage. S-IVB-505N installed in the Beta I stand 1.9.67.

68-70037 Loading the S-IVB-504N stage in the Super Guppy at Mather AFB for the flight to KSC 10.9.68.

The S-IVB-504N aft interstage undergoing weight and balance tests 1967 (Courtesy Phil Broad)

Installation of insulation, which had been in progress on the S-IVB-505 tankage throughout November, ended on 5 December 1966 at Huntington Beach.

LH2 installations began with the installation of the Cold Helium Bottle.

68-64834 Removal of the S-IVB-504N stage from the Beta I test stand at SACTO following post-firing checkout 11.3.68.

On 13 January 1967 DAC personnel completed tank installations in the S-IVB-505 stage, and moved the stage to Tower 4 of the Space Systems Center for structural assembly.

On 25 January the S-IVB-505 stage was re-identified as S-IVB-504N following the loss of the S-IVB-503 stage on 20 January 1967.

During January 1967 DAC completed structural assembly of the S-IVB-504N stage. The stage was installed in VCL checkout tower 5 on 8 February 1967. The engine was installed in the S-IVB-504N stage on 9 February 1967, following which stage checkout was initiated.

Stage testing

Post-manufacturing checkout of the S-IVB-504N stage was initiated on 27 February and completed on 12 April 1967 in the Space Systems Center and the stage was moved to the Manufacturing Tower 7 for installation of the dual re-pressurization system on 13 April. On 17 and 18 April, in tower 8, a production acceptance leak check was performed.

On 26 May the stage underwent painting following final inspection completed the day before, in Building 45 at Huntington Beach. On 9 June the stage was weighed as 26,397.8 lbs.

The S-IVB-504N stage was shipped from DAC to SACTO on 16 June 1967. Transportation was by Super Guppy from Los Alamitos Naval Air Station in California to Mather Air Force Base in Sacramento. At SACTO the stage was placed in Tower number 1 of the VCL for pre-firing modifications.

On 7 July the stage was placed in the Beta I Test Stand at SACTO to begin pre-static firing checkout. On 4 August an integrated systems check ended the pre-fire subsystem checkout of the stage.

Pre-static checkout was completed on 9 August 1967 with a successful simulated static firing countdown.

Problems with the Beckman Digital Data Acquisition System (DDAS) on 16 August caused rescheduling of the static firing to 23 August. Countdown had proceeded through LOX and LH2 loading, 68% levels for the O2-H2 burner firing, and Test Stand walk-around.

The first static firing of the S-IVB-504N stage at the Beta I Test Stand at SACTO took place on 23 August 1967 but was terminated after 51.23 seconds. The reason was that there was a fire indication on the J-2 engine. However, post-test investigation revealed that a section of the fire detection wiring set to note "fire" at 170-200 F was improperly touching a fuel turbine inlet duct having estimated external temperatures of 500-600 F.

The two-day countdown went smoothly and included a 204 second O2-H2 burner firing. The last of two static firings took place successfully on 26 August 1967. The firing duration was 438 seconds. Cutoff was initiated by the propellant utilization system at the planned level of 1% LOX. Liquid O2 pre-valve response was slow during critical components check. The valve was subsequently replaced.

McDonnell Douglas completed an abbreviated post-static firing checkout on the S-IVB-504N stage on 31 August 1967 and then transferred the stage from the Beta I Test Stand to the VCL for storage.

Post-storage checkout of the stage began on 25 September 1967 at SACTO.

On 8 January 1968 personnel at SACTO installed the S-IVB-504 stage in the Beta I Test Stand for post-firing checkout, following insulation stripping and re-priming. Post-static checkout of the stage in the Beta I Test Stand was completed on 8 March. The stage was removed from Beta I on 11 March. The stage was then transferred to the VCL Tower 1 for modification work.

Repeat post-static checkout was completed on 16 August 1968.

The stage departed SACTO on 10 September 1968, arriving at KSC on 12 September 1968. Transportation was via the Super Guppy aircraft that departed Mather Air Force Base near SACTO. The journey took two days because of problems with the aircraft.

The S-IVB-504N stage ultimately was launched on 3 March 1969 as the third stage for the Apollo 9 Saturn V. In free flight it was given the designation 1969-18B. It is still in free flight with an indefinite lifetime in a heliocentric orbit.

S-IVB-505N

Summary

Manufactured from the former S-IVB-506 stage. Still in orbit.

Engine

The initial and final engine configuration was:

Position 301: J-2091

Stage manufacture

On 25 January 1967, following the loss of the S-IVB-503 stage five days earlier, the S-IVB-506 stage was re-identified as S-IVB-505N.

Proof, leak and dye-penetrant checks were completed on the S-IVB-505N tankage on 31 January. The stage was then moved to the Space Systems Center Insulation Chamber for insulation and tank installations.

On 23 March 1967 DAC completed internal insulations in the S-IVB-505N stage.

On 14 April the tank installations, cleaning and closing were completed in Tower 4 of the Space Systems Center and the stage was moved to Tower 2 for skirt joining and installation.

The engine was installed in the stage on 4 May. Also in May personnel completed hydraulic hookup, and erected the stage in Tower 5 to begin systems checkout.

Stage testing

Power-on was achieved on 20 May.

Post-manufacturing checkout of the S-IVB-505N stage was completed on 29 June 1967. The stage was moved to Tower 8 for 10psi system leak checks that were completed during the second week of July. The stage was moved to Tower 7 on 25 July where the O2-H2 Burner System was installed. The stage was then painted and prepared for shipment. The stage turnover meeting was held on 15 August.

The S-IVB-505N stage was shipped from DAC to SACTO on 17 August 1967 by Super Guppy aircraft. The aircraft took off from Los Alamitos Naval Air Station near Huntington Beach and landed at Mather Air Force Base near SACTO. The stage was transferred to the South Tower of the VCL the following day, and mechanical and electrical modifications were accomplished.

On 1 September the stage was installed in the Beta I Test Stand at SACTO for pre-firing checkout. During the checkout, the O2-H2 Burner Propellant valves were tagged for replacement. Checkout at the sub-system level was completed with the Integrated Systems Test on 27 September. On 5 October a simulated static firing countdown on the stage was completed.

Pre-static checkout was performed at SACTO, being completed on 9 October 1967. The replacement of a defective hydraulic accumulator necessitated a 24 hour hold in the S-IVB-505N acceptance firing countdown.

67-58386 Mid-shipmen viewing loading of S-IVB-505N aboard Super Guppy at Los Alamitos 17.8.67.

67-58399 S-IVB-505N stage being transported past Administration building at SACTO en-route from Mather AFB 18.8.67.

S-IVB-505N being lowered from the Beta III test stand at SACTO onto the "birdcage" stand. 11.68 (Courtesy Phil Broad)

67-60461 Static firing of the S-IVB-505N stage in the Beta I test stand at SACTO 12.10.67.

69-71996 The S-IVB-505N stage en route from SACTO to Mather AFB for the flight to KSC 2.12.68.

68-61994 The S-IVB-505N stage being removed from the Beta I Test Stand at SACTO 28.11.67.

The S-IVB-505N being loaded onto the Super Guppy at Mather AFB. 2.12.68 (Courtesy Phil Broad)

The single static firing took place on 12 October 1967, in the Beta I stand at SACTO. The firing duration was 448.4 seconds with cutoff initiated by LOX depletion. Propellant utilization cutback time was 152 seconds.

Between 19 and 25 October McDonnell Douglas performed an abbreviated post-firing checkout of the S-IVB-505N stage and declared the stage "ready for storage". The stage would remain in storage until the S-IVB-206 stage was removed from the VCL. On 28 November 1967 the stage was removed from the Beta I Test Stand and placed in storage in the VCL birdcage.

On 24 January 1968 workmen at SACTO removed the S-IVB-505 stage from storage and placed it in the VCL Tower 2 for post-static checkout preparations.

On 19 February McDonnell Douglas workmen at SACTO closed the LH2 tank after completing internal inspection.

On 11 April 1968 post-static checkout of the stage began, eight days earlier than scheduled, following completion of stage modification work at SACTO.

68-69213 The S-IVB-505N stage approaching the Beta III test stand at SACTO for pre-shipment modifications 1.8.68.

Post-static checkout of the stage was completed on 11 June 1968. The stage was then removed from the test stand on 21 June.

On 1 August the stage was removed from storage and transferred to the Beta III Test Stand for pre-shipment modifications. During September these modifications included rework of the switch selector panel, installation of additional instrumentation, configuration upgrading of propulsion system components, and modification of the wiring harness. In addition a special cold helium test was performed in order to try and identify the cause of a helium leak detected on the S-IVB-502 flight vehicle.

Final post-static checkout was completed on 23 November 1968. Stage modification work would continue.

The stage departed SACTO on 2 December 1968, arriving at KSC on 3 December 1968. Transportation was by Super Guppy aircraft from the Mather Air Force Base near SACTO.

The S-IVB-505N stage ultimately was launched on 18 May 1969 as the third stage for the Apollo 10 Saturn V. In free flight it was given the designation 1969-43B. It is still in free flight with an indefinite lifetime in a heliocentric orbit.

S-IVB-506N

Summary

Stage made from former S-IVB-507 tankage. Stage still in orbit.

Engine

The initial and final engine configuration was:

Position 301: J-2101

Stage manufacture

On 25 January 1967, following the loss of the S-IVB-503 stage five days earlier, a replacement stage using an old S-IVB-507 tankage became S-IVB-506N.

Tank cylinder panels were received at the Space Systems Center and assembly of the S-IVB-506N LH2 tank was initiated in the Building 45 trim and weld fixture in February. The LOX tank was completed and shipped to DAC on 21 March. Fabrication of the forward dome was in process at Santa Monica during March 1967. The LH2 tank cylinder assembly was completed on 23 March and was moved to Tower 1 for joining with the LOX tank and forward dome.

Joining of the LOX tank and forward dome was completed on 14 April in Tower 1, followed by proof testing and cleaning operations. Leak checks and dye-penetrant checks were then conducted in Manufacturing Tower 8 on 2 May 1967.

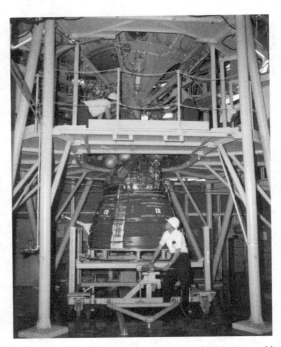

67-58379 J-2 engine being mated to S-IVB-506N in assembly tower 5 at Huntington Beach 21.8.67.

68-62966 The S-IVB-506N stage being installed in the Beta III test stand at SACTO 26.1.68.

The stage was placed in the Insulation Chamber in early May where clip sanding was initiated. Insulation was completed by 26 June and internal installations were initiated.

LOX tank installations began on 24 July. The stage was moved to Tower 2 for joining of the skirts to the tank assembly on 27 July. On 21 August structural assembly of the stage was completed and it was moved from Assembly Tower 2 to Assembly Tower 5 at Huntington Beach to accomplish engine installation. The engine was installed on 21 August.

conducted a simulated static firing of the stage in the Beta III Test Stand. Pre-static checkout was completed on 9 July 1968. A propellant loading test was conducted on 11 July 1968 to prove the operational compatibility of the stage mated to the rebuilt Beta III stand.

68-68410 S-IVB-506N static firing at the Beta III test site at SACTO 17.7.68.

68-68411 J-2 engine close up during the stagtic firing of the S-IVB-506N stage at the Beta III test site at SACTO 17.7.68.

Stage testing

On 6 September the stage entered systems checkout, which was completed by 3 November. Testing had been suspended for a week in the middle of September for rework of a multiplexer test cable.

Post-manufacturing checkout of the S-IVB-506N stage was completed on 3 November 1967. An insulation liner inspection was made the following week, and the stage was placed in Tower 8 for leak checks. Final inspection began in Tower 5 on 15 November 1967.

The S-IVB-506N stage was shipped from DAC to SACTO on 25 January 1968. It was transported by Super Guppy aircraft flying from Los Alamitos Naval Air Station in Los Angeles to Mather Air Force Base in Sacramento. On 26 January the stage was installed in the Beta III Test Stand at SACTO for installation of modifications, pre-static checkout and acceptance firing. Pre-static checkout was initiated on 20 March. Because of problems with the J-2 engine pre-static checkout was halted during April. On 26 April workmen completed rework of the engine's LOX dome.

Between 3 and 5 July McDonnell Douglas personnel

The S-IVB-506N stage being raised into the Beta III test stand at SACTO. 23.11.68 (Courtesy Phil Broad)

The firing countdown was # CD 614109. The single static firing took place at 1823:11 PDT on 17 July 1968. The test consisted of a two-burn oxygen-hydrogen burner firing (of durations 456 and 130 seconds, initiated at 1433:07 PDT) followed by a J-2 engine firing of 445.2 seconds duration. Abbreviated post-firing checkout followed immediately. Results indicated the stage was structurally adequate for the environment in which it was fired. Hairline cracks were found in the Korotherm coating on the forward skirt. Abbreviated post fire checkout on the stand was completed on 24 July.

The stage was removed from the test stand on 1 August 1968 and transported to the VCL for post-firing checks and modifications. When these were completed the stage was placed in storage in the VCL.

On 23 November S-IVB-506N was again installed in the Beta III Test Stand for further modification and checkout. The checkout on Beta III started on 25 November 1968 and ended on 3 January 1969.

On 14 January the stage was weighed at 26,857.2 lbs.

The stage departed SACTO on 17 January 1969, arriving at KSC on 18 January 1969. Transportation was via the Super Guppy aircraft flying from Mather Air Force Base.

The S-IVB-506N stage ultimately was launched on 16 July 1969 as the third stage for the Apollo 11 Saturn V. In free flight it was given the designation 1969-56B. It is still in free flight with an indefinite lifetime in a heliocentric orbit.

S-IVB-507

Summary

Launched Apollo12. Believed to have been sighted in 2002.

Engine

The initial and final engine configuration was:

Position 301: J-2119

Stage manufacturing

In early 1967 fabrication of the LOX tank was in process at Santa Monica. The forward dome was nearing completion by the end of March. The forward dome arrived at Huntington Beach from Santa Monica on 10 May.
On 10 May 1967 the LH2 tank and the LOX tank and forward dome were joined to complete the tank assembly, which was then placed in the Insulation Chamber of the Space Systems Center. The stage propellant tank assembly was subjected to a hydrostatic proof test on 13 and 14 June 1967. The propellant tanks leak check was conducted between 19 and 21 June and the dye-penetrant check performed on 26 June. Insulation and bonding was completed on 15 August 1967.

68-70106 Static firing of the S-IVB-507 stage in the Beta I test stand at SACTO 16.10.68.

The S-IVB-507 stage being removed from the Beta I test stand at SACTO. 30.10.68 (Courtesy Phil Broad)

67-58365 Repair work performed on insulation tiles in the S-IVB-507 LH2 tank 8.67.

67-58373 Assembly area Huntington Beach, left to right, S-IVB-507, 506N, 212 8.67.

On 16 August McDonnell Douglas placed the S-IVB-507 stage in the Assembly Tower 4 at Huntington Beach and began J-2 engine installation. Installation was complete by 11 November. During hardware installation in the LH2 tank a "blister" effect was noted on the liner of the internal insulation. It was determined that this was due to a number of tiles having passed inspection (in April) with an improper mix ratio of polyurethane ingredients. It was decided to replace the liner over the affected areas. Rework was completed on 30 September.

The stage was installed in the SSC VCL tower 6 on 8 November 1967. On 15 November the LH2 and LOX hardware installations in the stage were complete and systems checkout was begun on 20 November 1967.

Stage testing

On 24 January 1968 McDonnell Douglas workmen at Huntington Beach completed systems checkout (all-systems test) of the S-IVB-507 stage.

Post-manufacturing checkout of the S-IVB-507 stage was completed on 8 February 1968. Final inspection was completed on 28 February and Korotherm insulation rework and painting began. The stage was removed from the tower and placed in storage on 5 April 1968. Final inspection was re-performed between 30 July and 6 August 1968. 356 defects were noted at this time. On 2 August the weight and balance measurements were performed. The stage weight was 26,934.6 lbs.

The S-IVB-507 stage was shipped from DAC to SACTO on 7 August 1968. Transportation was by Super Guppy aircraft from Los Alamitos Naval Air Station near Huntington Beach to Mather Air Force Base near SACTO. On 9 August the stage was installed in the Beta I Test Stand in preparation for acceptance testing of the stage. Power was applied for the first time on 13 August 1968. Integrated systems checkout began on 23 September and ended on 15 October.

Countdown, number 614113, was initiated on 15 October 1968. The single static firing took place at 1607:57 PST on 16 October 1968 with a duration of 433.2 seconds. The acceptance firing was preceded by a simulated orbital coast, a cold helium leak check, two O2-H2 burner firings (of durations 455 and 130 seconds performed at 1255:43 PST), and an ambient re-pressurization test. All test objectives were achieved. The abbreviated post fire checkout was initiated on 17 October. The propulsion system leak check was conducted from 17 to 29 October.

The stage was moved from the Beta I Test Stand at SACTO on 30 October 1968 and transferred to the VCL for post-static modifications and checkout.

On 10 January 1969 an all-systems test (AST) was completed in the VCL. A further AST was completed in the South Tower at SACTO on 10 February.

Post-static checkout was completed on 10 February 1969 and preparations began for final tank purging. Subsequent to this the stage was stored temporarily before being removed on 25 February in preparation for shipment to KSC. A final weight and balance check was performed on 25 and 26 February 1969.

The stage departed SACTO on 6 March 1969, arriving at KSC on 10 March 1969. Transportation was from the Super Guppy aircraft flying out of Mather Air Force Base.

The S-IVB-507 stage ultimately was launched on 14 November 1969 as the third stage for the Apollo 12 Saturn V. In free flight it was given the designation 1969-99B. It is still in free flight with an indefinite lifetime in an initial orbit of 163,100 km by 861,800 km and a period of 60,480 minutes.

On 12 September 2002 it was reported that the S-IVB-

507 stage may have been sighted. Analysis of the orbital motion of the newly discovered object, J002E3, indicated that it had the right properties to be the old Apollo 12 third stage.

The new object was discovered on 3 September 2002 by Bill Yeung. It had a magnitude of 16.5, and orbital analysis indicated that the object was captured by the Earth from heliocentric orbit in April 2002. It was believed that the stage entered heliocentric orbit before being captured again by the Earth in 2002. The brightness of the object was consistent with that of the S-IVB stage, and investigative work was performed to determine the exact properties of the paint used on the S-IVB stage.

S-IVB-508

Summary

Launched Apollo 13. First S-IVB stage to be purposely crashed on the moon.

Engine

The initial and final engine configuration was:

Position 301: J-2122

Stage manufacture

Fabrication work on the S-IVB-508 LOX tank was in process during March 1967. The forward dome was completed and shipped to Huntington Beach during April for assembly. Assembly of the LH2 tank cylinder was completed at Huntington Beach by the end of June 1967.

In July 1967 the LH2 cylinder and the LOX tank were joined in Tower 1 at Huntington Beach. Forward dome-to-LH2 tank welding operations began the first week in August. Welding was accomplished and the tank assembly was placed in the Hydrostatic Test Tank in Tower 8 for leak checks. Hydrostatic testing was performed on 8 and 9 August 1967. The tanks were leak checked on 16 and 17 August.

The S-IVB-508 stage was then moved to the Insulation Chamber for internal insulation installations. The same blister effect on the tank internal insulation lining, as noted on the S-IVB-507 stage, was observed on this stage. Rework of 26 tiles was required.

On 15 October 1967 McDonnell Douglas began LH2 tank installations in the S-IVB-508 stage.

On 15 November 1967 LOX tank hardware installations were completed and stage joining operations were started in Tower 2. Stage joining was completed in mid-December 1967 and electrical wiring and plumbing installations were begun.

The stage was installed in the SSC VCL tower 5 on 22 January 1968.

The single J-2 engine was installed in the S-IVB-508 stage on 23 January 1968.

Stage testing

Post-manufacturing checkout of the S-IVB-508 stage, comprising 38 tests, was started on 30 January 1968 and completed on 29 May 1968. Power was first applied on 16 February 1968. The stage was removed from the VCL on 26 June 1968.

69-72068 Loading the S-IVB-508 stage onto the Super Guppy at Los Alamitos Naval Air Station for the flight to SACTO 30.12.68.

69-73064 Aerial view of the static firing of the S-IVB-508 stage at the Beta III test stand at SACTO 20.2.69.

On 8 August the stage was removed from a post-manufacturing checkout tower and towed to the manufacturing building for painting.

Following S-IVB-508 stage painting and installation of the LH2 probe on 10 September, the stage was moved from the Vertical Assembly and Checkout Building to Tower 7 at Huntington Beach for modifications. The stage was weighed on 18 December. The weight was 29,960.9 lbs. Final stage inspection was completed on 19 December 1968. A total of 127 defects were noted at this time.

The S-IVB-508 stage was shipped from DAC to SACTO on 30 December 1968. Transportation was by Super Guppy aircraft from Los Alamitos Naval Air Station in Southern Los Angeles to Mather Air Force Base in Sacramento.

The stage was installed in the Beta III test stand on 3 January 1969.

Pre-static checkout was performed at SACTO, beginning on 10 January 1969 and being completed on 15 February 1969. The integrated systems test had been completed on 8 February.

The firing countdown, # CD 614116, was started at 0800 PST on 18 February 1969. Computer problems forced a 24-hour hold in the count. Prior to the main stage firing a single O2-H2 burner firing of 460 seconds was initiated at 1118:19.000 PST on 20 February 1969.

The single static firing took place in the Beta III Test Stand at 1422:15.695 PST on 20 February 1969. All test objectives were achieved in the test that included an oxygen-hydrogen burner firing and a 457.0 second main-stage duration J-2 engine firing.

Post-static checkout was completed on 13 March 1969. On 25 March the post-firing inspection of the stage was completed and it was transferred from the Beta III Test Stand to the VCL at SACTO for LH2 tank inspection and temporary storage.

On 15 May McDonnell Douglas shipped the S-IVB-508 Aft Inter-stage aboard the S.S. Steel Seafarer to KSC, where it arrived on 28 May.

The stage departed SACTO on 12 June 1969, arriving at KSC on 13 June 1969. Transportation was via the Super Guppy aircraft that departed from Mather Air Force Base.

The S-IVB-508 stage ultimately was launched on 11 April 1970 as the third stage for the Apollo 13 Saturn V. In free flight it was given the designation 1970-29B. It had a flight lasting 3.25 days before crashing on the moon on 15 April 1970.

S-IVB-509

Summary

Launched Apollo 14. Crashed on the moon.

Engine

The initial and final engine configuration was:

Position 301: J-2124

Stage manufacturing

Fabrication of the LOX tank was in process at Santa Monica at the end of June 1967. Welding of segments of the LH2 tank cylinder was underway at the Space Systems Center.

A "growler" check for de-bonding of the honeycomb insulation of the Aft Common Dome of the S-IVB-509 stage was completed the first week of July 1967, and the Forward Dome/Aft Dome assembly was prepared for welding. Welding of the LH2 tank cylinder segments was also completed by the end of July. An out-of-contour area on the Aft LOX Dome was determined to be too serious to be reworked. The Aft Dome of the S-IVB-510 stage was selected as a replacement.

The LH2 Forward Dome was shipped to Huntington Beach 8 August, followed by the LOX tank assembly on 30 August.

69-73476 The S-IVB-509 stage being prepared for weight and balance at Huntington Beach Late 3.69.

69-75855 Static firing of the S-IVB-509 stage in the Beta III test stand at SACTO 14.5.69.

S-IVB-509 in the "Pandjuris" welding rig at Huntington Beach. 1967 (Courtesy Phil Broad)

During September 1967 structural assembly of the S-IVB-509 tank assembly was completed in Tower 1 at Huntington Beach. The hydrostatic proof test was performed on 30 September and 1 October 1967. The stage was then moved to Tower 8 for leak checks.

On 8 October 1967 leak checks on the tank assembly were completed and the stage was moved to the Insulation Chamber for insulation installations that were completed on 11 December 1967. In late December 1967 the stage was moved to Tower 4 for cleaning.

On 5 January 1968 McDonnell Douglas personnel at Huntington Beach completed liquid hydrogen (LH2) tank cleaning operations as well as insulation inspection of the stage. Only minor problems were found. LOX tank cleaning and inspection was completed on 15 February.

The J-2 engine was installed in the S-IVB-509 stage on 14 March 1968, after which the stage was installed in the SSC VCL tower 6 on 15 April.

Stage testing

Post-manufacturing checkout of the S-IVB-509 stage, involving 38 tests, was started on 23 April and completed on 18 September 1968.

On 8 October 1968 workers at Huntington Beach placed the S-IVB-509 stage on the paint dolly and moved it to the paint booth in the manufacturing building.

On 6 November 1968 the stage was placed into tower 8 for pre-shipment retention.

On 19 February 1969 the structural assembly activity and painting of the S-IVB-509 stage was completed at Huntington Beach. On 15 March McDonnell Douglas completed modifications to the stage and prepared it for shipment to Sacramento.

The S-IVB-509 stage was shipped from Huntington Beach to SACTO on 31 March 1969. Transportation was via the Super Guppy aircraft flying from Los Alamitos Naval Air Station to Mather Air Force Base near SACTO. On 1 April the stage was installed in the Beta III Test Stand. Pre-static checkout commenced the following day and was completed on 10 May 1969.

Countdown # CD 614120 was initiated at 0830 PDT on 13 May 1969. Prior to the firing the O2-H2 burner was fired at 1134:53.445 PDT on 14 May 1969 for 460 seconds. The LH2 tank was re-pressurized 177.3 seconds after burner start and the LOX tank re-pressurization was terminated approximately 0.2 seconds later.

The single static firing took place in the Beta III Test Stand at 1409:18.981 PDT on 14 May 1969. All test objectives were met in the 452.4 second main-stage firing.

Post-static checkout operations in the Beta III Test Stand were completed on 29 May 1969 with an AST being performed at the end of the sequence. On 4 June the stage was transferred from the Beta III Test Stand to the VCL for completion of post-static checkout activities. These were accomplished on 19 June and the stage was placed in storage awaiting shipment to KSC.

On 8 September 1969 McDonnell Douglas personnel completed shipping preparations for the S-IVB-509 stage and placed the stage in temporary horizontal storage in the VCL. On 8 October it was moved to position D-2 in the VCL for completion of final modifications. These were completed on 11 December 1969 and technicians began repainting the stage in preparation for shipment to KSC.

The stage departed SACTO on 17 January 1970, arriving at KSC on 20 January 1970. Transportation was with the Super Guppy aircraft that left Mather Air Force Base for the journey to KSC skid strip.

The S-IVB-509 stage ultimately was launched on 31 January 1971 as the third stage for the Apollo 14 Saturn V. In free flight it was given the designation 1971-08B. It had a flight lasting 3.44 days before crashing on the moon on 4 February 1971.

69-00082 The S-IVB-510 stage being loaded aboard the Super Guppy at Los Alamitos Naval Air Station for the flight to Mather AFB 19.6.69.

S-IVB-510

Summary

Launched Apollo 15. Crashed on the moon.

Engine

The initial and final engine configuration was:

Position 301: J-2079

Stage manufacture

The S-IVB-510 Common Bulkhead welding was accomplished at Santa Monica during July 1967. Welding of the LH2 Tank Cylinder quarter panels was completed at Huntington Beach at the same time.

Fabrication of the LH2 Forward Dome at Santa Monica was accomplished, and the assembly was shipped to Huntington Beach on 19 September.

On 3 October 1967 McDonnell Douglas completed fabrication of the S-IVB-510 stage LOX tank assembly at Santa Monica using the S-IVB-511 aft LOX dome as a replacement for the 510 dome reassigned to the S-IVB-509 stage. The contractor shipped the completed assembly to Huntington Beach to be joined with the LH2 tank assembly. Tank assembly was completed in mid-November and was moved to Tower 3 for proof testing. Hydrostatic proof testing took place on 28 and 29 November 1967.

69-01857 Static firing of the S-IVB-510 stage in the Beta III test stand at SACTO 14.8.69.

The S-IVB-510 stage being removed from tower 5 in the VA&C building at Huntington Beach. 1968 (Courtesy Phil Broad)

On 6 December 1967 McDonnell Douglas at Huntington Beach completed leak checks on the S-IVB-510 LOX and LH2 tank assembly and moved the stage to the insulation chamber for insulation installations.

On 14 March 1968 installation of the LH2 tank internal insulation, which had started on 7 February, was completed at Huntington Beach without significant problems. Final cleaning and closing of the LH2 tank was accomplished on 25 April. On 15 May workers began joining the various sections of the S-IVB-510 stage, a task that was completed during July.

The single J-2 engine was installed in the S-IVB-510 stage on 18 July 1968.

On 9 August the stage was moved from Tower 2 to Tower 5 in the Vehicle Assembly and Checkout Building at Huntington Beach in preparation for engine electrical and hydraulic installations.

Stage testing

Post-manufacturing checkout of the S-IVB-510 stage was initiated by McDonnell Douglas on 6 September 1968 and completed on 14 December 1968.

Following the VCL final acceptance checkout the stage was moved to tower 8 for a production acceptance leak test, which was accomplished between 30 December 1968 and 7 January 1969. Final inspection was performed between 10 and 14 January 1969. 139 defects were noted that were corrected by 10 February.

On 1 June post-manufacturing modifications were completed and preparations for shipment to SACTO were begun. The stage was weighed on 14 June 1969. The weight was 26,920.67 lbs.

The S-IVB-510 stage was shipped from Huntington Beach to SACTO on 19 June 1969. Transportation was via the Super Guppy aircraft flying from Los Alamitos Naval Air Station to Mather Air Force Base. Upon arrival at SACTO the stage was placed in the Beta III Test Stand.

Pre-static checkout was performed at SACTO, being completed on 31 July 1969. These checks included a power-on test sequence performed on 23 June.

The single static firing took place on 14 August 1969. The firing lasted 448.7 seconds. During the firing the fuel tank pressurization control module failed to operate satisfactorily. Failure analysis indicated the manufacturer had failed to process a static seal to specification and that this had generated the abnormal condition.

Post-static checkout was completed on 18 September 1969. The stage was then moved from the Beta III Test Stand to the VCL for additional inspection and temporary storage.

The stage departed SACTO on 11 June 1970, arriving at KSC on 12 June 1970. Transportation was via the Super Guppy aircraft flying from Mather Air Force Base in Sacramento.

The S-IVB-510 stage ultimately was launched on 26 July 1971 as the third stage for the Apollo 15 Saturn V. In free flight it was given the designation 1971-63B. It had a flight lasting 3.30 days before crashing on the moon on 29 July 1971.

S-IVB-511

Summary

Final S-IVB-500 stage to be static fired. Crashed on the moon.

Engine

The initial and final engine configuration was:

Position 301: J-2134

Stage manufacture

The aft LOX tank dome that was originally manufactured for the S-IVB-511 stage was transferred to the S-IVB-510 stage as a replacement on 3 October 1967.

Initial work on assembly of the S-IVB-511 stage at Huntington Beach began during January 1968.

On 29 February McDonnell Douglas completed washing and degreasing the tankage for the S-IVB-511 following completion of hydraulic proof testing on 27 February.

As a result of AS-502 vehicle flight anomalies, program officials diverted the S-IVB-511 forward skirt from the Huntington Beach assembly plant to the Manned Spacecraft Center, during June, to support longitudinal vibration tests.

On 11 July McDonnell Douglas completed and shipped from MSFC to Los Alamitos Naval Air Station the replacement forward skirt for S-IVB-511, which was designated –511A. The S-IVB-511A replacement skirt was airlifted via the Super Guppy. The replacement had been used to support the "short stack" POGO test at Wyle Laboratories, Huntsville.

68-64831 The S-IVB-511 tank assembly entering insulation chamber at Huntington Beach 3.68.

69-02593 The S-IVB-511 stage leaving McDonnell Douglas at Huntington Beach for Los Alamitos for the flight to Mather AFB 16.9.69.

Joining of the thrust structure and the aft skirt of the S-IVB-511 to the propellant tank assembly began at Huntington Beach on 26 August.

The single J-2 engine was installed in the S-IVB-511 stage on 11 November 1968.

Stage testing

Post-manufacturing checkout of the S-IVB-511 stage was completed on 27 February 1969.

S-IVB-511 stage modifications were completed on 24 April in Tower 6 at Huntington Beach. Personnel then placed the stage on the paint dolly. Painting operations were completed on 20 May. The stage was then moved to Tower 8 on 28 May for final modification and storage prior to shipment to SACTO. On 2 June the stage was moved to Tower 2 at Huntington Beach for additional modification by McDonnell Douglas personnel.

On 5 August the stage was moved from Tower 2 to Tower 6 to complete minor electrical and mechanical modifications. These were done by 21 August and the stage was prepared for shipment to SACTO. On 5 September the stage was repainted, an activity made necessary due to the discovery of peeling paint on the S-IVB-506N stage at KSC and a subsequent check of all S-IVB stages. All modifications to the stage were completed by 11 September.

The S-IVB-511 stage was shipped from Huntington Beach to SACTO on 16 September 1969. Transportation was via the Super Guppy aircraft that flew from Los Alamitos Naval Air Station to Mather Air Force Base. On 17 September the stage was installed in the Beta III Test Stand at SACTO and pre-static checkout was started. Electrical power turn on was initiated on 25 September and was followed by the initiation of mechanical, electrical and propulsion tests.

Pre-static checkout was completed on 25 November 1969.

The single static firing took place on 18 December 1969 in the Beta III Test Stand at SACTO. The full duration firing was 442.8 seconds long. The S-IVB-511 stage was the last stage to undergo a static acceptance firing. For cost reasons the final three stages that had engines, S-IVB-512, S-IVB-513 and S-IVB-514 were not hot fire tested at stage-level. One of these stages, S-IVB-512, became the only Saturn V stage to launch a manned crew and yet not have achieved a static firing.

Post-static checkout was completed on 28 January 1970.

The stage departed SACTO on 29 June 1970, arriving at KSC on 1 July 1970. Transport was via Super Guppy aircraft flying from Mather Air Force Base to the KSC Skid Strip.

The S-IVB-511 stage ultimately was launched on 16 April 1972 as the third stage for the Apollo 16 Saturn V. In free flight it was given the designation 1972-31B. It had a flight lasting 3.13 days before crashing on the moon on 19 April 1972.

S-IVB-512

Summary

The only Saturn V stage to be launched, never having been subjected to a stage static firing. Crashed on the moon.

Engine

The initial and final engine configuration was:

Position 301: J-2137

Stage manufacturing

McDonnell Douglas began fabricating components for the S-IVB-512 stage during February 1968. Hydrostatic testing of the LH2 tank was accomplished on 10 May, followed by X-ray and dye-penetrant inspection checks that were completed on 31 May.

On 23 September S-IVB-512 LH2 tank insulation installation was completed at Huntington Beach. This procedure had been in progress since 10 July.

On 2 October 1968 workmen at Huntington Beach placed the S-IVB-512 LH2 tank assembly in Tower 4 of the Vehicle Assembly and Checkout Building for washing, degreasing, and installation of fuel ducts.

On 15 January 1969 McDonnell Douglas completed structural fabrication and assembly of the S-IVB-512 stage at Huntington Beach.

The J-2 engine was installed in the S-IVB-512 stage on 17 February 1969. However, on 1 May, stage systems checkout was temporarily suspended to allow for J-2 engine modification.

Stage testing

Post-manufacturing checkout of the S-IVB-512 stage was completed on 9 July 1969. On 24 July the stage was moved to Tower 8 for post-checkout modifications. On 17 September the stage was moved from Tower 8 to Tower 6 for final installation of the tunnel cover, impingement curtain fit check, and temporary storage.

The stage was removed from temporary storage on 14 October 1969 in preparation for final inspection and painting. On 6 November painting was complete and the stage was again placed in storage.

No stage static firings were performed. Instead an expanded all systems test was completed on 29 September 1970.

69-03476 The S-IVB-512 stage in the Vehicle Checkout Assembly Building at Huntington Beach 11.69.

70-04827 The S-IVB-512 stage en route to the Manufacturing Building at Huntington Beach 1.70.

The stage departed Huntington Beach on 4 December 1970, sailing from Seal Beach Naval docks together with the S-IVB-F stage, aboard the Point Barrow. They arrived at Michoud on 18 December 1970, where the S-IVB-F stage was unloaded. At Michoud the S-II-15 stage was loaded onboard the Point Barrow. The S-IVB-512 stage carried on to KSC, aboard the Point Barrow, arriving on 21 December 1970. This was the first S-IVB-500 series stage to be shipped directly from Huntington Beach to KSC and the first to be transported to KSC by sea.

The S-IVB-512 stage ultimately was launched on 7 December 1972 as the third stage for the Apollo 17 Saturn V. In free flight it was given the designation 1972-96B. It had a flight lasting 3.63 days before crashing on the moon on 10 December 1972.

S-IVB-513

Summary

No stage static firings performed. Stage on display at JSC.

Engine

The initial and final engine configuration was:

Position 301: J-2140

Stage manufacture

The S-IVB-513 H2 tank cylinder fabrication and inspection ended on 18 June 1968 at the McDonnell Douglas Huntington Beach facility.

On 27 September the S-IVB-513 cylindrical tank section was hoisted into an assembly tower for joining to the forward dome and LOX tank assembly. The propellant tank assembly of the S-IVB-513 stage was moved from the Vehicle Assembly and Checkout Building to Tower 8 for gas leak checks on 12 November 1968.

On 21 February 1969 the stage was moved into the insulation chamber at McDonnell Douglas' Huntington Beach facility and LH2 tank installation was started. By 6 April the installation was complete and the stage was moved to Tower 2 for tank joining operations. This operation was completed on 20 April and the stage was moved to Tower 6 for modifications.

The J-2 engine was installed in the S-IVB-513 stage on 22 May 1969 and all vertical installation tasks were completed by 19 June.

Stage testing

On 4 August modifications to the stage in Tower 6 were completed and the stage was moved to Tower 5 for systems checkout. On 26 November the stage was moved to Tower 8 for x-ray and leak check.

Post-manufacturing checkout of the S-IVB-513 stage was completed on 4 December 1969.

No stage static firings were performed. Instead an expanded all systems test was completed on 8 December 1970.

The stage was transported by truck the short distance to NAR at Seal Beach on 18 January 1971 for long-term storage. It was returned to Huntington Beach by truck on 12 November 1971.

70-05608 S-IVB-513 stage en route to the Manufacturing Building at Huntington Beach for painting. In the foreground the S-IVB-515 tank assembly 2.70.

70-06056 Top left S-IVB-513 stage and centre S-IVB-515 propellant tank assembly. In the foreground S-IVB-212 forward skirt, aft skirt, and thrust structure at Huntington Beach 3.70.

The S-IVB-513 stage prior to raising into tower 8 of the VA&C building at Huntington Beach. 1969 (Courtesy Phil Broad)

The stage departed Huntington Beach on 25 May 1972, arriving at KSC on 26 May 1972. Transportation was by Super Guppy aircraft flying from Los Alamitos Naval Air Station to the KSC Skid Strip. The stage was then placed in long-term storage at KSC on 9 June 1972.

In 1977 the stage was moved from KSC to the Johnson Space Center (JSC) in Houston, Texas. S-IVB-513 left KSC on board a barge in late 1977 and arrived at JSC a few days later after traveling through the inter-coastal waterways. The stage has been on display to the public at JSC ever since. The ownership of the stage was transferred to the Smithsonian museum in 1978. In 2004 plans were announced to refurbish the JSC Saturn V stages (including S-IVB-513) and enclose them in a purpose built building.

S-IVB-514

Summary

No stage firings performed. Stage on display at KSC.

Engine

The initial and final engine configuration was:

Position 301: J-2143

Stage manufacture

During October 1968 McDonnell Douglas technicians at Huntington Beach finished the welding of the skin segments to form the cylindrical tank assembly of the S-IVB-514 stage.

Hydrostatic testing of the LH2 forward dome-to-LH2 cylinder weld of the S-IVB-514 stage was completed on 12 February 1969. On 13 May the stage was moved into the insulation chamber at Huntington Beach to begin installation of insulation. This activity was completed on 12 June. The stage was positioned horizontally for LH2 installations on 25 June. The installations were in place by 9 July and the stage was moved to Tower 4 for LOX tank cleaning. This was completed on 28 August and the stage was moved to Tower 2 for tank and skirt joining and installations.

On 18 September the stage was removed from temporary storage and positioned back in Tower 2 for operations leading to J-2 engine installation and checkout.

Installation of the thrust structure was achieved on 20 November 1969.

The J-2 engine was installed in the S-IVB-514 stage on 28 January 1970.

Stage testing

During March 1970 the stage was removed from Tower # 5, positioned on an A-frame in the Checkout Building and then lowered into Tower # 7. In April 1970 the stage was back in Tower # 5 of the vehicle assembly and checkout building.

Post-manufacturing checkout of the S-IVB-514 stage was completed on 20 July 1970.

No stage static firings were performed.

The stage was transferred by truck to NAR at Seal Beach for long-term storage on 28 December 1970. It was returned by truck to Huntington Beach on 20 December 1971. The stage was placed in long-term storage on 6 December 1972. It was removed from storage on 7 March 1973. Finally, the stage departed Huntington Beach on 27 March 1973, arriving at KSC on 28 March 1973. It was transported by Super Guppy aircraft flying from Los Alamitos Naval Air Station to the KSC Skid Strip. The stage was placed in KSC long-term storage on 4 April 1973.

In April 1976 it was transferred outside the VAB to form a Saturn V display. It was on display in an unmated horizontal configuration together with other Saturn V stages. In 1979 the title was turned over to the Smithsonian. The stage remained there until 1996, when on 23 April it was transferred 2 miles to its new home. It was refurbished, and located in the new indoor Saturn V center at KSC that opened its doors to the public on 5 December 1996.

S-IVB-515

Summary

Final S-IVB stage manufactured. Converted and used as the Skylab back-up OWS and now on display at the NASM, Washington.

Engine

No engine was ever installed in this stage.

Stage manufacturing

On 19 March 1969 the S-IVB-515 LH2 tank cylinder was placed in Tower 1 at Huntington Beach and preparations for welding began. By 31 March fabrication of the stage had progressed to welding of the aft dome to the tank cylinder at McDonnell Douglas' Huntington Beach facility.

McDonnell Douglas completed the S-IVB-515 tank assembly operation on 22 May at Huntington Beach.

69-03409 The S-IVB-515 and 212 stages (back up and flight Skylab OWSs) in isolation chambers at Huntington Beach 10.69.

The S-IVB-515 stage (Skylab OWS back-up) on display at NASM in Washington DC in 1977

On 12 June hydrostatic pressure tests of the S-IVB-515 stage were completed. On 30 June cleaning, leak, and dye-penetrant checks of the LH2 tank were completed and the stage was moved to the insulation chamber number 2 for installation of insulation.

On 7 November 1969 McDonnell Douglas personnel at Huntington Beach completed installation of LH2 tank insulation in the S-IVB-515 stage.

In January 1970 the stage was removed from Tower # 4. By April 1970 the stage was in storage in the manufacturing building.

Prior to installation of a J-2 engine on the stage, the stage was allocated as the back-up Skylab vehicle and final construction was to the Skylab configuration.

The stage was transferred from Huntington Beach to KSC, arriving on 25 May 1972.

Following the successful launch and operation of the primary Skylab vehicle (S-IVB-212) the S-IVB-515 stage was provided to the Smithsonian Institute.

In July 1976 the Smithsonian Air and Space museum in Washington DC opened its doors. The S-IVB-515 stage has been displayed in a Skylab configuration ever since. Modifications for display include a door allowing visitor entry to the interior of the stage.

69-75006 The S-IVB-515 tank being lowered onto the forward dome for mating at Huntington Beach 4.69.

S-IVB-212 (Skylab Orbital Workshop)

Summary

S-IVB-200 stage (final one) converted for use and flown as Skylab OWS. Originally built with a J-2 engine but never test fired at stage level. Engine removed as part of the conversion. Re-entered the atmosphere in 1979.

Engines

The initial engine configuration was:

Position 301: J-2103

Stage manufacture

During the first quarter of 1967 internal insulation installation and clip bonding of the S-IVB-212 stage tankage was in process. The stage propellant tank hydrostatic proof testing was performed successfully on 8 and 9 March 1967. The propellant tank leak check was conducted between 14 and 17 March. Aft skirt and forward skirt assembly was in progress at the end of March 1967.

Stage testing

The stage was installed in the SSC VCL tower 6 on 29 June 1967. On 25 July 1967 the S-IVB-212 stage entered systems checkout at Huntington Beach. Power was first applied to the stage on 1 August 1967.

69-71990 The S-IVB-212 stage being hoisted into Tower # 8 at Huntington Beach during storage 12.68.

69-74993 The J-2 engine being disconnected from the thrust structure of the S-IVB-212 stage at Huntington Beach 4.69.

68-66302 Left to right, S-IVB-211, 212, 210, 505 aft inter stage in background. At right 510 forward skirt, aft skirt and thrust structure 4.68.

69-75845 S-IVB-212 stage being dissassembled at Huntington Beach 5.69.

70-06083 S-IVB-212 (Skylab OWS) being removed from the insulation chamber # 1 at Huntington Beach 3.70.

On 14 September 1967 McDonnell Douglas concluded the All Systems Test that ended systems checkout of the S-IVB-212 stage. This test demonstrated the combined operation of the stage electrical, hydraulic, propulsion, instrumentation, and telemetry systems under simulated flight conditions. A total of 33 checkout procedures involving the stage systems were accomplished during this period. The stage was removed from the VCL on 18 September 1967.

Following the VCL final acceptance checkout, the stage was moved to tower 8 for a production acceptance leak test of the propellant tanks assembly. This was accomplished between 25 and 27 September 1967.

On 18 October 1967 McDonnell Douglas completed painting and final inspection of the S-IVB-212 stage and prepared it for storage at the Space Systems Center at Huntington Beach. The stage remained in storage at Huntington Beach from 3 November 1967 to March 1969.

On 26 March 1969 McDonnell Douglas shipped the inter-stage for S-IVB-212 to MAF.

In April 1969 S-IVB-212 was moved from the vertical checkout tower # 8 to tower # 6, where the J-2 engine was disconnected as part of the refurbishment of the stage for use as the Skylab workshop. In May 1969 the forward skirt was removed from the stage. In June 1969 the stage was removed from the insulation chamber and transported to the vertical assembly and checkout building.

During November 1969 S-IVB components were removed from the thrust structure aft skirt. McDonnell Douglas completed the conversion of the S-IVB-212 stage to the Skylab Orbital Workshop in the summer of 1972.

The Skylab Orbital Workshop (former S-IVB-212 stage), together with the payload shroud, spent two weeks traveling on the AKD Point Barrow from Seal Beach Naval docks to KSC, leaving on 8 September 1972 and arriving on 22 September 1972.

The S-IVB-212 stage was eventually launched in the modified Skylab configuration on 14 May 1973. It was host to three visiting astronaut crews before reentering on 11 July 1979 after 2248 days in orbit. It had been given the orbital designation of 1973-27A.

References

1. History of the George C Marshall Space Flight Center from January 1 through June 30 1961 volume 1. NASA MHM-3. November 1961.
2. History of the George C Marshall Space Flight Center from July 1 through December 31 1961 volume 1. NASA MHM-4. March 1962.
3. History of the George C Marshall Space Flight Center from January 1 through June 30 1962 volume 1. NASA MHM-5. September 1962.
4. History of the George C Marshall Space Flight Center from July 1 through December 31 1962 volume 1. NASA MHM-6. May 1963.
5. History of the George C Marshall Space Flight Center from January 1 through June 30 1963 volume 1. NASA MHM-7. November 1963.
6. History of the George C Marshall Space Flight Center from January 1 through June 30 1963 volume 2. NASA MHM-7. November 1963.
7. History of the George C Marshall Space Flight Center from July 1 through December 31 1963 volume 1. NASA MHM-8. July 1964.
8. History of the George C Marshall Space Flight Center from January 1 through June 30 1964 volume 1. NASA MHM-9. May 1965.
9. History of the George C Marshall Space Flight Center from July 1 through December 31 1964 volume 1. NASA MHM-10.
10. History of the George C Marshall Space Flight Center from January 1 through December 31 1965 volume 1. NASA MHM-11. April 1968.
11. History of the George C Marshall Space Flight Center from January 1 through December 31 1965 volume 2. NASA MHM-11. April 1968.
12. A chronology of the George C Marshall Space Flight Center from January 1 through December 31 1966. NASA MHR-6. November 1969.
13. A chronology of the George C Marshall Space Flight Center from January 1 through December 31 1967. NASA MHR-7. April 1970.
14. A chronology of the George C Marshall Space Flight Center from January 1 through December 31 1968. NASA MHR-8. February 1971.
15. A chronology of the George C Marshall Space Flight Center from January 1 through December 31 1969. NASA MHR-9 (draft never issued). June 1972.
16. An illustrated chronology of the NASA Marshall Center and MSFC programs 1960-1973. NASA MHR-10. May 1974.
17. Saturn illustrated chronology. April 1957 to April 1968. NASA MHR-5. January 20 1971.
18. Saturn V quarterly progress report, October 1 1965 to December 31 1965. NASA MPR-SAT V. 65-4.
19. Saturn V quarterly progress report, January 1 1966 to March 31 1966. NASA MPR-SAT V 66-1.
20. Saturn V semi-annual progress report, July 1 1966 to December 31 1966. NASA MPR-SAT V 66-3.
21. Saturn V semi-annual progress report, January 1 1967 to June 30 1967. NASA MPR-SAT-67-3.
22. Saturn V semi-annual progress report, July 1 1967 to December 31 1967. NASA MPR-SAT-67-4.
23. Saturn V annual progress report, fiscal year 1967, July 1 1966 through June 29 1967. The Boeing Company.
24. Test program summary for Saturn V/Apollo program S-IC stage. NASA MSFC. October 29, 1970.
25. S-II MTF stage history summary (extract). MSFC.11 December 1970.
26. S-IC MTF stage history summary (extract). MSFC. 11 December 1970.
27. Untitled datasheets – listings of all firings at MSFC. December 29 1970.
28. History of static firings conducted at the Saturn static test facility. MSFC internal memorandum. December 20 1966.
29. Saturn S-II stage monthly progress report July 1966. NAA SID 63-266-42. 28 August 1966.
30. Saturn S-II stage quarterly progress report, first quarter 1968. NAR SID 63-266-50. April 1968.
31. Saturn S-II stage quarterly progress report, second quarter 1968. NAR SID 63-266-51. July 1968.
32. Static firing test report: Saturn S-II all-systems test stage (S-II-T). NAA SID 66-967. 12 July 1966.
33. S-II-8 static firing final test report, volume 2. NAR SD 69-329. 20 June 1969.
34. Saturn S-II chronology of events. NAA. 1971.
35. Historical summary – S&ID Apollo program. NAA January 20 1966.
36. Industrial security report. Explosion of Saturn S-II-T at Mississippi Test Facility on May 28, 1966. NAA M-14-0-3. July 18 1966.
37. Mississippi Test Operations. NAR. 15 January 1971.
38. Saturn S-IVB quarterly technical progress report. Douglas DAC-56533. March 1967.
39. Transportation of Douglas Saturn S-IVB stages. Douglas 3688. November 1965.
40. Report to the Congress. Review of the Saturn S-IVB-503 stage accident under the Apollo program. B-156556. April 15 1969.
41. Narrative end item report on Saturn S-IVB-501. Douglas SM-53184. August 1966.
42. Saturn S-IVB-502 stage acceptance firing report. Douglas DAC-56357. September 1966.
43. Narrative end item report on Saturn S-IVB-503. Douglas DAC-56505. May 1967.
44. Narrative end item report Saturn S-IVB-504N, volume 1, SSC. Douglas DAC-56561. July 1967.
45. Saturn S-IVB-505N stage acceptance firing report (special cold helium test). Douglas SM-47461A. September 1968.
46. Sacramento Narrative end item report Saturn S-IVB-506N. McDonnell Douglas DAC-56608. February 1969.
47. Saturn S-IVB-506N stage acceptance firing report. McDonnell Douglas SM-47462. September 1968.

48. Narrative end item report Saturn S-IVB-507. McDonnell Douglas DAC-56622. August 1968.
49. Sacramento Narrative end item report Saturn S-IVB-507. McDonnell Douglas DAC-56623. April 1969.
50. Saturn S-IVB-507 stage acceptance firing report. McDonnell Douglas DAC-56727. December 1967.
51. Huntington Beach Narrative end item report Saturn S-IVB-508. McDonnell Douglas DAC-56639. January 1969.
52. Saturn S-IVB-508 stage acceptance firing report. McDonnell Douglas DAC-56757. April 1969.
53. Huntington Beach Narrative end item report Saturn S-IVB-509. McDonnell Douglas DAC-56662. April 1969.
54. Saturn S-IVB-509 stage acceptance firing report. McDonnell Douglas DAC-56799. July 1969.
55. Huntington Beach Narrative end item report Saturn S-IVB-510. McDonnell Douglas DAC-56725. July 1969.
56. Narrative end item report Saturn S-IVB-212. McDonnell Douglas DAC-56581. September 1968.
S57. -IVB/IB and S-IVB/V common battleship report. Douglas SM-47012. 21 February 1966.
58. Telex from NASA to Aero Spacelines, 12 December 1966.
59. J-2 engine performance analysis, flight AS-501. Rocketdyne R-7450-1. 17 May 1968.
60. J-2 engine AS-502 flight report. Volume 2 S-II stage failure analysis. Rocketdyne R-7450-2. 17 June 1968.
61. J-2 engine AS-502 flight report. Volume 3 S-IVB stage failure analysis. 17 June 1968.
62. Saturn F-1 configuration identification & status report. Rocketdyne R-5857. 16 July 1975.
63. Saturn J-2 configuration identification & status report. Rocketdyne R-5788. 16 July 1975.
64. J-2 Propulsion Production System Data. Rocketdyne. No refs.
65. F-1 major configuration change points. Internal letter. NAR. 9 March 1971.
66. J-2 engine significant configuration change points. Internal letter, NAR. 9 March 1971.
67. MSFC test laboratory, monthly progress report, February 1, 1967 through February 28, 1967.
68. Contractor photographic coverage still and motion picture. Monthly reports from August 1967 to January 1971. MSFC.
69. Skylab a chronology. NASA SP-4011. 1977.
70. Saturn V news reference. December 1968.
71. MSFC – Mississippi Test Facility general data. Datasheet 1969.
72. From Michoud to the moon. NASA brochure. Cica 1966.
73. Liquid Propellant Rocket Propulsion Systems, Rocketdyne brochure. 1999.
74. Advanced Engine Test Facility. The American Society of Mechanical Engineers. 1993.
75. Marshall Space Flight Center. Historic Aerospace Site. AIAA. 2002.
76. Air Force Research Lab. Historic Aerospace site. AIAA. 2000.
77. The Rocketdyne Santa Susana Field Lab. Historic Aerospace site. AIAA. 2001.
78. NASA JSC Roundup. 16 September 1977.
79. Newly found object could be leftover Apollo rocket stage. Near-Earth object program office news release. 12 September 2002.
80. Rocket test stand gets facelift. Air force news release. January 19 2004.
81. Air Force Rocket Laboratory. Web site.
82. NavSource Online: Service Ship Photo Archive. Web site.
83. All about Guppys. Daren Savage. Web site.
84. Cloudster. S-IVB photographs. Phil Broad. Web site
Rocket propulsion testing. Arnold engineering development center. Web site.
85. J-4 rocket test cell reaches 400th test firing milestone. Air force press release. 23 September 1997.
86. Comprehensive Field data review #1. SSFL. California Dept of Toxic Substances control.
87. SSFL preliminary site evaluation. Agency for toxic substances and disease registry. 1999.
88. Saturn V. KSC spaceport news. February 2 1996.
89. Efforts underway to relocate KSC's Saturn V components. KSC release 43-96. April 18 1996.
90. Smithsonian works to save Saturn V moon rocket at JSC. NASA news release J04-022. April 5 2004.
91. Saturn Rising. Air and Space magazine. December 1996/January 1997.
92. MSFC Marshall Star. October 16 1968.
93. MSFC Marshall Star. May 4 1966.
94. MSFC Marshall Star. January 11 1967.
95. MSFC Marshall Star. April 12 1967.
96. MSFC Marshall Star. February 14 1968.
97. MSFC Marshall Star. May 1 1968.
98. MSFC Marshall Star. August 14 1968.
99. MSFC Marshall Star. January 22 1969.
100. MSFC Marshall Star. February 5 1969.
101. MSFC Marshall Star. April 9 1969.
102. MSFC Marshall Star. June 25 1969.
103. MSFC Marshall Star. April 29 1970.
104. MSFC Marshall Star. May 6 1970.
105. MSFC Marshall Star. May 13 1970.
106. MSFC Marshall Star. November 4 1970.

Stage Allocations

	Apollo	First stage	Second stage	Third stage
SA-501	Apollo 4	S-IC-1	S-II-1	S-IVB-501
SA-502	Apollo 6	S-IC-2	S-II-2	S-IVB-502
SA-503	Apollo 8	S-IC-3	S-II-3	S-IVB-503N
SA-504	Apollo 9	S-IC-4	S-II-4	S-IVB-504N
SA-505	Apollo 10	S-IC-5	S-II-5	S-IVB-505N
SA-506	Apollo 11	S-IC-6	S-II-6	S-IVB-506N
SA-507	Apollo 12	S-IC-7	S-II-7	S-IVB-507
SA-508	Apollo 13	S-IC-8	S-II-8	S-IVB-508
SA-509	Apollo 14	S-IC-9	S-II-9	S-IVB-509
SA-510	Apollo 15	S-IC-10	S-II-10	S-IVB-510
SA-511	Apollo 16	S-IC-11	S-II-11	S-IVB-511
SA-512	Apollo 17	S-IC-12	S-II-12	S-IVB-512
SA-513	Skylab	S-IC-13	S-II-13	S-IVB-212
SA-514	N/A	S-IC-14	S-II-14	S-IVB-513
SA-515	N/A	S-IC-15	S-II-15	S-IVB-514

Engine Allocations

	First stage					Second stage					Third
	101	102	103	104	105	201	202	203	204	205	301
SA-501	F-3013	F-3015	F-3016	F-3012	F-3011	J-2026	J-2043	J-2030	J-2035	J-2028	J-2031
SA-502	F-4017	F-4018	F-4019	F-4021	F-4020	J-2057	J-2044	J-2058	J-2040	J-2041	J-2042
SA-503	F-4024	F-4022	F-4025	F-4026	F-4027	J-2051	J-2053	J-2059	J-2045	J-2055	J-2071
SA-504	F-5029	F-5032	F-5031	F-5033	F-5030	J-2067	J-2068	J-2069	J-2070	J-2066	J-2094
SA-505	F-5035	F-5041	F-5040	F-5042	F-5034	J-2075	J-2077	J-2080	J-2081	J-2076	J-2091
SA-506	F-6043	F-6046	F-6051	F-6054	F-6044	J-2089	J-2086	J-2088	J-2084	J-2085	J-2101
SA-507	F-6048	F-6052	F-6047	F-6053	F-6050	J-2090	J-2092	J-2093	J-2096	J-2097	J-2119
SA-508	F-6055	F-6058	F-6057	F-6078	F-6056	J-2082	J-2099	J-2102	J-2098	J-2100	J-2122
SA-509	F-6061	F-6064	F-6063	F-6065	F-6062	J-2106	J-2110	J-2108	J-2109	J-2105	J-2124
SA-510	F-6088	F-6069	F-6068	F-6071	F-6073	J-2112	J-2113	J-2114	J-2115	J-2116	J-2079
SA-511	F-6095	F-6096	F-6087	F-6094	F-6059	J-2117	J-2125	J-2121	J-2123	J-2118	J-2134
SA-512	F-6084	F-6076	F-6075	F-6083	F-6074	J-2130	J-2126	J-2127	J-2129	J-2128	J-2137
SA-513	F-6079	F-6080	F-6082	F-6077	F-6081	J-2104	J-2132	J-2135	J-2136	J-2138	J-2140
SA-514	F-6089	F-6093	F-6085	F-6086	F-6092	J-2139	J-2141	J-2142	J-2144	J-2145	J-2143
SA-515	F-6066	F-6091	F-6097	F-6060	F-6098	J-2147	J-2152	J-2149	J-2150	J-2151	N/A

S-IC-T Tests

Stage	Firing no.	Test stand	Date	Time	Duration Planned	Actual	101	102	103	104	105	Comments
S-IC-T	S-IC-01	MSFC/S-IC	9.4.65	1620 CDT	7s	3s	F-2005	F-2007	F-2008	F-2010	F-2003	Single engine. Unintentional observer cut-off
S-IC-T	S-IC-02	MSFC/S-IC	9.4.65	1845 CDT	7s	2.5s	F-2005	F-2007	F-2008	F-2010	F-2003	Single engine. Automatic safety cut-off
S-IC-T	S-IC-03	MSFC/S-IC	10.4.65	1710 CDT	15s	16.73s	F-2005	F-2007	F-2008	F-2010	F-2003	Single engine. Successful
S-IC-T	S-IC-04	MSFC/S-IC	16.4.65	1458 CDT	7s	6.5s	F-2005	F-2007	F-2008	F-2010	F-2003	Five engine. Successful
S-IC-T	S-IC-05	MSFC/S-IC	6.5.65	1510 CDT	15s	15.55s	F-2005	F-2007	F-2008	F-2010	F-2003	Successful
S-IC-T	S-IC-06	MSFC/S-IC	20.5.65	1458 CDT	40s	40.84s	F-2005	F-2007	F-2008	F-2010	F-2003	Successful
S-IC-T	S-IC-07	MSFC/S-IC	8.6.65	1608 CDT	90s	41.1s	F-2005	F-2007	F-2008	F-2010	F-2003	Terminated early by observer
S-IC-T	S-IC-08	MSFC/S-IC	11.6.65	1459 CDT	90s	90.9s	F-2005	F-2007	F-2008	F-2010	F-2003	Successful
S-IC-T	S-IC-09	MSFC/S-IC	29.7.65	1756 CDT	40s	17.6s	F-2005	F-2007	F-2008	F-2007	F-2003	Terminated early by observer
S-IC-T	S-IC-10	MSFC/S-IC	5.8.65	1602 CDT	LOX depletion	143.65s ib/147.63s ob	F-2005	F-2007	F-2008	F-2007	F-2003	Successful
S-IC-T	S-IC-11	MSFC/S-IC	8.10.65	1641 CDT	LOX depletion	42.38s ib/47.80s ob	F-4T2	F-2003	F-2008	F-2007	F-3T1	Automatic configuration. Successful
S-IC-T	S-IC-12	MSFC/S-IC	3.11.65	1640 CST	90.5s		F-4T2	F-2003	F-2008	F-2007	F-3T1	Terminated early by observer in error
S-IC-T	S-IC-13	MSFC/S-IC	24.11.65	1307 CDT	LOX depletion	148.4s ib/153.4s ob	F-4T2	F-2003	F-2008	F-2007	F-3T1	Successful
S-IC-T	S-IC-14	MSFC/S-IC	9.12.65	1609 CST	150s	146.07s ib/150.02s ob	F-4T2	F-2003	F-2008	F-2007	F-3T1	Successful
S-IC-T	S-IC-15	MSFC/S-IC	16.12.65	1500 CST	40s	40.96s ib/45.96s ob	F-4T2	F-2003	F-2008	F-2007	F-3T1	Successful
S-IC-T	S-IC-T-1	MTF/B-2	3.3.67	1721 CST	15s	15.2s	F-2003	F-2007	F-2008	F-2010	F-3T1	Successful
S-IC-T	S-IC-T-2	MTF/B-2	17.3.67	1516 CST	60s	60.184s	F-2003	F-2007	F-2008	F-2010	F-3T1	Successful
S-IC-T	S-IC-20	MSFC/S-IC	1.8.67	1500 CDT	40s	2.15s	F-2003	F-2007	F-2008	F-2010	F-3T1	Terminated early by observer in error
S-IC-T	S-IC-21	MSFC/S-IC	3.8.67	1500 CDT	40s	3.60s	F-2003	F-2007	F-2008	F-2010	F-3T1	Terminated early by observer in error
S-IC-T	S-IC-22	MSFC/S-IC	3.8.67	1923 CDT	40s	41.74s	F-2003	F-2007	F-2008	F-2010	F-3T1	Successful

S-IC Tests

Stage	Firing no.	Test stand	Date	Time	Duration Planned	Duration Actual	Engine 101	Engine 102	Engine 103	Engine 104	Engine 105	Comments
S-IC-1	S-IC-16	MSFC/S-IC	17.2.66.	1518 CST	40s	40.79s	F-3013	F-3015	F-3016	F-3012	F-3011	Successful
S-IC-1	S-IC-17	MSFC/S-IC	25.2.66.	1459 CST	125s	83.2s	F-3013	F-3015	F-3016	F-3012	F-3011	Pc measurement failure
S-IC-2	S-IC-18	MSFC/S-IC	7.6.66.	1843 CDT	125s	126.3s	F-4017	F-4018	F-4019	F-4021	F-4020	Successful
S-IC-3	S-IC-19	MSFC/S-IC	15.11.66.	1538 CDT	125s	121.7s	F-4023	F-4022	F-4025	F-4026	F-4027	Successful
S-IC-4	S-IC-4-1	MTF/B-2	16.5.67.	1520 CDT	125s	125.096s	F-5029	F-5032	F-5031	F-5033	F-5030	Successful
S-IC-5	S-IC-5-1	MTF/B-2	25.8.67.	1814 CDT	125s	125.096s	F-5035	F-5041	F-5040	F-5042	F-5034	Successful
S-IC-6	S-IC-6-1	MTF/B-2	13.8.68.	1734 CDT	125s	126.504s	F-6043	F-6046	F-6051	F-6054	F-6044	Successful
S-IC-7	S-IC-7-1	MTF/B-2	30.10.68.	1516 CDT	125s	126.464s	F-6048	F-6052	F-6047	F-6053	F-6050	Successful
S-IC-8	S-IC-8-1	MTF/B-2	18.12.68.	1639 CDT	125s	126.688s	F-6055	F-6058	F-6057	F-6059	F-6056	Successful
S-IC-9	S-IC-9-1	MTF/B-2	19.2.69.	1516 CST	125s	126.640s	F-6061	F-6064	F-6063	F-6065	F-6062	Successful
S-IC-10	S-IC-10-1	MTF/B-2	16.4.69.	1433 CST	125s	126.372s	F-6066	F-6069	F-6068	F-6071	F-6073	Successful
S-IC-11	S-IC-11-1	MTF/B-2	26.6.69.	1722 CDT	125s	96.8s	F-6049	F-6045	F-6072	F-6060	F-6070	Abort due to engine fire
S-IC-11	S-IC-11R-1	MTF/B-2	25.6.70.	1523 CDT	125s	70.628s	F-6095	F-6096	F-6087	F-6094	F-6059	Abort due to LOX pressure redline exceedance
S-IC-12	S-IC-12-1	MTF/B-2	3.11.69.	1512 CST	125s	126.328s	F-6084	F-6076	F-6075	F-6083	F-6074	Successful
S-IC-13	S-IC-13-1	MTF/B-2	6.2.70.	1547 CST	125s	126.432s	F-6079	F-6080	F-6082	F-6077	F-6081	Successful
S-IC-14	S-IC-14-1	MTF/B-2	16.4.70.	2015 CST	125s	126.364s	F-6089	F-6093	F-6085	F-6086	F-6092	Successful
S-IC-15	S-IC-15-1	MTF/B-2	30.9.70.	1817 CDT	125s	126.672s	F-6066	F-6091	F-6097	F-6060	F-6098	Successful

S-II Tests

Stage	Test stand	Date	Time	Firing duration	Engine 201	Engine 202	Engine 203	Engine 204	Engine 205	Comments
S-II-1	MTF/A-2	1.12.66.		384s	J-2026	J-2043	J-2030	J-2035	J-2028	Successful
S-II-1	MTF/A-2	22.12.66.	2200 CST	2s	J-2026	J-2043	J-2030	J-2035	J-2028	Aborted due to faulty probe
S-II-1	MTF/A-2	30.12.66.		363s	J-2026	J-2043	J-2030	J-2035	J-2028	Successful
S-II-2	MTF/A-2	6.4.67.	1608 CST	363s	J-2057	J-2044	J-2058	J-2040	J-2041	Successful
S-II-2	MTF/A-2	15.4.67.		367s	J-2057	J-2044	J-2058	J-2040	J-2041	Successful
S-II-3	MTF/A-1	19.9.67.	2010 CDT	65s	J-2051	J-2053	J-2059	J-2045	J-2055	Successful
S-II-3	MTF/A-1	27.9.67.	1523 CDT	358s	J-2051	J-2053	J-2059	J-2045	J-2055	Successful
S-II-4	MTF/A-2	30.1.68.		17s	J-2067	J-2068	J-2069	J-2070	J-2066	Aborted by false reading from observer
S-II-4	MTF/A-2	10.2.68.	1439 CST	347s	J-2067	J-2068	J-2069	J-2070	J-2066	Successful
S-II-5	MTF/A-1	1.8.68.		7.54s	J-2075	J-2077	J-2080	J-2081	J-2076	Aborted by false reading from observer
S-II-5	MTF/A-1	9.8.68.		367s	J-2075	J-2077	J-2080	J-2081	J-2076	Successful
S-II-6	MTF/A-2	3.10.68.	1522 CDT	371s	J-2089	J-2086	J-2088	J-2084	J-2085	Successful
S-II-7	MTF/A-2	22.1.69.		365s	J-2090	J-2093	J-2092	J-2096	J-2097	Successful
S-II-8	MTF/A-2	8.4.69.	1345 CDT	381s	J-2082	J-2099	J-2102	J-2098	J-2100	Centre engine purposely shut down early. Successful
S-II-9	MTF/A-1	20.6.69.		349s	J-2106	J-2110	J-2108	J-2109	J-2105	Successful
S-II-10	MTF/A-1	1.10.69.		368s	J-2112	J-2113	J-2114	J-2115	J-2116	Successful
S-II-11	MTF/A-1	14.11.69.		371.6s	J-2117	J-2125	J-2121	J-2123	J-2118	Successful
S-II-12	MTF/A-2	26.2.70.		2s	J-2130	J-2126	J-2127	J-2129	J-2128	Aborted due to faulty engine ox valve
S-II-12	MTF/A-2	4.3.70.		376.2s	J-2130	J-2126	J-2127	J-2129	J-2128	Successful
S-II-13	MTF/A-2	30.4.70.		364.9s	J-2104	J-2132	J-2135	J-2136	J-2138	Successful
S-II-14	MTF/A-2	31.7.70.		373.2s	J-2139	J-2141	J-2142	J-2144	J-2145	Successful
S-II-15	MTF/A-2	30.10.70.	1515 CST	373.0s	J-2147	J-2152	J-2149	J-2150	J-2151	Successful

S-IVB Tests

Stage	Countdown number	Test stand	Date	Time	Duration	Engine 301	Comments
S-IVB-501	CD 614061	SACTO/Beta I	20.5.66.		50s	J-2031	Automatic cutoff
S-IVB-501	CD 614063	SACTO/Beta I	26.5.66.		151s	J-2031	Successful
S-IVB-501	CD 614063	SACTO/Beta I	26.5.66.		301s	J-2031	Successful
S-IVB-502	CD 614067	SACTO/Beta I	28.7.66.	1420 PDT	150.7s	J-2042	Successful
S-IVB-502	CD 614067	SACTO/Beta I	28.7.66.	1554 PDT	291.2s	J-2042	Successful
S-IVB-503N		SACTO/Beta I	3.5.67.		446.9s	J-2071	Successful
S-IVB-504N		SACTO/Beta I	23.8.67.		51.23s	J-2094	Terminated due to fire indication
S-IVB-504N		SACTO/Beta I	26.8.67.		438s	J-2094	Successful
S-IVB-505N		SACTO/Beta I	12.10.67.		448.4s	J-2091	Successful
S-IVB-506N	CD 614109	SACTO/Beta III	17.7.68.	1823 PDT	445.2s	J-2101	Successful
S-IVB-507	CD 614113	SACTO/Beta I	16.10.68.	1608 PST	433.2s	J-2119	Successful
S-IVB-508	CD 614116	SACTO/Beta III	20.2.69.	1422 PST	457.0s	J-2122	Successful
S-IVB-509	CD 614120	SACTO/Beta III	14.5.69.	1409 PDT	452.4s	J-2124	Successful
S-IVB-510		SACTO/Beta III	14.8.69.		448.7s	J-2079	Successful
S-IVB-511		SACTO/Beta III	18.12.69.		442.8s	J-2134	Successful

Santa Susana Field Laboratory (SSFL)

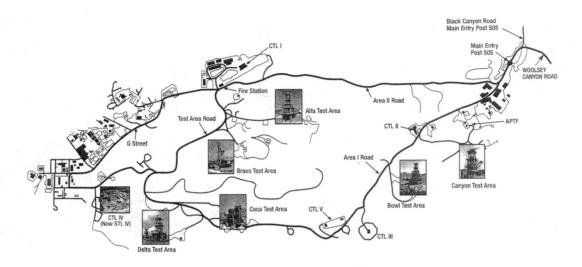

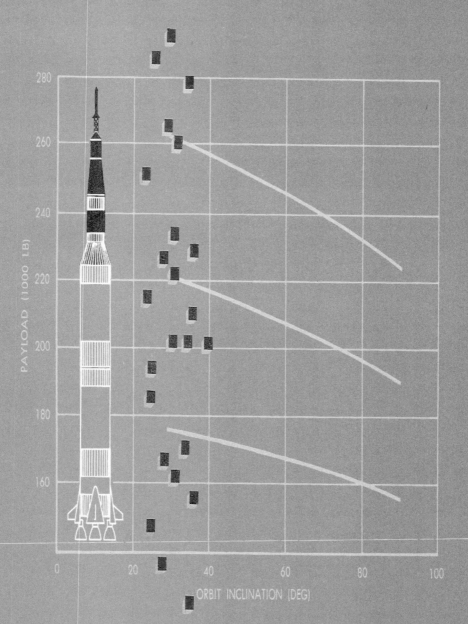

SATURN V PAYLOAD PLANNER'S GUIDE

November 1965
Douglas Report SM-47274
Prepared By: L.O. Schulte Development Planning - Saturn Payload Applications
Approved By: F. C. Runge Program Manager - Saturn Payload Applications
Approved By: T. J. Gordon Director of Advance Saturn and Large Launch Vehicles

DOUGLAS MISSILE & SPACE SYSTEMS DIVISION HUNTINGTON BEACH, CALIFORNIA

This guide has been prepared by Douglas to acquaint payload planners with the capability of the Saturn V Launch Vehicle and to assist them in their initial payload/launch vehicle planning. This guide is not an offer of space aboard Saturn. Only NASA can commit experiments to this vehicle. This book attempts to show methods by which Saturn could accommodate payloads of various weights, volumes and missions. You will see that the capabilities of this vehicle permit a wide spectrum of assignments, including scientific, technological as well as operational type payloads. A similar guide has been prepared on the capabilities of the Saturn IB Launch Vehicle. This book, called the Saturn IB Payload Planner's Guide, Douglas Report No. SM-47010 is available upon request. Requests for additional information may be addressed to:

Mr. Fritz Runge, Program Manager Saturn Payload Applications
Douglas Missile and Space Systems Division 5301 Bolsa Avenue
Huntington Beach, California 92646 Telephone 714-897-0311
SM – 47274

I - INTRODUCTION

The Saturn V is a three stage launch vehicle under development by the NASA to support the Apollo Lunar Landing mission.

The Saturn V Vehicle will also be used to achieve many other objectives related to the national goal of lunar exploration and space flight. Certainly the development of future space stations and interplanetary spacecraft will rely heavily upon hardware and techniques developed in the Apollo.

The Saturn V Vehicle is designed to launch very large manned and unmanned payloads into space.

Each of the stages are now on the production line and progressing on schedule. The initial flight tests for the Saturn V Vehicles will be in early 1967 and will be capable of injecting over 261,000 pounds of payload into a 100 nautical mile circular earth orbit. Since the S-IVB third stage actually goes into orbit along with the payload, the total weight in orbit is nearly 300,000 pounds. In the future, a Saturn V with a high-energy fourth stage could provide an effective configuration for high velocity missions; for example, approximately 12,000 pounds could be accelerated to a hyperbolic excess velocity of 45,000 feet/second. The Saturn class of vehicles thus constitutes a great national resource which is destined to serve the launch vehicle needs of a wide variety of future manned and unmanned space missions.

This Payload Planner's Guide is intended as a starting point for engineers, scientists, and executives who are planning to conduct engineering tests, space science experiments, or, operational missions. It outlines, for the payload planner, the technical information and procedures with which large prime, or small auxiliary payloads can be effectively integrated and flown on the vehicle. The payload planner will find here the characteristics of the Saturn V launch vehicle, its performance, the accommodations it offers to potential experimenters, suggested procedures to be followed in obtaining support for the experiment, approximate flight schedules and engineering data needed to initiate the design of a payload. To planners of prime payloads, the guide offers four protective shroud designs. To auxiliary payload planners, it presents several payload accommodation concepts for identifying and describing volumes in the Saturn V where such payloads could be installed.

Environmental data and payload weight limitations for each payload volume are provided.

The Saturn V performance capabilities are included for payload flight planning. The major subsystems of the launch vehicle and their relation to the payloads are described. A concept-to-flight chronology of events is presented to support payload/launch vehicle system planning on the part of prospective users.

The Douglas Missile and Space Systems Division will be pleased to discuss the planning, support, operation, and data evaluation involved in the flight of any payload on Saturn V.

SATURN V CONFIGURATIONS

The three-stage configuration of the Saturn V, depicted in Figure 1-1, is the basis for the data presented in this guide. The three stage Saturn V with the Apollo spacecraft is about 363 feet tall and weighs nearly 3200 tons. A possible four-stage configuration is described in Section V.

FIRST STAGE (S-IC)

The S-IC stage is 138 feet tall and 33 feet in diameter. The propellants, liquid oxygen (LOX) and RP-1 (special kerosene fuel), are stored in two separate tanks with the fuel in the lower tank. Five 1.5 million pound thrust F-1 engines are used to generate a total of 7.5 million pounds of thrust at lift off, and propel the vehicle to an altitude of 30 nm in 150 seconds. Four of the engines are hydraulically gimballed to provide thrust vector control in response to steering commands from the guidance system located in the Instrument Unit. The first stage is separated from the second by eight 80,000 pound thrust solid rocket motors.

SECOND STAGE (S-II)

The S-II stage is 81.5 feet tall and 33 feet in diameter. The propellants, liquid oxygen (LOX) and liquid hydrogen (LH2), are stored in two tanks separated by a common bulkhead with the LOX in the lower tank. Five 205,000 pound thrust J-2 engines propel the vehicles to a burnout altitude of 90 to 100 nautical miles depending upon the mission. Four of the engines are gimbaled for control during flight, similar to the S-IC. Eight 22,900 pound thrust solid motors are fired to ullage the propellants for engine start. The second stage is separated from the third by four solid rocket motors, each of which produces 35,000 pounds of thrust for 1,5 seconds. The interstage which mates the S-IVB to the S-II remains with the S-II.

THIRD STAGE (S-IVB)

The S-IVB stage is 58.5 feet tall and is about 22 feet in diameter. The S-IVB is powered by a single Rocketdyne J-2 engine that burns liquid oxygen and liquid hydrogen to provide a thrust of 205,000 lb. at an engine mixture ratio of 5/1. During flight, the main engine is hydraulically gimbaled in pitch and yaw to provide thrust vector control in response to commands from the instrument unit. Powered roll control is provided by fixed position hypergolic engines located in two auxiliary propulsion system (APS) modules mounted 180° apart on the aft skirt. Three axis (roll, pitch and yaw) attitude control during coast is also provided by the APS. Two solid-propellant ullage rockets are fired at stage separation, just prior to ignition of the J-2 engine, to assure that the propellant is settled in the bottom of the tanks during the start phase. After second stage separation, the J-2 engine on the S-IVB stage ignites and propels the payload to the desired altitude. The S-IVB as presently designed has a 4-1/2 hour orbital coast plus a 2 hour translunar coast capability. Two 72 lb-thrust hypergolic engines in the APS modules, are fired during the first shutdown of the J-2 engine to control the position of the propellants and again, prior to the second J-2 start, to position the propellants during chilldown and restart of the main engine.

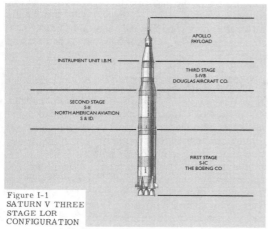

Figure I-1
SATURN V THREE STAGE LOR CONFIGURATION

INSTRUMENT UNIT (I.U.)

The Instrument Unit houses the guidance and control systems and the flight instrumentation systems for the Saturn V launch vehicle. Specifically, the I.U. contains electrical, guidance and control, instrumentation, measuring, telemetry, radio frequency, environmental control, and emergency detection systems.

SATURN V CAPABILITY

Saturn V has the capability to perform a broad spectra of manned and unmanned space missions and can carry large prime and auxiliary payloads as summarized in Figure I-2 and presented in detail in Sections IV and V.

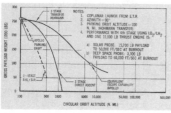

Figure I-2
SATURN V MISSION POTENTIAL

The major advantages of utilizing the Saturn V are:

- Largest orbital payload capability of any vehicle in the world (about 261,000 lb. to 100 n. mi.).
- Large diameter payload volume.
- Large escape payload capability (about 98,000 lb.).
- Large synchronous orbit payload capability (over 72,000 lb. to a 20,000 n. mi. orbit with a 28.5° inclination and about 62,000 lb. to an equatorial synchronous orbit).
- Low transportation costs per pound of payload in orbit. Flight proven systems and subsystems.
- Man-rated systems. Production stages. Complete and existing manufacturing, test and launch facilities. Flexibility of planning two, three or four stage configurations. Associated NASA data acquisition and tracking networks are operational.
- Auxiliary payload volumes, weight, power, and data channels may be available.
- Growth potential of the vehicle is considerable.

II PAYLOAD CONSIDERATION

II-1 Planning and Schedules

Technical assistance is available at Douglas to aid the experimenter or payload originator in planning, flying and evaluating a payload on the Saturn V. The three-stage Saturn V vehicles can carry prime or auxiliary payloads on a great variety of manned or unmanned missions. Since it is beyond the scope of this guide to include all the data on each payload volume, some of the significant examples are shown in Figure II-1. Space, power and weight carrying capability is available in almost every part of the vehicle. Depending on specific mission requirements, auxiliary payloads may be carried in the:

(a) Apollo Command Module
(b) Apollo Service Module
(c) Lunar Excursion Module (LEM) Ascent or Descent Modules
(d) LEM Adapter
(e) Instrument Unit (I.U.)
(f) S-IVB Stage
(g) Fourth Stage (in four-stage, high energy mission configurations)

A summary of Saturn V payload potentials is presented in Figure II-1. New prime payloads may use either existing or special shroud designs.

Information in this guide is primarily associated with the prime payload carrying ability of the Saturn vehicle, auxiliary payloads within the S-IVB and payloads supported by the S-IVB and extending above it. Experimenters desiring more information on the other stages or modules should contact the appropriate agency as listed in Figure II-1.

The general steps normally required to bring a prime or auxiliary payload from concept, through integration and flight with the launch vehicle system, to final evaluation, are presented in the flow-diagram shown in Figure II-2.

Payloads and experiments may be conceived by any organization or individual in the government, universities or industry. In some cases, in order to be effective, payload proposals must include certain launch vehicle and program interface data. Douglas will assist experiment originators in the definition of payload/launch vehicle concepts. The payload/experiment proposals submitted by the originating organization are evaluated by NASA experiment review

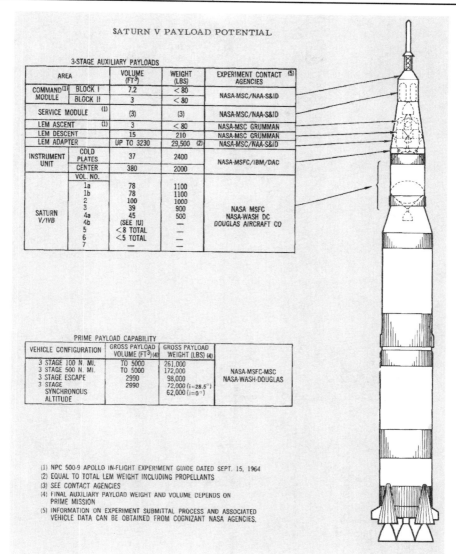

SATURN V PAYLOAD POTENTIAL

3-STAGE AUXILIARY PAYLOADS

AREA		VOLUME (FT^3)	WEIGHT (LBS)	EXPERIMENT CONTACT AGENCIES (5)
COMMAND [1] MODULE	BLOCK I	7.2	< 80	NASA-MSC/NAA-S&ID
	BLOCK II	3	< 80	
SERVICE MODULE [1]		(3)	(3)	NASA-MSC/NAA-S&ID
LEM ASCENT [1]		3	< 80	NASA-MSC GRUMMAN
LEM DESCENT		15	210	NASA-MSC GRUMMAN
LEM ADAPTER		UP TO 3230	29,500 (2)	NASA-MSC/NAA-S&ID
INSTRUMENT UNIT	COLD PLATES	37	2400	NASA-MSFC/IBM/DAC
	CENTER	380	2000	
SATURN V/IVB	VOL. NO. 1a	78	1100	NASA MSFC NASA-WASH DC DOUGLAS AIRCRAFT CO
	1b	78	1100	
	2	100	1000	
	3	39	900	
	4a	45	500	
	4b	(SEE IU)	—	
	5	< 8 TOTAL	—	
	6	< 5 TOTAL	—	
	7	—	—	

PRIME PAYLOAD CAPABILITY

VEHICLE CONFIGURATION	GROSS PAYLOAD VOLUME (FT^3) (4)	GROSS PAYLOAD WEIGHT (LBS) (4)	
3 STAGE 100 N. MI.	TO 5000	261,000	NASA-MSFC-MSC NASA-WASH-DOUGLAS
3 STAGE 500 N. MI.	TO 5000	172,000	
3 STAGE ESCAPE	2990	98,000	
3 STAGE SYNCHRONOUS ALTITUDE	2990	72,000 (i=28.5°) 62,000 (i=0°)	

(1) NPC 500-9 APOLLO IN-FLIGHT EXPERIMENT GUIDE DATED SEPT. 15, 1964
(2) EQUAL TO TOTAL LEM WEIGHT INCLUDING PROPELLANTS
(3) SEE CONTACT AGENCIES
(4) FINAL AUXILIARY PAYLOAD WEIGHT AND VOLUME DEPENDS ON PRIME MISSION
(5) INFORMATION ON EXPERIMENT SUBMITTAL PROCESS AND ASSOCIATED VEHICLE DATA CAN BE OBTAINED FROM COGNIZANT NASA AGENCIES.

boards to determine the concept's priority in meeting national objectives. With mission objectives approved, budget and vehicle allocations can be made and the concept can be processed through normal procurement channels to obtain the final contractual authority.

Upon receipt of payload contractual authority, more detailed mission planning will be accomplished among NASA, the Saturn V stage contractors, and the payload originator. NASA acts as overall program integration manager.

Development and qualification of payloads proceed in parallel with launch vehicle production.

Peculiar payload requirements may necessitate accomplishment of detailed testing, test support planning and test documentation. These must be accomplished at the beginning of final checkout of the Saturn vehicle to ensure compatibility of payloads and launch vehicle.

A typical schedule of S-IVB stage production and critical auxiliary payload integration periods is shown in Figure II-3. Also shown are typical delivery dates that an S-IVB mounted auxiliary payload might have to meet to minimize interference with the delivery schedule of the stages. The complexity of the payload and the nature of its integration will establish the lead time for a particular flight.

A schedule indicating the type of operations that must be accomplished at Kennedy Space Center (KSC) to prepare a prime payload is presented in Figure 11-4.

Figure 11-5 indicates typical delivery dates to KSC for Saturn V vehicles SA-501 through SA-515. Deliveries beginning with SA-516 may be estimated at a rate of six per year. While most of these currently have prime payload assignments, some are not expected to be fully loaded and may have room for auxiliary payloads.

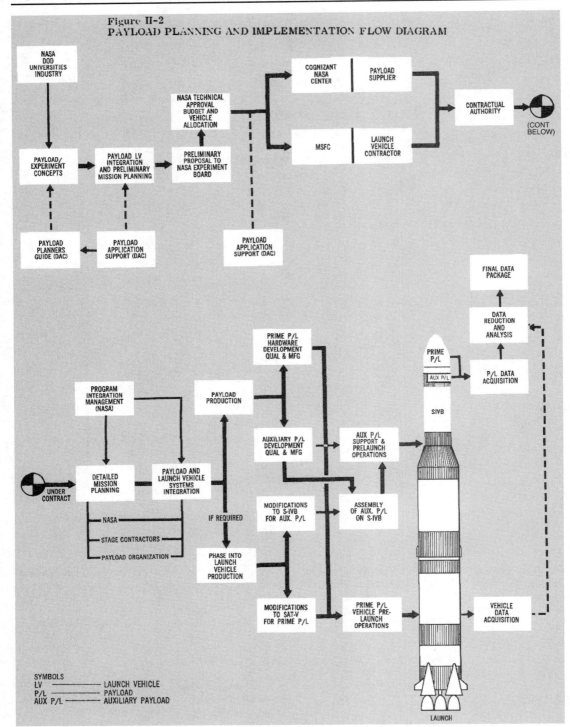

Figure II-2
PAYLOAD PLANNING AND IMPLEMENTATION FLOW DIAGRAM

II-2. Launch Vehicle Accommodations

Since auxiliary payloads can vary widely in size, shape and weight, the portion of the stage and in the I.U. The envelopes of available space within the forward skirt and I.U. extend from the electrical/electronic units mounted on the skirt to the forward dome of the S-IVB tank as shown in Figure II-6. Also, additional space is available in pods mounted externally on the forward skirt. Many combinations of space, power, data, and environmental systems can be furnished to meet the needs of auxiliary payloads. These systems do not exist in the present vehicle. This discussion is intended to illustrate feasible techniques which could be employed to accommodate auxiliary payloads.

The possible experimental payload volumes within the Saturn S-IVB stage are listed below:

experimental volumes are shown in Figure II-7. The first concept shows a payload incorporating a solid propellant motor (ABL-X-258), an optional second motor (ARC-XM-85), and a satellite payload. The assembly is mounted in a support cradle that also provides means of ejection from the S-IVB stage. The ABL-X-258 motor is ignited by a signal from a timer keyed to the ejection sequence. The payload assembly has an attitude control system referenced to the S-IVB stage attitude at payload separation. The payload is protected through the boost phase of the trajectory by a fairing that is jettisoned just before payload ejection. The support cradle is attached to a honeycomb mounting plate that in turn is attached to the S-IVB stage structure through the forward skirt frames and to pads on the S-IVB Liquid Hydrogen (LH_2) tank skin. Some performance figures for this type of installation are shown in Configuration B Figure II-S. Other payloads with different requirements that could also be accommodated are indicated by concepts 2, 3, and 4 of Figure II-7.

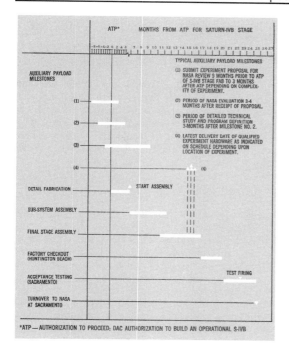

Figure II-3 TYPICAL S-IVB INTEGRATION SCHEDULE FOR AUXILIARY PAYLOADS

a. Experiment Volume No. 1a and 1b. (Figure II-6)

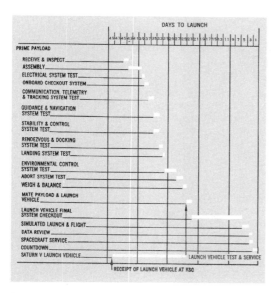

Figure II-4 TYPICAL KSC SATURN PREPARATION SCHEDULE FOR PRIME PAYLOADS

About 78 cubic feet could be provided external to the forward interstage in each of two pods as shown. Since these pods have not been designed, it may be possible to include provisions for certain unique payload requirements in the basic layout of these volumes. Approximately 1,100 lb. of payload maybe carried in each location. Some modification to the forward skirt for structural support and rerouting of some electrical cables will be required. Mounting concepts for these

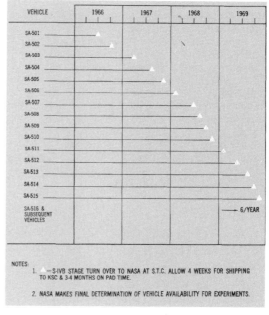

Figure II-5 SATURN S-IVB DELIVERY SCHEDULE

b. Experiment Volume No. 2

Variation in the shape of Volume 2 is possible depending on the payload configuration. However, some limitations on the use of this space are set by checkout requirements on equipment and wiring in the interstage, I.U., and the LEM descent module. Accessibility to these areas requires the use of a vertical access kit that restricts the available volume to that under the access kit platform. This volume consists of approximately 109 cubic feet. The experiment modules can be mounted on a lightweight structural cone supported by one of the forward skirt frames. A total payload weight of about 1,000 lb. can be carried in this location. Weight limitations on a specific experiment module must be controlled by prime mission requirements as well as by structural design factors.

c. Experiment Volume No. 3

Experimental modules can be mounted directly on the thermal conditioning panels in the forward skirt. See Figure II-9. On all vehicles beginning with SA-504, six or more of the sixteen panels will be available for mounting experiments because of a simplified telemetry system. A volume of at least 39 cubic feet with a maximum weight of 900 lb. is available. This weight and volume indicated may be increased if the accessibility, the payload center of gravity location and the mounting method permit.

d. Experiment Volume No. 4

For some missions in which the LEM descent stage is not carried, an additional large volume may be available above the access kit platform. This space extends over the S-IVB forward dome and into the instrument unit to Station 3258.5. The volume available within the forward skirt is about 45 cubic feet. Approximately 380 cubic feet is available in the I.U. The experiment modules can be mounted on an auxiliary payload adapter. This payload adapter consists of a 'spider' structure supported from the S-IVB forward skirt frames as shown in Figure II-10. The adapter would also serve as an access kit when removable work platforms are inserted as shown. The experimental modules are mounted on honeycomb panels attached to the adapter, The adapter accompanying the modules must be removable for access to the liquid hydrogen tank. A total payload weight of up to 2,500 lb. may be carried in this location, prime payload weight permitting.

Other auxiliary payloads, such as the Delta third stage, may be carried as shown in Figure II-8. The payloads are mounted on the auxiliary payload adapter through additional supporting structure. The internal mounting depicted shows the Delta third stage including separation and spin-up mechanism. The payload is carried in a horizontal or stowed position during the boost phase until the separation of the S-IVB from the prime payload. At this time the Delta third stage is erected, spun-up, and separated at a signal in the S-IVB stage separation sequence. Ignition of the ABL-X-258 motor is triggered by a timer after an appropriate separation distance is achieved. Separation forces can be generated by small solid propellant motors similar to the spin rockets. In the stowed position the Delta third stage projects approximately 33 inches above the S-IVB/I.U. interface at Station 3222.6. Some representative performance figures for two possible configurations are shown (Figure 11-8).

e. Experiment Volume No. 5

A small amount of usable volume may be available in the aft skirt area of the operational configuration. Certain modules (five volumes of about 1.5 cubic feet each) may be mounted directly on the existing mounting plates in place of R&D equipment not required on operational flights.

f. Experiment Volume No. 6

Experimental modules of light weight may be mounted directly on the thrust structure. Precise locations and volumes available cannot be defined at present, but small modules of the proper size (about one cubic foot each), shape, and weight could be accommodated depending on the mounting requirements of the payloads.

g. Experiment Volume No. 7

Volume 7 is within the hydrogen tank itself. Any experiment placed in this volume would, of course, displace the LH_2 and be subjected to the temperature and pressure conditions of the LH_2. Some experimenters may want to take advantage of these conditions to study a system under cryogenic and space environment, or to study the fluid behavior of the liquid or gaseous hydrogen. There are eight cold helium spheres in the LH_2 tank to pressurize the liquid oxygen (LOX) tank during powered flight. There are four additional flanged connections on which spheres could be installed to hold experiments at liquid hydrogen temperatures while protecting them from direct contact with the hydrogen. Each of the spheres has a volume of 3.5 cubic feet, The entrance to the sphere is only 1.44 inches in diameter. However, this could be increased to about 4 inches in diameter.

II-2-2. Prime Payloads Above the S-IVB Stage

Minimum vehicle changes will be required if the volume normally occupied by the Lunar Excursion Module (LEM) were to be utilized for other payloads (Figure II-11). This volume within the LEM adapter might be used on future flights if prime mission objectives permit.

Figures 11-12, II-13 and II-14 illustrate three other possible configurations of payloads and payload fairings. The fairings protect the payloads from aerodynamic loads and temperatures while in flight. They may also be used for payload thermal conditioning on the launch pad, if such conditioning is required and provisions for it are included. The fairings may be made of aluminum honeycomb or of fiberglass, if RF transparency is a requirement. The fairings are jettisoned during second-stage operation when atmospheric effects are negligible.

A typical adapter which supports the prime payload, Figure II-15, can be designed to the diameter dictated by payload requirements. The adapter is mounted directly on the instrument unit and is a conical frustrum of semi-monocoque construction, The adapter includes a structural ring to bear the lateral components of the loads imposed by the payload and to provide clearance for the end frame of the fairing at Station 3264.6. The height of the adapter above the mating plane at Station

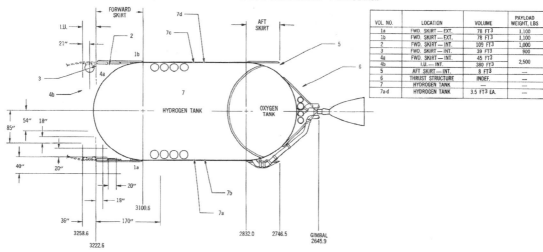

Figure II-6
PROPOSED CONCEPTS FOR S-IVB AUXILIARY PAYLOAD VOLUMES

VOL. NO.	LOCATION	VOLUME	PAYLOAD WEIGHT, LBS
1a	FWD. SKIRT — EXT.	78 FT³	1,100
1b	FWD. SKIRT — EXT.	78 FT³	1,100
2	FWD. SKIRT — INT.	109 FT³	1,000
3	FWD. SKIRT — INT.	39 FT³	900
4a	FWD. SKIRT — INT.	45 FT³	2,500
4b	I.U. — INT.	380 FT³	
5	AFT SKIRT — INT.	8 FT³	—
6	THRUST STRUCTURE	INDEF.	—
7	HYDROGEN TANK	—	—
7a-d	HYDROGEN TANK	3.5 FT³ EA.	—

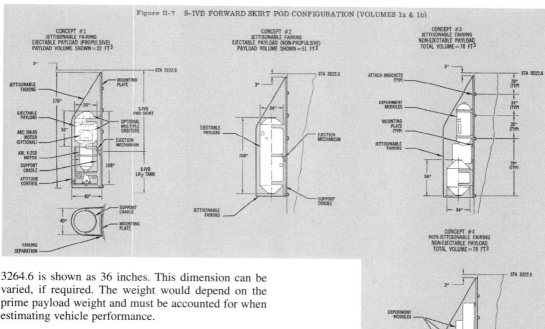

Figure II-7 S-IVB FORWARD SKIRT POD CONFIGURATION (VOLUMES 1a & 1b)

3264.6 is shown as 36 inches. This dimension can be varied, if required. The weight would depend on the prime payload weight and must be accounted for when estimating vehicle performance.

Configuration "A" shown in Figure II-11 utilizes the volume that would be available if mission objectives are such that the LEM is not used. A payload volume of about 3,230 cubic feet and a weight of 29,500 lb. could be accommodated in this space. The payload could remain with the S-IVB in orbit, or be ejected after separation of the forward section of the spacecraft/LEM adapter. The LEM adapter incorporates four panels that are unfolded at the time of command module separation.

Configuration "B" shown in Figure II-12 is the shape originally designed for the Voyager spacecraft and encompasses a volume of approximately 2,990 cubic feet. These dimensions are approximate and should be used for preliminary layout only. The final dimensions depend on payload configuration and adapter height requirements. The approximate weight of the fairing is 2,500 lb. if made of aluminum honeycomb.

Configuration "C" shown in Figure II-13 is for a modified LEM adapter and encompasses a volume of about 5,000 cubic feet with the approximate dimensions shown. The weight of the fairing is about 3,000 lb.

Configuration "D" shown in Figure II-14 combines the Voyager nose fairing with an S-IVB stage forward skirt. It encompasses a usable volume of approximately 6,000

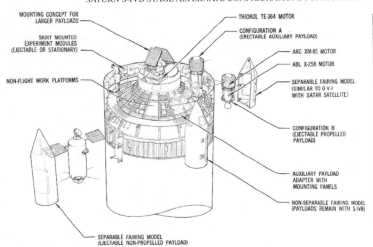

SATURN S-IVB STAGE ALTERNATE CONFIGURATIONS FOR AUXILIARY PAYLOADS

CONFIGURATION A (ABL X-258 MOTOR + SATELLITE)		
ELLIPTICAL ORBIT CAPABILITIES (INITIAL ORBIT ALTITUDE = 100 N. MILES)		
SATELLITE WEIGHT (LBS)*	APOGEE (N. MI.)	PERIGEE (N. MI.)
157	LUNAR MISSION	—
200	44,000	100
300	15,300	100
400	9,500	100
500	7,000	100

*SATELLITE WEIGHT INCLUDES GUIDANCE SYSTEM

CONFIGURATION B (ABL X-258 MOTOR + ARC XM-85 MOTOR + SATELLITE)	
CIRCULAR ORBIT CAPABILITIES** (INITIAL ORBIT ALTITUDE = 100 N. MI.)	
SATELLITE WEIGHT (LBS)*	ORBITAL ALTITUDE (N. MI.)
150	7,700
200	4,700
300	2,850

ELLIPTICAL ORBIT CAPABILITIES (INITIAL ORBIT ALTITUDE = 100 N. MI.)		
SATELLITE WEIGHT (LBS)*	APOGEE (N. MI.)	PERIGEE (N. MI.)
100	24,700	17,000
200	9,800	6,600
300	7,000	3,500

**REQUIRES OFF-LOADING X-258 MOTOR

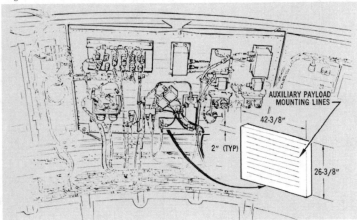

Figure II-9 S-IVB FORWARD SKIRT THERMAL CONDITIONED PANELS

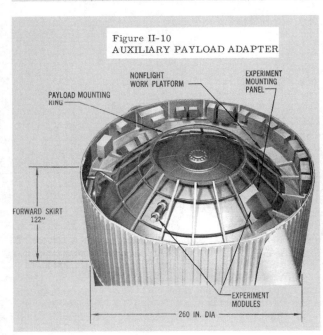

Figure II-10 AUXILIARY PAYLOAD ADAPTER

cubic feet and weighs about 3,600 lb.

a. Prime Payload Attitude Control Systems

Prime payloads may require their own attitude control systems if they are to be separated from the S-IVB stage during orbital coast. A concept using existing Saturn IB/S-IVB(A) or Saturn V/S-IVB(B) Auxiliary Propulsion System (APS) modules on prime payloads is presented in Figure II-16. The APS modules are presently designed for 4-1/2 and 6-1/2 hours coast. respectively, when mounted on the S-IVB stage used in the Saturn IB and V missions. Much longer coast periods can be achieved if these units are used for payload attitude control. The duration will be a function of payload moment of inertia and required operating cycle.

The APS modules are self-contained propulsion units which require electrical power, vehicle attitude sensors, control circuitry and guidance signals. The guidance and attitude sensing signals are provided by the I.U. The electrical power requirements for either the Saturn V or IB modules are 28 volts at a maximum of 26.5 amp for operating valves and switches. The attitude control band requirements of the payload, moments of inertia, center of gravity, location of the payload, and environmental disturbances dictate the total propellant needed for a given mission. The 150 lb. thrust is, perhaps, larger than necessary but there are techniques available for reducing it by about 50% for better propellant economy. Smaller engines from other programs could also be used. However, with the payloads indicated in Figure II-16, for the Saturn V module, control periods in excess of 6-1/2 hours with a deadband of ±1° in all three axes are possible, Reducing the control accuracy requirements extends the operating duration. Detailed descriptions of the APS modules and techniques for extending their operational life are presented in Section II-5.

II-3. Payload Thermal Environment and Control

Accurate determination of the payload temperature control requirement demands realistic thermodynamic models. Douglas employs a series of heat transfer

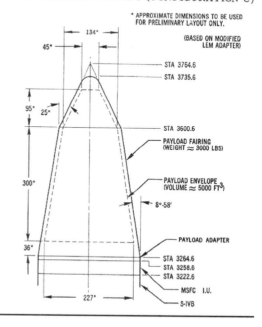

Figure II-13
PRIME PAYLOAD FAIRING (CONFIGURATION C)

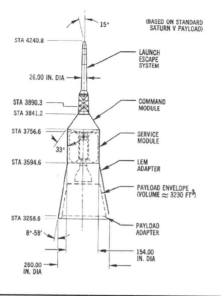

Figure II-11
PRIME PAYLOAD FAIRING (CONFIGURATION A)

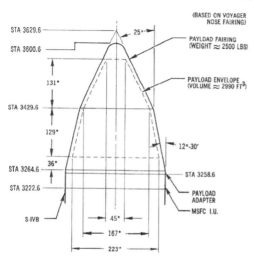

Figure II-12
PRIME PAYLOAD FAIRING (CONFIGURATION B)

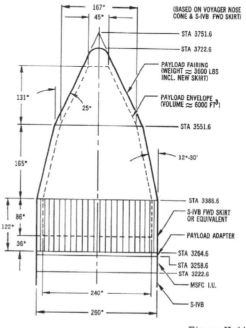

Figure II-14
PRIME PAYLOAD FAIRING (CONFIGURATION D)

computer programs (both 1 and 3 dimensional) which indicate the temperature history that can be expected at any point in the vehicle for a multitude of thermodynamic environments. Should the temperature of the volume be critical for a particular experiment, a thermal protection system can then be designed.

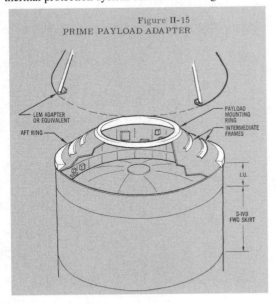

Figure II-15
PRIME PAYLOAD ADAPTER

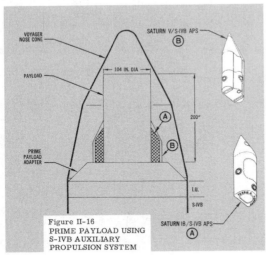

Figure II-16
PRIME PAYLOAD USING
S-IVB AUXILIARY
PROPULSION SYSTEM

The forward skirt thermal control systems for cooling electronic equipment differ from those used in the aft skirt. The equipment mounted on the forward skirt is conditioned actively and that mounted aft is conditioned passively.

In the active system, electronic components are mounted on 16 thermal conditioned panels (cold plates, Figure II-9) which transfer heat to a coolant (60% methanol and 40% water) flowing through the panel. For the present S-IVB and I.U. flight plans, the coolant will enter the cold plates at 60°F maximum and leave at 70°F maximum. A coolant flow rate of approximately 0.5 gpm per panel is used at present. Units generating high heat loads should not be mounted close together since the coolant may not be capable of removing the required heat and excessive temperatures could result. The total allowable heat load per panel is 500 watts.

The mounting methods and the vibration levels predicted during launch allow 150 lb. of equipment to be carried on each plate. Concentrated loads should be avoided. Experiments must be designed so that they can be mounted without interference to the coolant channels.

No cooling is available from the end of the prelaunch phase until approximately 130 sec after lift-off. A pre-launch purge gas system, utilizing air and gaseous nitrogen, provides the forward skirt area with a warming medium. It operates only up to the time of launch and provides no thermal control after that time. This system protects the electronic components and reduces oxygen present to 4% by volume. The total flow rate in the forward area is about 275 lb./min. The purge gas surrounding the components located in the I.U. and S-IVB forward skirt will be at a temperature of 35°F to 75°F.

In the passive system, electronic components are mounted in the aft skirt area on 18 fiberglass panels. No fluid thermo-conditioning system is used. Temperature is controlled through the proper surface finish of each electronic package and by providing conduction paths and insulation. Appropriate coatings are added when a special heating or cooling problem is revealed by calculation or test.

As how designed, each fiberglass panel in the aft skirt area is capable of supporting 100 lb. of electronic packages. Four panels will be available for auxiliary payloads.

A separate pre-launch purge gas system maintains the equipment mounted on the aft skirt at a temperature of 20°F to 70°F during prelaunch procedures. Dry air at a flow of 300lb per minute to the S-IVB stage is provided from a ground source. Gaseous nitrogen purge of approximately the same flow rate is initiated about 30 minutes before LH_2 loading. During flight, heat is radiated to space and to local sinks such as the LOX tank. If possible, high heat dissipating components or temperature-sensitive components should be mounted on the cold plates in the forward skirt.

If the above systems do not meet the needs of an experiment, modifications can be made to the thermal conditioning system or the purge gas system. Example of such changes are:

(a) Coolant flow rate in the thermal conditioning system can be changed to control the temperature of the experiment equipment.
(b) A space radiator could be installed to cool electronic equipment for long periods of time.
(c) Insulation and thermal control coatings may be engineered.
(d) Mounting procedures and requirements can be altered to vary heat conduction paths.

(e) Flow rate and temperature of the purge gas system could be varied.

(f) Purge gas could also be ducted directly to the experiment equipment.

II-4. Payload Acoustics and Vibration Environment

Acoustic and vibration phenomena have a similar time-history during a flight. A time-history of the former is shown in Figure II-17. The acoustic noise level inside the vehicle at three auxiliary and prime payload locations are shown in the figure. At lift-off, the exhaust of the first stage engines generates high frequency noise in the high shear mixing region close to the nozzles and lower frequency noise in the fully turbulent cores of the exhaust jets. This is transmitted through the air to the spacecraft and vehicle.

Following lift-off, the acoustic noise decreases as the exhaust pattern straightens out and as the distance between the vehicle and the ground reflecting surface increases. A further reduction occurs as the vehicle reaches supersonic speeds because the sound generated aft of the vehicle is left behind the vehicle. However, turbulent pressure fluctuations in the aerodynamic boundary layer intensify as free stream dynamic pressure increases. The maximum noise from this source occurs at the time of maximum dynamic pressure or shortly thereafter and it decreases as the dynamic pressure is reduced. The remaining excitation is structurally transmitted from engines, pumps, etc. These are of lower intensity and remain relatively constant until engine cut-off, Brief periods of vibration occur during retro rocket and ullage firings and stage separation.

To provide design criteria for payloads, design specifications have been developed which cover all of these environments. Figure II-18 is a broad-band acoustic specification for three payload locations, and

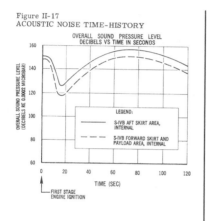

Figure II-17
ACOUSTIC NOISE TIME-HISTORY

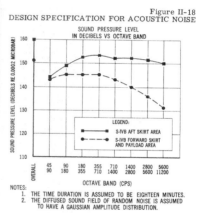

Figure II-18
DESIGN SPECIFICATION FOR ACOUSTIC NOISE

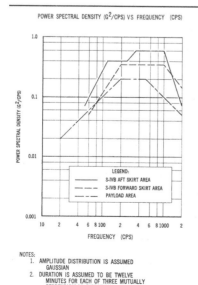

Figure II-19
DESIGN SPECIFICATION FOR RANDOM VIBRATION

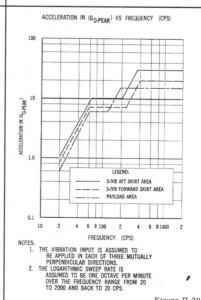

Figure II-20
DESIGN SPECIFICATION FOR SINUSOIDAL VIBRATION

represents an acoustical environment to which an item may be designed and ground tested to ensure satisfactory operation during an actual flight. All frequencies are assumed to be excited at the same time and at the appropriate level in each octave band. Figure II-19 is a broadband random vibration specification for the same purpose. The duration of these qualification tests is longer than the duration of the significant environment for an actual flight to allow for exposure during static firings and to increase the reliability of the items. Figure II-20 is a sinusoidal vibration specification. The purpose of the sinusoidal sweep test requirement is to provide assurance that the item has adequate strength for transitory or unsteady phenomena that could occur in a flight. There is also a shock specification for each location but it is not included in this brief discussion.

II-5. S-IVB Stage Subsystem Information

There are at least four major subsystems of the S-IVB stage that may influence, or be of benefit to an auxiliary payload. They include the auxiliary propulsion, electrical power, thermal conditioning, and data acquisition systems.

II-5-1. Auxiliary Propulsion System

The Auxiliary Propulsion System (APS) modules for the S-IVB stage have two basic configurations as indicated in Figure II-21. The two are necessary to meet the mission requirements of the Saturn V vehicle and the Saturn IB vehicle. The major differences between the two are in propellant capacity and degrees of freedom.

The Saturn V/S-IVB APS is sized to provide roll control during powered flight, three axis attitude control during a 4-1/2 hour earth orbital coast and a two hour translunar coast, and propellant settling for continuous vent initiation and main engine restart. The attitude control function is provided by three 150 lb. thrust engines in each module and propellant settling by a 72 lb. thrust engine in each module. A mock-up of the Saturn V/S-IVB module is shown in Figure II-22. The Saturn IB/ S-IVB APS does not provide for the translunar coast period nor does it have the 72 lb. thrust engines.

The attitude control system is a pulse-modulated on-off system. The system is based on the minimum impulse capability of the 150-lb. thrust engine, which has a minimum impulse bit capability of 7.5 lb.-sec with an electrical input pulse width of approximately 65 milliseconds. The attitude control system is designed to operate with an attitude dead zone of ±1 degree in all axes. The undisturbed limit cycle rates of the Saturn V/S-IVB with payload in a 100 n. mi. circular orbit are approximately 0.02 deg/sec in roll and 0.001 deg/sec in pitch and yaw (0.003 deg/sec during translunar coast).

The auxiliary propulsion system is a completely self-contained modular propulsion sub-system. The modules require electrical power and command signals to provide the necessary stage functions. They are mounted on the aft skirt 180° apart. The equipment for loading propellants to the modules is a semi-automatic system, with individual umbilical connectors in each module.

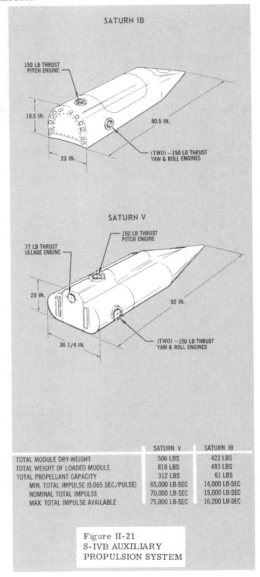

	SATURN V	SATURN IB
TOTAL MODULE DRY-WEIGHT	506 LBS	422 LBS
TOTAL WEIGHT OF LOADED MODULE	818 LBS	483 LBS
TOTAL PROPELLANT CAPACITY	312 LBS	61 LBS
MIN. TOTAL IMPULSE (0.065 SEC/PULSE)	65,000 LB-SEC	14,000 LB-SEC
NOMINAL TOTAL IMPULSE	70,000 LB-SEC	15,000 LB-SEC
MAX. TOTAL IMPULSE AVAILABLE	75,000 LB-SEC	16,200 LB-SEC

Figure II-21
S-IVB AUXILIARY PROPULSION SYSTEM

Each module contains one 72 lb. thrust and three 150-lb. thrust ablatively-cooled liquid bi-propellant hypergolic engines, a positive expulsion (Teflon bladder) propellant feed system for zero gravity operations and a helium pressurization system. Each Saturn V APS module contains 119.4 lb. of MMH (Monomethylhydrazine) fuel and 192.6 lb. of N_2O_4 (nitrogen tetroxide) oxidizer. The nominal oxidizer to fuel mixture ratio is 1.65/1.

The total firing time for the engines is 7 minutes for steady state operations. Pulse operation at a pulse frequency of up to 10 pulses per second is possible.

Testing has demonstrated a pulse mode capability of over 20 minutes accumulated burn time.

Many attitude control and maneuvering functions, other than those now required for the 6-1/2 hour mission, could be performed by extending the S-IVB attitude stabilization capability. An increase in the S-IVB stabilization capability would require design or operational changes to the subsystems to overcome the limitations of the present S-IVB. The major items involved in extended coast characteristics include: (a) the available mass of propellants and pressurization gases, (b) the engine life expectancy, (c) the propellant conditioning requirements to avoid freezing, (d) the attitude control dead bands, (e) the S-IVB electrical power supply, and (f) the IU electrical power supply. All of these items are closely inter-related and affect the coast time capability. If the dead band control zones were to be relaxed to ±2° in pitch and ±10° in yaw and roll, and electrical power added in the form of a fuel cell, then under certain conditions there could be sufficient propellant on-board for controlled coast times up to 30 days. Of course payload, orbit, FPR and orientation also affect coast times and must be studied for a specific mission. Consideration has been given to the above items and preliminary design concepts have confirmed that the necessary modifications can readily be made if the mission requires longer attitude-stabilized coast periods.

II-5-2. Electrical Power System

The S-IVB has four independent electrical systems with 56- and 28-volt silver-oxide primary batteries. Forward system #1 (350 ampere hours, 28 vdc) supplies power to the data acquisition system which produces low-level, high-frequency signals that must be isolated from other systems. Forward system #2 (15 ampere hours, 28 vdc) supplies power to systems which cannot tolerate switching transients or high frequency interference, such as the propellant utilization system and inverter converter. Both batteries for the forward systems are mounted in the forward skirt.

Aft system #1 (270 ampere hours, 28 vdc) supplies power to valves, heaters and relays in the main propulsion engine, pressurization system, stage sequencer, APS modules and ullage rockets which generate switching transients, that must be isolated from other systems. Aft system #2 (70 ampere hours, 56 vdc) supplies power for an auxiliary hydraulic pump, LOX chilldown inverter and LH2 chilldown inverter. Both batteries for aft system #1 and #2 are located in the aft skirt. Both the aft and the forward systems are wired through distribution boxes located in their respective areas.

The batteries are sized to handle the stage load requirements for 6-1/2 hours. If additional power is required for the planned 6-1/2 hours or for longer periods, additional batteries could perhaps be used for as many as 72 hours which is the wet life of the batteries. If power is required for even longer periods in orbit, other batteries or fuel cells could be used.

II-5-3. Data Acquisition System

The early Saturn V R&D vehicles have five telemetry systems; one single sideband/frequency modulation (FM) system, one pulse code modulation (PCM)/FM system, and three FM/FM systems. One channel of each FM/FM system will be used for sampled pulse amplitude modulation data. Pertinent data on these systems are shown in Table II-I.

Vehicle SA-504, to be delivered in mid-1967, and all subsequent vehicles will have only one telemetry system (PCM/FM). The capability of this operational telemetry system is:

TABLE II-I
SATURN V R & D TELEMETRY SYSTEMS
(SA-501, 502 & 503)

T/M System	Frequency (MC/S)	Prime Channels	Prime Sampling Rate (Per sec)
1. SS/FM	226.2	15	Continuous
2. PCM/FM	232.9	0-100 Bi-level + Parallel Acceptance of 3 PAM Multiplexers at	120 120, 40
3. FM/FM	246.3	15	Continuous
PAM/FM/FM		30	120
4. FM/FM	253.8	15	Continuous
PAM/FM/FM		30	120
5. FM/FM	258.5	15	Continuous
PAM/FM/FM		30	120

Total Measurement Capability

1. SS/FM		15 prime channels possible to sub-multiplex by 5 =	75
2. PCM/FM		100 Bi-level channels + 30 prime channels on checkout multiplexer: 3 prime channels for frame sync & calibration; 23 prime channels possible to sub-multiplex by 10 =	234
3. Three-5 FM/FM		15 prime channels possible to sub-multiplex by 3 =	45
PAM/FM/FM		30 prime channels per multiplexer: 3 prime channels for frame sync & calibration; 23 prime channels possible to sub-multiplex by 10 = 234 x 3	702
			1,056

8 channels at 120 samples/second
360 channels at 12 samples/second
<u>190</u> bi-level using remote digital sub-multiplexer
558 total measurement capability

or

44 channels at 120 samples/second
<u>190</u> bi-level using remote digital sub-multiplexer
234 total measurement capability

The 558 or the 234 measurement capability is based on the utilization of two multiplexers. If increased to four multiplexers, the capability becomes:

130 channels at 12 samples/second
4 channels at 120 samples/second
690 channels at 4 samples/second
12 channels at 40 samples/second
<u>190</u> bi-level using remote digital sub-multiplexer
1026 total measurement capability

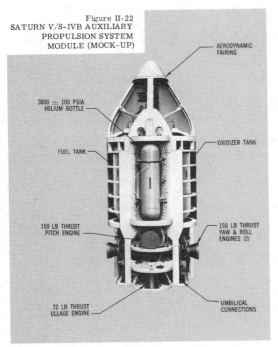

Figure II-22
SATURN V/S-IVB AUXILIARY PROPULSION SYSTEM MODULE (MOCK-UP)

Additional combinations of sampling rates are obtainable. It is estimated that about 50 channels would be available for auxiliary payloads. The exact number of telemetry channels available can only be determined after the vehicle is selected because instrumentation varies from one vehicle to another. Once a payload application is established and scheduled, and bandwidth, accuracy, etc. are known, a determination of available channels can be made.

The operational vehicles (all vehicles after SA-503) also have provisions for mounting one complete set of modified R&D FM/FM systems in kit form. Its capability is 18 channels of continuous data, or with sub-multiplexing, 34, 66 or 82 channels depending upon the sampling rate.

Of course the payload originator may wish to furnish part or all of the Data Acquisition System associated with the experiment.

II-6. Orbital and Deep Space Tracking, Data and Control Stations

Requests for payload data must be integrated into the overall mission plan and approved by the appropriate NASA office. Orbital or space tracking and control functions required by a payload after separation from the Saturn must also be specifically approved.

II-7. Launch Support Facilities

Launch operations for the Saturn V vehicle will be conducted at Complex 39 and will utilize the mobile, or off-pad-assembly, concept. This concept, which provides for a greater flexibility and launch rate than on pad assembly, employs four basic operations: (1) vertical assembly and checkout of the Saturn V on a mobile launcher in a controlled environment, (2) transfer of the assembled and checked-out vehicle to the launch pad on a mobile launcher, (3) automatic checkout at the launch pad, and (4) launch operations by remote control from a distant launch control center. The major units involved in this concept are the Vertical Assembly Building (VAB), Launch Control Center (LCC), Mobile Launcher (ML), Mobile Service Structure, Crawler-Transporter, Launch Pads, and High Pressure Gas Facility. Figure II-23 shows an artist's conception of the Complex 39 area. Figure II-24 shows a schematic illustration of the complex.

The Vertical Assembly Building (VAB) has two major operating areas - High Bay and Low Bay. The High Bay provides the facilities and services to assemble the complete launch vehicle in a controlled environment, and to conduct pre-launch preparations. This building is 524 feet high, 513 feet wide, and 432 feet long, and has four vehicle assembly bays and supporting facilities. Each bay is equipped with extendable platforms which are designed to permit access to the vehicle as it is assembled vertically on a mobile launcher. When the assembly of the vehicle is completed in the High Bay, pre-launch system and subsystem checks are conducted before it is moved to a launch pad.

The Low Bay area has two pairs of bays for performing continuity checks on the S-II and the S-IVB stages, engineering shops, offices, and storage space for stage pre-assembly. The S-IVB area in the Low Bay has two active stage preparation and checkout cells. This building, with its two pairs of assembly bays arranged similarly to those of the High Bay, is 118 feet high, 437 feet wide, and 256 feet long. The transfer aisle portion of the Low Bay, which connects with the High Bay transfer aisle, is 210 feet high.

The Launch Control Center, (LCC) a four-story rectangular building adjacent to the High Bay, is 76 feet high, 378 feet wide, and 181 feet long. The LCC contains offices, a cafeteria, a Complex control center, telemetry and data processing equipment for use during stage and vehicle checkouts. It also houses the firing and computer rooms which contain the control and monitoring equipment required for automatic vehicle checkout and launch.

The Mobile-Launcher (ML), upon which the Saturn V is assembled and launched, can be divided into four major elements—structure, umbilical service arms, firing accessories, and operations test and launch equipment. The structure consists of the two-story launch platform, 25 feet high, 160 feet long, and 135 feet wide, and the umbilical tower. The umbilical tower, mounted on one end of the launch platform, extends 380 feet above the deck of the structure and has eight umbilical swing arms. The arms vary in length from 35 feet to 45 feet and carry electrical, pneumatic, and propellant lines to the space vehicle. The firing accessories installed on, and considered part of the ML, include fuel fill and drain umbilicals, electrical and pneumatic umbilicals, cable masts, pneumatic-valve panels, deluge, flushing, and fire fighting systems, access platforms and ladders, and a heating and ventilating system. The ML operation test and launch equipment includes a ground power system, test sets, and a computer complex.

The Mobile Service Structure is an open-frame steel truss tower designed to perform some functions at a parked position and also to be moved to the pad by the Crawler-Transporter for servicing and arming of the space vehicle. The structure is 402 feet high and is 135 feet by 132 feet at the base.

A Crawler-Transporter is used to position the Mobile Launcher (ML) in the Vertical Assembly Building (VAB), to move the ML space vehicle configurations from the VAB to the launch pad, and to move the Mobile Service Structure from its parked position to the launch pad. It incorporates a large platform and four tractor units as a self-contained vehicle. This equipment, 131 feet long, 114 feet wide, and weighing 5.5 million pounds, is powered by diesel generators developing a total of over 7000 kilowatts for motivation, leveling, and the steering system. The leveling system keeps the ML and the Saturn V Vehicle within one-sixth of a degree of true vertical while negotiating a curve of up to five hundred feet in radius and a 5 per cent grade. The Crawler Transporter, which can be steered from either end, has a normal loaded speed of one mph maximum and an unloaded speed of two mph maximum.

Each launch pad of Complex 39 is in the shape of an eight-sided polygon, with a distance across (perpendicular to crawlerway) of 3000 feet. The overall hardstand area of 390 feet by 325 feet has a center portion elevated 42 feet to allow sufficient clearance for positioning a two-way flame deflector beneath the ML after it is anchored for launch.

Installed at the launch pad at varying distances from the hardstand area are the propellant storage and transfer facilities for LO_2, LH_2, and RP-1 (900,000 gallons for LO_2, 850,000 gallons for LH_2 and 250,000 gallons for RP-1).

Those accommodations which are required to support the launch of an auxiliary payload must be arranged for, or provided by the experiment sponsor, unless available equipment can be used without conflict. It is imperative that the auxiliary payload planner evaluate, at an early date, the ground support equipment and range support required to launch his experiment. Small payloads can often be accommodated within the existing equipment and facilities. Special calibration, checkout, alignment, and handling equipments which are peculiar to the auxiliary payload will require early planning and arrangements. Small payloads can often be handled with relative ease with mobile trailers which are brought in by the payload agency. In the event that building space is required, special arrangements must be made in advance.

One of the earliest actions involved in staging the launch of any payload is the preparation of the Range Data and Support Requirements Document. If separate auxiliary and primary payload documents are required, they must be coordinated.

In the Block House, specific control of the experiment countdown may be handled in one of several ways, namely:

(a) Automatic countdown control of auxiliary payload functions with manual override from payload console.
(b) Manual checkout and control of auxiliary payload functions from payload console.
(c) Automatic integrated countdown of auxiliary payload functions with launch vehicle or primary payload with no separate payload console.

Obviously, the specific requirements and objectives of the auxiliary payload will dictate which of the above modes of operation will be used. Again, early coordination with the launch vehicle is important.

Space can be provided near the Launch Complex for payload originators to park instrumented trailers of their own for remote radio-line checkout if available checkout facilities are not adequate.

II-8. Data Reduction and Evaluation

Prime payloads (mounted above the S-IVB stage and the I.U.) generally will have self-contained telemetry systems; whereas, auxiliary payloads could use available S-IVB or I.U. Telemetry systems. The following information is needed to insure proper use of vehicle telemetry:

(a) Type of measurement (pressure, temperature, signal, vibration, strain, etc.)
(b) Range and accuracy of measurement needed
(c) Type of monitoring (continuous, sampling, real time)
(d) Type of presentation (punch tape, magnetic tape or strip chart)

The interconnect between the test article and the stage will be made by standard flight proven methods. Where appropriate, transducers and signal conditioning devices will be existing flight qualified items.

The worldwide tracking network is utilized for orbital and deep space operations. The tracking network is composed of ground stations around the world which have been established for various space missions. These stations provide the capability of tracking, data acquisition, and communications for space programs. For manned Earth orbital and Lunar operations, the Manned Spaceflight Network or the Space Tracking and Data Acquisition Network would normally be used. For planetary missions, the Deep Space Instrumentation Facility is used. The use of these networks must be arranged in advance with NASA. Procedures exist which will permit data to be recorded in the form of tapes and strip charts from NASA, USAF, and other organizations.

Figure II-23
SATURN V ON PAD 39-KSC

A limited amount of data may be received prior to lift-off via stage umbilicals and could be recorded on strip charts, sequence recorders or magnetic tape as required. All data telemetered from the vehicle is received and recorded during pre-launch, launch, and orbit. Vehicle data may aid reduction and interpretation of payload data. Quick look and in some cases, real time data can be provided by the ETR Facilities.

Douglas can give experimenters partial or complete reduction and evaluation of data in a final report. Douglas has data reduction facilities at the Douglas Huntington Beach Data Laboratory which can reduce data to the following forms:

(a) Analog strip charts
(b) Tabulated digital readouts in engineering units
(c) Plots of digital data in engineering units
(d) Analog oscillograph plots
(e) Digital magnetic tapes in engineering units

The Huntington Beach Computer Facility is equipped with IBM 7094 computers for lengthy, iterative, computation processes.

II-9. Checklist for Auxiliary Payload/Vehicle Interface Requirements

The first important step in planning an auxiliary or prime payload for Saturn V is to document information on the physical and operating characteristics of the payload along with the required launch vehicle accommodations and ground support. With such information, it will be possible for you to discuss the various aspects of Saturn V flight accommodations with appropriate planning agencies. A typical check list is given below of the items of information required to properly consider and define Launch Vehicle accommodations for your payload. Information on the experiment submittal process and associated vehicle data can also be obtained from cognizant NASA Agencies.

A. General Information Required

Experiment title
Proposal Originator
Purpose and application of experiment
Relationship to Apollo or other national goals
Description of experimental procedures
Present status of experimental equipment
Scope of budget or available funding

B. Experiment Mission Requirements

Orbital Altitude; circular, elliptical (apogee-perigee)

Synchronous orbit/Hohmann transfer
(1) circularization by S-IVB
(2) circularization by payload

Suborbital or orbital flight durations, minimum, maximum
Desired launch azimuth
Desired launch inclination
Desired date of launch (year)

Astronauts' time required, pre-flight, inflight, post-flight

C. Experimental Equipment Capability or Requirements
Envelope description or volume requirements
Weight

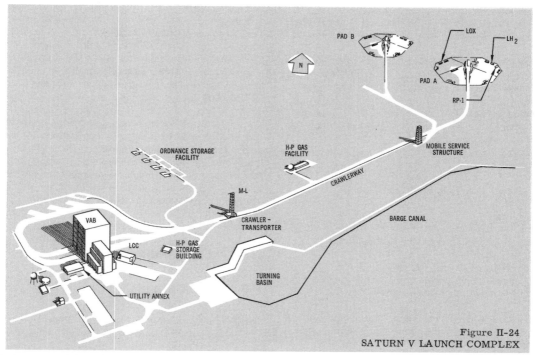

Figure II-24
SATURN V LAUNCH COMPLEX

Environmental Limitations or Capability

(1) temperature
(2) acoustics
(3) vibration
(4) shock
(5) acceleration
(6) humidity and free moisture
(7) atmosphere and pressure
(8) sand and dust
(9) meteoroids
(10) fungus
(11) salt spray
(12) ozone
(13) hazardous
(14) particle radiation
(15) electromagnetic radiation
(16) electromagnetic compatibility
(17) explosion proofing
(18) sterilization requirements
(19) special environmental Interface control

Electrical Power Loads; voltage, current, duration, AC-DC
(1) steady state
(2) intermittent
(3) peak
(4) desired interface locations

Vehicle gas requirements; flow rates, pressures, temperatures
(1) helium
(2) nitrogen
(3) oxygen
(4) hydrogen
(5) others

Jettison Requirements

Special Attitude Control Requirements
(1) stabilization-control precision
(2) angular acceleration and velocity in pitch, yaw and roll

Schedule Information

Range Safety Requirements

D. Instrumentation Requirements
Type and Numbers:
Pressure, temperature, signal, vibration, strain, special

Range and Accuracies

Type of Monitoring
(1) continuous
(2) sampling
(3) real time

Duration or Time period of monitoring.

Interface

(1) transducer part of experimental package
(2) transducer part of stage contractor responsibility
(3) location

E. Final Data

Raw data desired
Reduced data desired
Evaluated data desired
Final data package, reports, tapes, graphs, etc.

F. Shroud Design for Prime Pay loads
Configuration A, B, C, D, or special (Figures II-11, 12, 13 and 14)

G. Suggested Mounting Location

H. Ground Support Equipment Location and Type

Electrical checkout
Pneumatic
Mechanical
Handling
Servicers

I. Tracking, Data Acquisition and Command

J. Facilities
Facilities needed by payload originator at Douglas Space Systems Center or Kennedy Space Center.

K. Special
Any special requirements that affect the integration of the payload with the launch vehicle

If you desire help in integrating your experiment with the Saturn V vehicle, please forward your request to the address shown in the foreword of this guide.

III SATURN V CONFIGURATION

SATURN V

The Saturn V Vehicle is a three-stage configuration designed primarily for accomplishing the Apollo Lunar Landing Mission. This vehicle, shown in Figure III-1, consists of (a) the S-IC first stage, built by The Boeing Company, (b) the S-II second stage, built by North American Aviation, (c) the S-IVB third stage, built by the Douglas Missile and Space Systems Division and (d) an Instrument Unit (IU) built by International Business Machine Corporation. The IU is mounted above the S-IVB and houses the guidance, control and non-stage oriented flight instrumentation. The payload shown above the IU is the Apollo system for the Lunar Landing mission. The payload is made up of a Command Module (CM), a launch escape system, a service module, a Lunar Excursion Module (LEM) and a LEM Adapter which houses the ascent and descent stages of the LEM.

The Saturn V Vehicles represent the largest launch vehicle under development in the United States. Its relationship to the earlier Saturn configurations is shown in Figure III-2. The first Saturn V launch vehicle will be flown in the later-half of 1966. The three stage standard Saturn V, with payload, weighs approximately 3200 tons at liftoff, can place up to 261,000 pounds of payload into a 100 nautical mile circular earth orbit and can accelerate 98,000 pounds to escape velocity. The first-stage engines generate a total of 7,500,000 pounds of thrust at sea level.

The Saturn V Vehicles stand alone in their payload class and have practically unlimited applications to both manned and unmanned earth orbital, Lunar or interplanetary missions. Although a single large payload may be the primary purpose for launching the Saturn V Vehicle, many auxiliary scientific or engineering payloads can also be carried into space economically aboard the Saturn upper (S-IVB) stage of the Saturn.

III-1. LOR/Apollo Configuration

III-1-1. First Stage (S-IC)

The S-IC stage, shown in Figures III-3 and III-4, the first stage of the Saturn V launch vehicle, is manufactured by The Boeing Company at the Michoud Plant near New Orleans, Louisiana. The first four development stages were built by MSFC at Huntsville, Alabama. The stage uses five Rocketdyne F-1 engines, each of which produces a nominal thrust of 1.5 million pounds and uses a mixture of Liquid Oxygen (LOX) and RP-1 (special kerosene fuel) as a propellant. The five engines burn far 150 seconds and lift the vehicle to an altitude of approximately 30 nautical miles before burnout occurs. Four of the engines are gimbal-mounted on a 364 inch diameter circle and are hydraulically gimbaled to provide thrust vector control in response to steering commands from the guidance system located in the Instrument Unit.

The stage utilizes separate propellant tanks that are all welded assemblies of cylindrical ring segments with dome-shaped end bulkheads. Each tank has slosh baffles over the full depth of the liquid. The LOX tank is pressurized by gaseous oxygen while the RP-1 tank uses stored helium.

Eight 80,000 pound-thrust retro-rockets provide separation of the S-IC from the S-II Stage. These solid propellant motors are mounted in pairs under each engine fairing.

III-1-2. Second Stage (S-II)

The second stage of the Saturn V Vehicle is the S-II (Figures III-5 and III-6) which is being developed by North American Aviation's Space and Information System Division at Downey, California. The stage uses five Rocketdyne J-2 engines, each rated at a nominal 205,000 pound-thrust, and burns a mixture of liquid oxygen and liquid hydrogen The propellants are contained in a cylindrical tank with domes at each end and an insulated common bulkhead to separate the

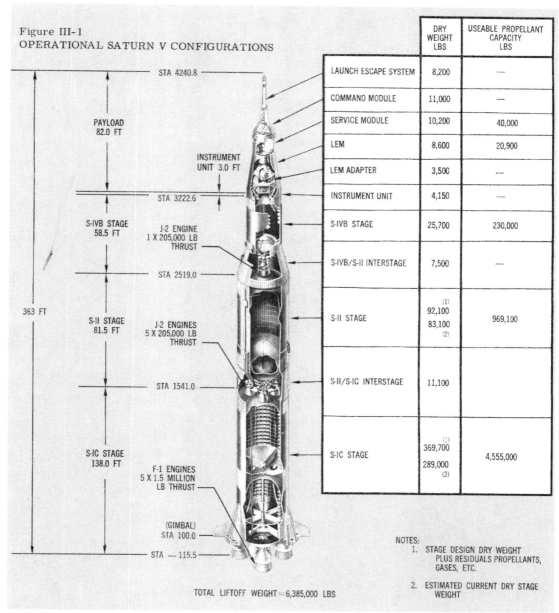

Figure III-1
OPERATIONAL SATURN V CONFIGURATIONS

	DRY WEIGHT LBS	USEABLE PROPELLANT CAPACITY LBS
LAUNCH ESCAPE SYSTEM	8,200	---
COMMAND MODULE	11,000	---
SERVICE MODULE	10,200	40,000
LEM	8,600	20,900
LEM ADAPTER	3,500	---
INSTRUMENT UNIT	4,150	---
S-IVB STAGE	25,700	230,000
S-IVB/S-II INTERSTAGE	7,500	---
S-II STAGE	(1) 92,100 / 83,100 (2)	969,100
S-II/S-IC INTERSTAGE	11,100	
S-IC STAGE	(1) 369,700 / 289,000 (2)	4,555,000

TOTAL LIFTOFF WEIGHT = 6,385,000 LBS

NOTES:
1. STAGE DESIGN DRY WEIGHT PLUS RESIDUALS PROPELLANTS, GASES, ETC.
2. ESTIMATED CURRENT DRY STAGE WEIGHT

upper LH_2 tank from the LOX tank. Each tank contains slosh baffles to minimize propellant slosh. Eight solid propellant 22,900 pound-thrust rocket motors burning for 3.74 seconds are used to ullage the propellants for engine start.

The five S-II engines burn for about 375 seconds and boost the vehicle to an altitude of approximately 100 nautical miles. Four of the engines are gimbal-mounted and are hydraulically gimbaled to provide thrust vector control in response to steering commands from the guidance system located in the Instrument Unit.

III-1-3. Third Stage (S-IVB)

The third stage of the Saturn V is the S-IVB (Figures III-7 and III-8) which is being developed by the Douglas Missile and Space Systems Division at Huntington Beach, California.

The S-IVB has a single 205,000 pound thrust Rocketdyne J-2 engine that burns liquid oxygen (LOX) and liquid hydrogen (LH_2). The Saturn V/S-IVB as presently designed has a 4-1/2 hour orbital plus a 2 hour translunar coast capability. The tankage contains 230,000 pounds of usable propellant at a LOX to LH_2 mass ratio of 5 to 1.

The thrust is transmitted to the stage through a skin and stringer structure shaped in the form of a truncated cone that attaches tangentially to the aft liquid oxygen dome. The hydrogen tank is internally insulated with reinforced polyurethane foam and contains a series of

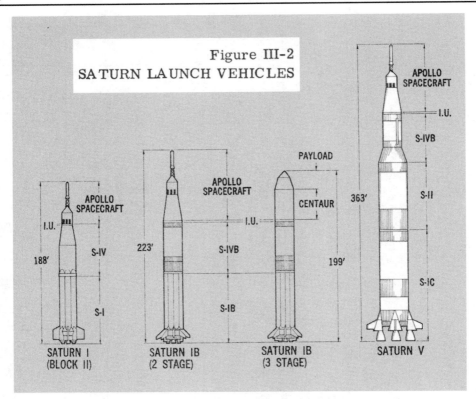

Figure III-2
SATURN LAUNCH VEHICLES

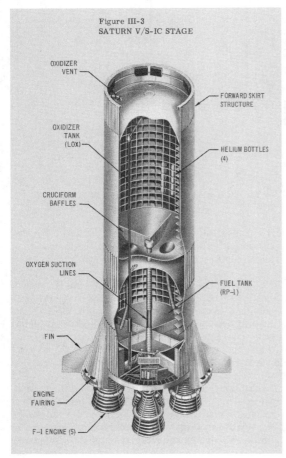

Figure III-3
SATURN V/S-IC STAGE

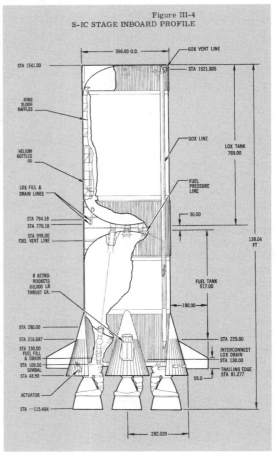

Figure III-4
S-IC STAGE INBOARD PROFILE

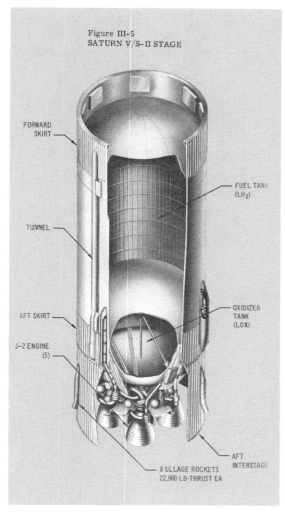

Figure III-5
SATURN V/S-II STAGE

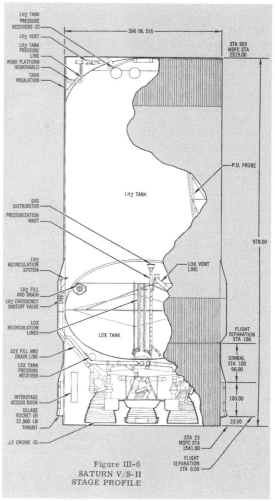

Figure III-6
SATURN V/S-II
STAGE PROFILE

high pressure spheres, storing gaseous helium, for liquid oxygen tank pressurization. Adapter structures, referred to as the forward and aft skirt and the aft interstage, provide the necessary interfaces for mating with the payload and the lower stages. The tank structure features a waffle-like pattern on the hydrogen tank sidewall to act as a semi-monocoque load bearing member. A double walled composite structure with an insulating fiberglass honeycomb core forms the common bulkhead which separates the hydrogen and oxygen tanks. The propellant tanks have spherical end domes. Skirt and interstage structures are composed of conventional skin, external stringers and internal frames.

Pitch and yaw attitude are controlled during powered flight by gimbaling the main engine. Roll control is provided by 150-pound thrust engines located in the Auxiliary Propulsion System (APS) modules. Three axis (roll, pitch, and yaw) attitude control during orbital and translunar coast or unpowered flight is provided entirely by the APS. The signals for vehicle attitude control originate in the guidance and control system located in the instrument unit.

The APS modules are located on the aft skirt assembly of the S-IVB, 180° apart from each other and utilize nitrogen tetroxide (N_2O_4) and monomethylhydrazine (MMH) as the propellant. Each Saturn V/S-IVB module has two 150-pound thrust roll/yaw engines, one 150-pound thrust pitch control engine and one 72-pound thrust ullage engine.

The separation of the S-IVB from the S-II is initiated by an explosive charge which parts the aft skirt from the aft interstage. Two 3400 pound thrust solid propellant ullage rockets mounted on the S-IVB are then ignited and burn for approximately 4 seconds to settle the propellant in the tanks by maintaining a positive acceleration. Four 35,000 pound thrust solid retro-rockets located on the aft interstage are fired simultaneously for 1.5 seconds to decelerate the first stage. The J-2 is ignited 1.6 seconds after separation signal and is at full thrust within 5 seconds of the ignition signal. The J-2 burn-into-orbit is about 152 seconds in duration. The two 72-pound ullage APS motors are burned for about 90 seconds during the J-2 shutdown to maintain propellant control. The ullage engines are fired again for 327 seconds during J-2 engine chilldown prior to the second J-2 start. Ten ambient-temperature helium spheres provide gas for repressurizing the LH_2 and LOX tanks for the restart. A weight saving modification to the stage is in progress which will utilize cold helium stored in bottles within the hydrogen tank and an oxygen-hydrogen burner for the repressurization.

The J-2 engine burns a second time for about 339 seconds to put the payload and S-IVB stage into a translunar trajectory. Engine shutdown is triggered by the guidance system when orbit insertion velocity is achieved, then three axis attitude stabilization is maintained as described above.

III-1-4. Instrument Unit (I.U.)

The instrument unit, fabricated by International Business Machines Corporation, is a 260-in. diameter by 36-in. high cylindrical section located forward of the S-IVB (Figure III-9). The I.U. is designed for a 6-1/2 hr. orbital and translunar coast capability but could be modified for longer durations. The electrical and environmental control systems are the limiting systems. This 3990 lb. unit, which is the "nerve center" of the launch vehicle, contains the guidance system, the control systems and the flight instrumentation systems for the launch vehicle. Access to the inside of the S-IVB forward skirt area is provided through an I.U. door. Electrical switch selectors provide the communications link between the I.U. computer and each stage. The computer controls the mode and sequence of functions in all stages. The I. U. consists of six major subsystems as listed below in (a) through (f).

(a) The structural system or the aluminum cylindrical body of the unit which carries the payload and supports the various systems.
(b) The environmental control system provides electronic equipment cooling during ground operations and throughout flight. The coolant is a 60%-40% methanol-water mixture which circulates through a series of cold plates. In flight, the absorbed heat is removed through a heat exchanger that vents boiled-off water to space. For ground operation, the system rejects heat to a thermo-conditioning servicer.
(c) The guidance and control systems provide guidance and control sensing, guidance steering computations, and control system signal shaping and summing. The shaped control signals are fed to the appropriate actuating devices on the S-IC, S-II and the S-IVB stages.
(d) The measuring and telemetry system transmits signals from the vehicle or experiment transducers during ground check-out and flight to ground command stations by various frequency bands and modulating techniques.
(e) Radio Frequency (RF) systems maintain contact between the vehicle and ground stations for tracking and command purposes. They consist of Azusa and C-band transponders, and an S-band command receiver and transmitter.
(f) A separate electrical system generates and distributes 28 vdc power required for operation of all of the above systems. Some of this power may be available to experimenters.

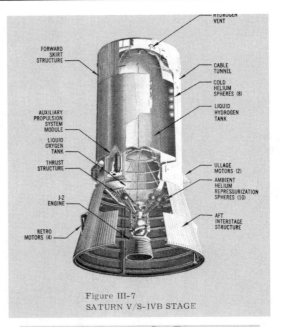

Figure III-7
SATURN V/S-IVB STAGE

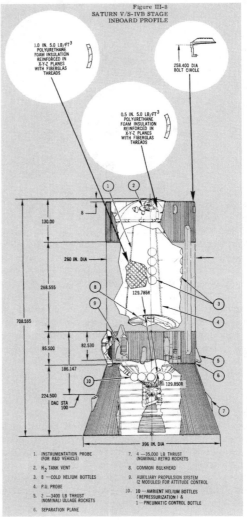

Figure III-8
SATURN V/S-IVB STAGE
INBOARD PROFILE

III-2. Man Rating, Reliability and Quality Control

One prime objective in the design of the Saturn S-V vehicle is to provide a safe, reliable vehicle able to carry a variety of manned and unmanned spacecraft into many different orbits and space trajectories. The approach used to achieve this objective is to impose stringent manrating, reliability, and quality control procedures throughout the design, production and checkout of the vehicle.

III-2-1. Man Rating

The S-V vehicle provides an Emergency Detection System (EDS) for automatic failure warning and also automatic mission abort capability should the crew have insufficient time to react to the failure. When sufficient time exists, the EDS provides the crew with data displays enabling them to decide whether or not to abort. It is designed to minimize the possibility of automatically aborting because of a false signal. The EDS and abort procedures are closely integrated with range safety procedures to ensure that the crew can escape safely. In addition, a Malfunction Detection System (MDS) is being developed for use oh the S-V vehicle.

The principle of the EDS is being expanded into an MDS so that more parameters are monitored, and the crew is provided with additional data displays. The mission go/no-go decision capability of the crew is increased. The safety features of the MDS include those of the EDS, in addition to a greater number of data displays providing the capability of verifying the out-of-tolerance condition of a parameter. This latter capability is added protection against the false abort mode.

III-2-2. Reliability

High reliability of components, subsystems, and systems are a basic design parameter of the Saturn Program. The program for attaining and maintaining high reliability consists of the following elements:

(a) Failure effect analysis of the design.
(b) A thorough and complete test program.
(c) Imposition of stringent reliability and quality control procedures.

The failure effect analysis consists of a detailed technical analysis of the design of the system to identify all the possible significant failure modes, categorizing the effects of each failure mode, and elimination of failure modes.

Thorough component, subsystem and system tests are conducted in the laboratory, at the Static Test Facility and during prelaunch checkout. Post-flight data evaluation of vehicle systems serve as a tool to assess reliability for future missions. Carefully planned procedures and controls used in these tests and data

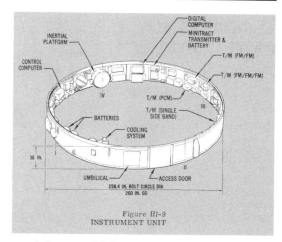

Figure III-9
INSTRUMENT UNIT

correlations establish a measure of reliability and determine the level of confidence in the measure. All these factors help meet the Saturn V reliability goal of 0.90. The reliability goal of the S-IVB stage is 0.95 at a 90% confidence level. Experimental payloads will be given the same attention to ensure the same high probability of success.

Stringent quality control standards in manufacture, fabrication, and testing ensure that reliability will not be degraded by human error or by manufacturing techniques. NASA documents of the NPC 200 series, (Quality Program Provisions) and NPC 250-1, (Reliability Provisions for Space System Contractors), contain the reliability requirements and quality control standards which guide payload planners.

III-2-3. Quality Control

Strict quality control standards including a comprehensive failure reporting system are implemented to assure non-degradation of reliability during manufacturing phases. Failure data is used to update reliability estimates and improve design. In addition, the failure data is automated and used throughout the test and checkout program to provide rapid, comprehensive reporting of the failure history of the vehicle.

A traceability program has also been implemented for the vehicle on all items which could cause an aborted mission during final countdown or a flight failure. Traceability is also imposed on those items utilizing new processes, new or exotic materials, or new and unique applications of old materials and processes. These items are serialized, lot coded or date coded and evidence that all inspections and test operations have been performed is retrievable. Thus, traceability requirements provide evidence of proper configuration of critical components.

The high reliability of the booster vehicle should be matched by high payload reliability. Therefore to provide assurance of the experiment's success, the reliability requirements imposed on the experiment and/or the payload must be at least equal to that of the S-IVB Stage.

IV SATURN V PERFORMANCE

IV-1. LOR/Apollo Mission Profile

The Saturn V three stage launch vehicle is being developed to provide a booster system for the Apollo Lunar Orbital Rendezvous (LOR) and Landing mission. That is, the Saturn V is designed to fulfill the mission velocity requirements of the Apollo mission up to injection of the Apollo spacecraft and Lunar Excursion Module (LEM) into a 72-hour lunar transfer trajectory. This mission requires two stages and a portion of the propellant from the third stage to achieve a one-hundred nautical mile circular orbit. A coast of up to three revolutions in this orbit (4-1/2 hours) is allowed for vehicle checkout and to determine the precise achieved orbit for third-stage engine restart operations. At the proper time, the third (S-IVB) stage re-ignites and boosts the spacecraft into the lunar transfer trajectory. During the translunar flight following S-IVB main engine shutdown, the spacecraft/launch vehicle separation occurs, followed by spacecraft transposition, docking, and midcourse orbit corrections. A sequence of events summary for the LOR/Apollo mission is illustrated in Figure IV-1. Upon reaching the vicinity of the moon, the spacecraft is injected into a parking orbit about the moon and the LEM separates, and descends to the lunar surface.

IV-2. General Three-Stage Mission Profile

The Saturn V three-stage launch vehicle is capable of placing heavy payloads in various circular, elliptical, or hyperbolic orbits. The vehicle capabilities and flight profiles for such missions are described in the following paragraphs:

For three-stage missions, the launch sequence is initiated with the ignition of the five first-stage engines. The vehicle rises vertically to clear the umbilical tower and rolls to align the pitch plane with the desired launch azimuth. Initiation of a pitch program then starts the vehicle down range. The pre-programmed pitch attitude history is designed to follow a ballistic trajectory (zero angle of attack) under no-wind conditions, At 154.6 seconds after liftoff, the center engine is shut down and four seconds later the remaining four control engines are shut down. Three and eight-tenths seconds are then required to separate the empty S-IC stage and reach 90% thrust on the S-II stage. After approximately thirty seconds, to ensure vehicle attitude stabilization, the forward section of the two-piece S-IC/S-II interstage is jettisoned. At approximately this same time, the launch escape system will be jettisoned on manned flights.

The S-II stage, which uses a programmed propellant mixture ratio to optimize the engine thrust/specific impulse history, reaches propellant depletion and is separated from the S-IVB stage approximately 536 seconds after liftoff. The second and third stage attitude history is determined by an iterative guidance scheme based on the calculus of-variations which minimizes the propellant burned in reaching the desired burnout velocity and position. A command cutoff occurs upon injection of the S-IVB/Payload into the desired orbit. If this orbit is a parking orbit (intermediate to a terminal point), the S-IVB will be reignited at some time later and will propel the payload to the desired final conditions of velocity and position.

Table IV-1 gives the weight breakdown of the Saturn V stages for the three-stage launch vehicle. The S-IVB propellant figure shown represents the total tank capacity. The actual amount consumed may be less and is dependent upon the specific mission.

Figure IV-2 presents the circular orbit capabilities for the three-stage vehicle for direct ascent missions to various orbit altitudes and inclinations when launched from the Eastern Test Range (ETR). Orbital payload capability, via Hohmann transfer, is shown in Figure

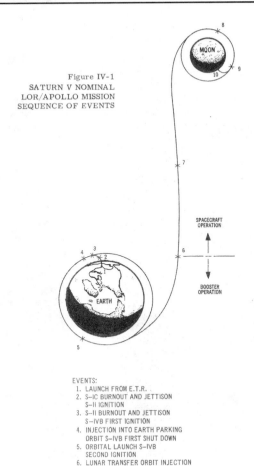

Figure IV-1
SATURN V NOMINAL
LOR/APOLLO MISSION
SEQUENCE OF EVENTS

EVENTS:
1. LAUNCH FROM E.T.R.
2. S-IC BURNOUT AND JETTISON
   S-II IGNITION
3. S-II BURNOUT AND JETTISON
   S-IVB FIRST IGNITION
4. INJECTION INTO EARTH PARKING
   ORBIT S-IVB FIRST SHUT DOWN
5. ORBITAL LAUNCH S-IVB
   SECOND IGNITION
6. LUNAR TRANSFER ORBIT INJECTION
   S-IVB SECOND SHUTDOWN, APOLLO
   TRANSPOSITION, DOCKING, AND S-IVB
   JETTISON
7. MID-COURSE CORRECTIONS
8. INJECTION INTO LUNAR PARKING
   ORBIT
9. DE-ORBIT LUNAR EXCURSION MODULE
10. LUNAR EXCURSION MODULE TOUCHES
    DOWN

TABLE IV-I SATURN V VEHICLE WEIGHT SUMMARY

Weights	lb.
S-IC at Separation	381,645
S-IC Stage/Residuals	(369,700)
S-IC/S-II Aft Interstage [1]	(1,330)
S-IC/S-II Separation/Start Losses	(10,615)
S-IC Propellant	4,555,003
S-IC/S-II Forward Interstage [1]	9,770
S-II at Separation	100,664
S-II Stage/Residuals	(92,139)
S-II/S-IVB Interstage	(7,468)
S-II/S-IVB Separation/Start Losses	(1,057)
S-II Propellant	969,078
S-IVB at Separation	28,549
S-IVB Dry Stage [2]	(25,708)
S-IVB Residuals	(2,841)
S-IVB Total Usable Propellant Capability	230,000
Flight Performance Reserves [3]	(2,907)
S-IVB Weight Loss in Parking Orbit [4]	3,495
Instrument Unit	4,150
	6,282,354 [5]

Note:
(1) Two plane separation - 1,330 pounds separates with S-IC. 9,770 pounds is carried with S-II for 30 seconds before jettison.
(2) Includes 204 pounds of jettisoned weight (ullage cases, etc.).
(3) Typical - 0.75 per cent of vehicle characteristic velocity, included in S-IVB propellant usable capacity.
(4) Typical for 4-1/2 hour coast.
(5) Total weights do not include payload, payload adapter, shroud, or launch escape system. Weights based on projected data for Vehicle SA-504.

IV-3 for the case of a due east launch. These data are based on the assumptions that the launch site is in the plane of the desired orbit and no trajectory plane-changing ("dog-leg") maneuvers are performed.

The approximate sector of allowable launch azimuths without requiring a "dog-leg" maneuver is between 40 to 140 degrees (launch orbit inclinations greater than approximately 55 degrees, or launch azimuths less than 40 degrees or greater than 140 degrees, would require special range safety waivers or a "dog-leg" ascent with a resultant decrease in payload).

Figure IV-4 shows the payload capabilities to a synchronous orbit (24-hour period) via Hohmann transfer as a function of orbit inclination for selected launch azimuths. The required plane rotation is accomplished coincident with circularization at apogee. Capabilities for a 60 degree inclined synchronous orbit are shown in Figure IV-5 as a function of launch azimuth. These missions require a third start capability which the S-IVB presently does not have. Weight penalties associated with a third start have been included.

Figure IV-6 shows the payload capability of the three-stage vehicle to various elliptical orbits for a due east launch. The interplanetary mission capability of the Saturn V Vehicle is shown in Figure IV-7. These data are based on a due east launch to a 100 nautical mile parking orbit and an orbit launch by the S-IVB. Some representative trajectory parameter histories for a high energy mission are also shown in Figure IV-8 through IV-14.

V SATURN V GROWTH POTENTIAL

V-1. Growth Configurations of the Saturn V

The three-stage configuration and performance capabilities described in this document are based on present estimates of a standard operational Saturn V Vehicle. Means of improving the basic vehicle to achieve higher performance are continually being studied. The various performance improvement techniques include such items as (a) adding engines to the S-IC and S-II, (b) increasing the thrust of the F-1 and J-2 engines, (c) increasing the propellant capacities of each of the three stages, (d) using improved engines on, the S-II and S-IVB with higher thrust and performance and (e) using solid motor strap-ons on the S-IC.

Other ways to improve the payload delivery capabilities and the versatility of the Saturn V will undoubtedly be studied for some time. The present operational vehicle can launch a payload weighing over 261,000 pounds to a 100 n. mi. circular orbit and 98,000 pounds to escape velocity. Improved performance studies indicate that payloads of 450,000 pounds to a 100 n. mi. circular orbit and 170,000 pounds to escape velocity are possible. The results of Saturn V improvement studies have shown that the Saturn V class of vehicles has substantial growth potential for future missions.

V-2. High Energy Mission Vehicles

The exploration of the solar system and the space beyond presents an exciting challenge to the scientific and engineering community. High spacecraft velocities or energies are required to explore the solar system or to perform deep space missions. The present Saturn V, even with the performance improvement techniques described above, is not capable of performing certain high energy missions; therefore, a high energy upper stage must be added. The Saturn V, with an additional upper stage, can offer the highest payload potential of any vehicle under development for these high energy missions. The Saturn V/Centaur is an example of a vehicle representing near term availability. The Centaur would serve as a high energy fourth stage on the Saturn V as depicted in Figure V-1. The four-stage mission profile is similar to the three-stage mission up to the one hundred nautical mile parking orbit. However, the shroud enclosing the Centaur is jettisoned at an altitude of approximately 340,000 feet, 214.5 seconds after liftoff. The S-IVB re-ignites in orbit, expends its propellant and is jettisoned. The Centaur stage is ignited and propels the payload to the desired final conditions.

Table V-I gives the weight breakdown of the Saturn V stages of the four-stage launch vehicle. The S-IVB and Centaur propellant data shown represent tank capacities. The actual amount needed would depend on the specific mission. In determining the performance of the four-stage vehicle, an S-IVB propellant loading of 230,000 pounds was used. This is not optimum for all missions and some small gain in performance may be achieved through optimization (within the tank capacity limits) of this parameter for a specific mission. The Saturn V/Centaur Vehicle has the capability to perform the types of missions as listed in Table V-II and illustrated in Figure V-2. Representative performance curves and trajectory parameter histories for a high energy mission are shown in Figures V-3 through V-8 for the four-stage Saturn V/Centaur Vehicle launched due east from the Eastern Test Range through a 100 nautical mile parking orbit.

TABLE V-I VEHICLE WEIGHT SUMMARIES

	Weights Lb.
S-IC at Separation	381,645
S-IC Stage/Residuals	(369,700)
S-IC/S-II Aft Interstage [1]	(1,330)
S-IC/S-II Separation/Start Losses	(10,615)
S-IC Propellant	4,555,003
S-IC/S-II Forward Interstage	9,770
S-II at Separation	100,664
S-II Stage/Residuals	(92,139)
S-II/S-IVB Interstage	(7,468)
S-II/S-IVB Separation/Start Losses	(1,057)
S-II Propellant	969,078
S-IVB at Separation	28,549
S-IVB Dry Stage [2]	(25,708)
S-IVB Residuals	(2,841)
S-IVB Total Usable Propellant Capacity	230,000
Flight Performance Reserve [3]	-
S-IVB Weight Loss in Parking Orbit [4]	3,495
Instrument Unit [8]	3,100
Centaur Shroud	5,933
Centaur Separation Weight [5]	4,564
Centaur Insulation Panels [6]	1,258
S-IVB/Centaur Interstage	630
S-IVB/Centaur Separation/Start Losses	146
Centaur Usable Propellant	29,900
Centaur Weight Loss in Parking Orbit [4]	475
	6,324,210 [7][8]

Note:
(1) Two plane separation - 1330 pounds separates with S-IC 9770 pounds is carried with S-II for 30 seconds before jettison.
(2) Includes 204 pounds of jettisoned weight (ullage cases etc.).
(3) Normally computed on a basis using a percentage of vehicle characteristic velocity.
(4) Typical for 4-1/2 hours coast.
(5) Centaur weight includes its own guidance for use during Centaur flight.
(6) Jettisoned after Centaur ignition.
(7) Total weights do not include payload, payload adapter, shroud, or launch escape system.
(8) Based on projected data for SA-513 and subsequent vehicles.

Saturn V can carry:
Experiments in the space sciences
Engineering tests that require actual environment of space
Prime missions that require a launch system to provide great weight-lifting ability or high velocity.

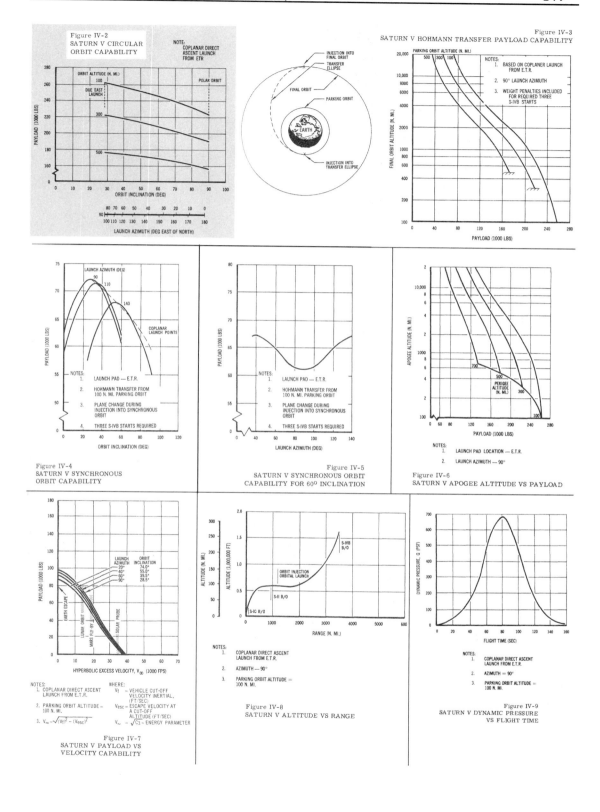

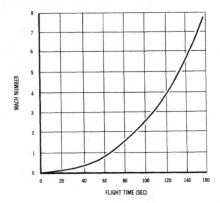

Figure IV-10
SATURN V MACH NUMBER
VS FLIGHT TIME

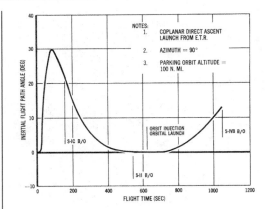

Figure IV-11
SATURN V INERTIAL FLIGHT
PATH ANGLE VS FLIGHT TIME

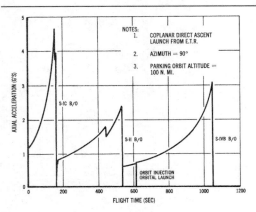

Figure IV-12
SATURN V AXIAL
ACCELERATION VS
FLIGHT TIME

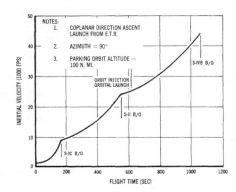

Figure IV-13
SATURN V INERTIAL
VELOCITY VS FLIGHT TIME

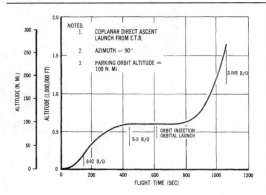

Figure IV-14
SATURN V ALTITUDE
VS FLIGHT TIME

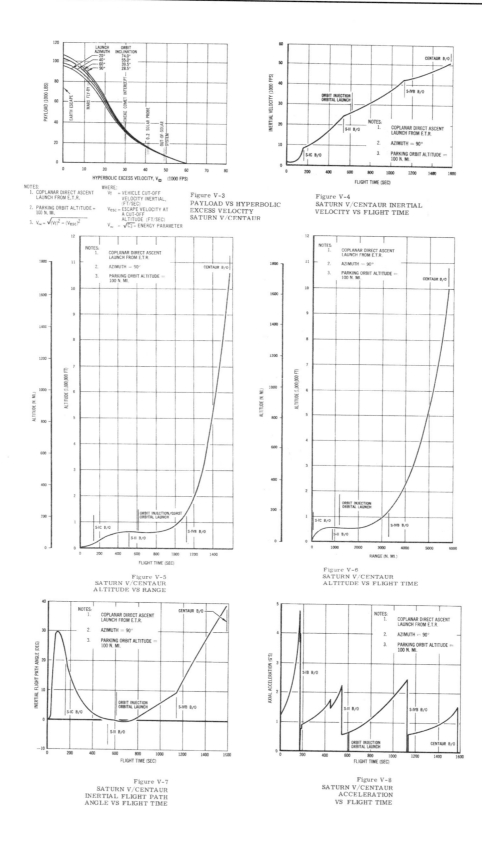

Figure V-3
PAYLOAD VS HYPERBOLIC EXCESS VELOCITY
SATURN V/CENTAUR

Figure V-4
SATURN V/CENTAUR INERTIAL VELOCITY VS FLIGHT TIME

Figure V-5
SATURN V/CENTAUR
ALTITUDE VS RANGE

Figure V-6
SATURN V/CENTAUR
ALTITUDE VS FLIGHT TIME

Figure V-7
SATURN V/CENTAUR
INERTIAL FLIGHT PATH ANGLE VS FLIGHT TIME

Figure V-8
SATURN V/CENTAUR
ACCELERATION VS FLIGHT TIME

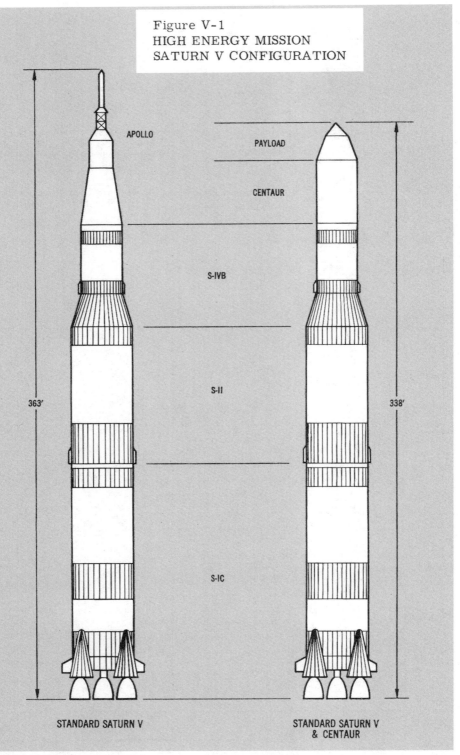

Figure V-1
HIGH ENERGY MISSION
SATURN V CONFIGURATION

Uprating of the basic Saturn V, as previously described in combination with the Centaur, would result in corresponding payload increases as shown in Table V-II. The basic Saturn V LOR Vehicle thus provides a base for substantial growth potential for accomplishing the high energy missions.

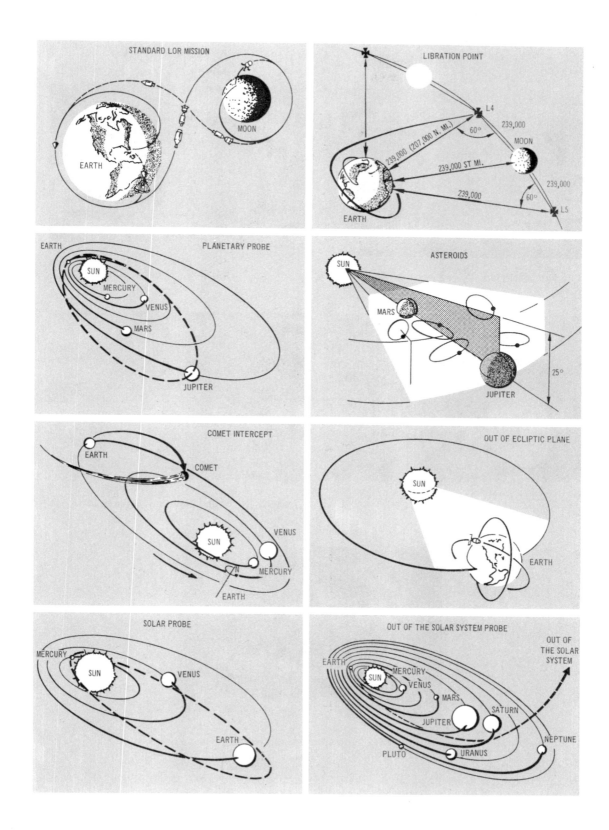

The Authors would like to acknowledge the invaluable assistance of the following people who made this book possible.

Olwyn Lawrie with Alan Lawrie at Rocketdyne

Bob Bell at Michoud

Pauline Roe

Arlene Royer and Charlie Reeves at NARA

Bob Jaques with Alan Lawrie
at the S-IC test stand MSFC

Carol Castle with Alan Lawrie
at the B Test Stand SSC

Paul Coffman at Rocketdyne

Tori Nichols

Dennis Wingo

Anne Coleman at the Salmon Library UAH

INDEX

A-1 Test Stand	173, 211, 216-219, 224-226, 231, 235-236, 239-241, 243-244
A-2 Test Stand	137, 173, 208, 210-211, 213, 215, 219-222, 227-228, 232-234, 237-239, 242, 244, 246-247
AEDC	140, 149, 253
Aft Skirt	49-53, 55, 60-65, 72, 75-76, 195-196, 199, 201-203, 205, 208, 212, 214, 217, 223, 241, 249-250, 255, 258, 260-261, 263-264, 279, 281, 284-285, 293, 298, 302, 304-305, 313
Alma (hurricane)	152, 206, 256
Aloha State	257
ASRC	142, 146, 152-153, 205, 258
Assembly and Checkout	67, 95-96, 101, 109, 136, 275, 278, 280-282, 285, 306
Automatic Checkout	75, 96, 103, 105, 108, 202, 235, 306
B-1 Test Stand	192
B-2 Test Stand	153-154, 160-161, 167-171, 173-192, 194
Battleship	19, 94-95, 99, 108, 135, 138-139, 149, 201, 203-205, 210, 250-253, 287
Beta I test stand	253, 260-261, 264-269, 272-273
Beta III test stand	255, 262, 264, 268-272, 274-279
Betsy (hurricane)	151, 153, 165, 167, 256
Boeing	11, 18, 21, 26, 93, 100, 123, 135-136, 138, 141, 149-155, 160, 162-163, 165-168, 170-182, 186, 188, 190, 192, 194, 196, 286, 310
Bowl SSFL test area	138, 146
Bravo SSFL test site	139
Camille (hurricane)	179, 181, 186, 188, 190, 236, 238
Canoga Park (City of)	11, 97, 136, 138, 140-141, 143, 145-149, 203
Canyon SSFL test area	138
Coca SSFL test site	94, 139, 197-199, 203-205
Control Pressure System	27, 34-35, 119
Courtland docks	260
Delta	139, 147
Douglas	11, 18, 67, 95-96, 124, 136-137, 249, 252, 255-257, 259-264, 267, 269, 271, 273-283, 285-287, 292-294, 301, 308, 310-311
Douglas Aircraft Company	136, 252, 257
Downey (City of)	94, 197, 201-202, 238, 310
Dynamic Test Stand MSFC	135, 153-154, 207, 257-258
EAFB	139, 142, 145
East Test Area MSFC	135, 142-143
Edwards Air Force Base	139, 142-143, 145
Edwards Field Laboratory	136
Electrical System	32-33, 39, 45, 61, 85, 92, 314
Engine Interface	44-45, 129
Environmental Control System	26, 35, 67, 71, 75, 85, 87-88, 103, 314
F-1 engine	13-14, 17-20, 22, 26-30, 34, 37-41, 47, 50, 97-99, 104, 106, 111, 135-136, 138-145, 149, 151, 155-159, 163-165, 167, 170, 172-174, 176, 178, 180-182, 186, 188, 190, 192-194, 228, 237, 287, 293, 310, 317
Fairings	24-25, 76, 109, 150, 154, 298
Fins	21, 24, 53, 109, 150, 154
First Stage	12-14, 16-19, 21-27, 31-37, 39, 50-51, 57, 59, 61, 80, 83, 89, 93, 98-101, 104, 108-111, 120-121, 125, 135, 139, 141, 150, 155, 163, 165-166, 168, 171, 174-175, 177, 179, 181-182, 184-185, 187, 189, 192, 194, 207, 218-220, 233, 255-256, 258-259, 266, 293, 303, 310, 313
Flight Control System	60, 67, 71-72
Flight Testing	103, 106-108
Forward Skirt	20-21, 24-26, 33, 35-36, 49, 51-52, 54-55, 60, 63-65, 70, 74-76, 151, 155-156, 158, 163-165, 167, 170, 176-177, 180, 182, 188, 190, 197-199, 201-202, 207-208, 212, 217, 221, 227, 230, 232, 235, 237-238, 240, 247, 250, 259-260, 278, 284-285, 296-299, 302, 305, 314
Fuel tank	20-23, 25-30, 32, 37, 42, 45, 69-71, 123, 149-151, 153-155, 157-158, 162-165, 167, 170, 173-183, 185, 187-189, 191, 193-194, 218-219, 251, 278
Gas Generator System	39, 42, 68, 129
Gimbal Bearing	39, 78, 129
Green Mountain State, ship	256
Ground Support	28, 31, 33, 58, 62, 67, 87, 98-99, 109, 112, 119, 122, 125, 135, 184, 260, 307-308, 310
Houston (City of)	14, 103, 110, 189, 192, 244, 248, 255-256, 282
Huntington Beach (City of)	11, 66-67, 95, 136-137, 141, 149, 249, 255-257, 259-260, 263-266, 268, 270-285, 287
Huntsville (City of)	11, 15, 21, 86-87, 93, 96-98, 108, 124-127, 135, 140, 142, 145-146, 152-155, 158, 161, 163, 196-197, 199, 201-202, 205, 207, 211, 248-250, 257-259, 278
Hydraulic Control System	39, 42, 45, 129, 144
Hypergol Cartridge	40, 45, 47, 129-130
IBM	11, 16, 19, 86-91, 96-97, 125, 248-249, 308
Instrument Unit	293-294, 298, 310-311, 313-314, 317-318
Instrumentation System	27, 33, 39, 46-47, 60, 68, 74, 78, 83, 85, 91, 129-132
Interstage	12, 49-52, 55, 59-66, 75-76, 95, 241, 254, 266, 293, 297, 313, 316-318
Intertank	20, 23-26
J-2 engine	12-15, 17-18, 49-52, 56, 58, 60-65, 67-74, 77-80, 98, 100, 106, 121, 135-136, 138-140, 145-149, 202-204, 206-210, 212, 214, 217, 219, 221, 224-227, 230, 232-233, 235, 237, 239-240, 242-243, 246-247, 250-253, 256-257, 259-260, 262, 265, 267, 270-276, 278-285, 287, 293-294, 310-311, 313-314, 317
Johnson Space Center	189, 192, 248, 282
Kennedy Space Center	11, 13-15, 18-19, 21, 50, 86, 93-94, 99, 101-103, 108-109, 111, 119-120, 122, 141, 155, 295, 310
Kerosene	141, 293, 310
KSC (Kennedy Space Center)	15, 19, 102-104, 122, 136, 140-141, 151-152, 156, 161-163, 165-166, 168, 171, 173-175, 177, 179, 181, 184-187, 189, 192, 194, 199, 205-207, 211, 213-216, 219, 223, 226, 228-229, 231, 233-234, 236, 239, 241, 243-246, 248, 255-257, 260-262, 264-267, 269-270, 272-273, 275-280, 282-283, 285, 287
Laurie (hurricane)	187
LH2 (liquid hydrogen)	12, 49, 54-55, 57-59, 63, 66-71, 73, 84, 122, 195-197, 200-203, 206-207, 210, 212-215, 217-219, 221-235, 238, 240, 242, 244-246, 250, 252, 259, 261, 263-264, 269-270, 272-278, 280-283
Liquid Hydrogen Tank	50-54, 56-58, 62, 119, 131, 298
Liquid Oxygen Tank	313
Liquid Oxygen Tank	23, 50-53, 55-56, 58, 62, 95
Little Lake, barge	141, 180, 208, 217, 219, 222, 230
Los Alamitos (City of)	123, 136-137, 141, 256, 258, 261, 263, 265, 267-268, 271, 273-279, 282
Los Alamitos Naval Air Station	136-137, 141, 256, 258, 261, 263, 265, 267-268, 271, 273-279, 282
Los Angeles (City of)	94, 97, 123-128, 136-139, 141, 145-146, 149, 200, 212, 229, 240, 243, 263, 271, 275

INDEX

LOX (liquid oxygen)	12-13, 20-32, 34, 36-37, 39, 41, 43-49, 52-53, 55, 57, 59, 63, 65-69, 71, 73, 83-84, 103-104, 121-122, 149, 151, 153-160, 162-165, 167-168, 170, 174, 176, 178, 180, 182, 185, 188, 190, 194, 196, 201, 203-204, 206, 212-213, 215-218, 220, 222-225, 227-235, 237-238, 246, 249-251, 257, 259-261, 264, 266, 270, 272, 274-278, 281-282, 293, 298, 302, 305, 310-311, 314
Marshall	11, 13, 15, 18-19, 21, 93, 98-100, 108, 111-112, 122, 125, 135, 137, 140, 162-163, 165, 286-287
Marshall Space Flight Center	11, 13, 15, 18, 21, 93, 98-100, 108, 111, 122, 131, 135, 140, 286
Mather Air Force base	137, 141, 260-263, 265-273, 275-279
McDonnell Douglas	136, 256, 267, 269, 271, 273-283, 285-287
Measurement System	32-33, 60, 131
Michoud Assembly Facility	18, 21, 25, 93-94, 101, 141, 145
Mississippi Test Facility (MTF)	18, 21, 93-95, 100, 286-287
Mississippi Test Operations	137, 286
MSFC	18-19, 86, 99, 112, 122, 135, 140-143, 145, 149-166, 168, 170, 173-174, 196-197, 199-202, 205-208, 210-212, 219, 222, 228, 233, 238, 240, 242-243, 245, 249-252, 256-259, 264, 278, 286-287, 310
NASA	11, 13-18, 25, 86, 92-93, 97, 100-101, 111-112, 119-120, 122, 135-140, 142, 144-145, 148-149, 152, 160, 162-163, 165-166, 168, 170-171, 173, 175, 179, 181, 184, 187, 189, 192, 195-197, 199, 201-202, 204-206, 208, 210, 223, 226-228, 231-232, 234, 236-240, 242-243, 245, 247, 249, 260, 264, 286-287, 292-295, 306-308, 315
NASM (Space Museum)	282-283
New Orleans (City of)	11, 18, 21, 93, 101, 141, 145, 163, 192, 257
North American	11, 17-18, 50, 94-95, 97, 100-101, 126-128, 136, 138, 142, 202-203, 212, 221, 224
North American Aviation	11, 17-18, 50, 94, 100-101, 126-127, 136, 138, 202-203, 212, 221, 310
Ordnance System	61
Orion, barge	141, 168, 171, 174-175, 177, 189, 200, 223, 229, 255-257
Oxidizer Dome	39, 43-45, 78
Oxidizer tank	41, 68, 82, 150, 170
Panama Canal	208, 211, 221, 255, 257
Payload	292-301, 303-308, 310, 313-318
Pearl River, barge	101, 161, 167, 170-171, 173, 175, 208, 211, 213
Point Barrow, ship	141, 143, 196-197, 201, 206, 208, 211, 213, 215, 218-219, 221, 224-226, 228, 230, 233, 235, 237, 240, 242, 244, 246-248, 255-256, 258, 280, 285
Port Huemene	197-198, 200
Poseidon, barge	94, 151-152, 154, 160-161, 163, 165-166, 214-216, 244
Pressurization System	27-29, 31, 39, 44, 58, 68, 70, 73, 119, 151, 159, 267, 304-305
Propellant Feed Control System	43
Propellant System	34, 56, 67-68
Propulsion System	27, 32, 47, 65-67, 72-73, 78, 83, 122, 173, 203, 260, 270, 273, 293, 300, 304, 313
Pyrotechnic Igniter	40
Range Safety System	32, 34, 76, 171
Redstone Airfield, Huntsville	141, 145, 256
Rocketdyne	11, 16-18, 78, 94, 97, 127-128, 136, 138-139, 141-146, 148-149, 171, 200-201, 203-204, 226, 257, 287, 293, 310-311
RP-1 (kerosene)	12-13, 20, 27, 31-32, 37, 39, 103-104, 141, 293, 307, 310
Sacramento (City of)	19, 67, 95-96, 136-137, 249, 255, 260-264, 286
SACTO	137-138, 141, 149, 249-250, 252-253, 255, 260-262, 264-279
Santa Monica (City of)	95, 125, 127-128, 136, 249-250, 255, 259, 261-262, 264, 266, 270, 272, 275, 277, 282
Santa Susana Field Laboratory	97-98, 141
Seal Beach (City of)	11, 51, 54-55, 94-95, 136, 141, 145, 149, 195-197, 199-201, 203, 205-206, 208, 211-215, 217-221, 223-225, 228-233, 235-237, 239-240, 242-248, 255-258, 280-282, 285
Seal Beach Naval Docks	141, 196-197, 200, 280, 285
Second Stage	12-14, 16-19, 24, 34, 37, 49-56, 59-62, 65, 76, 78, 80-83, 89-90, 94-95, 99-101, 110, 120-121, 136, 145, 186, 207, 211, 214, 216, 219, 223, 226, 229, 231, 234, 236, 239, 241, 243-244, 248, 256, 293, 310
Skid strip	141, 262, 277, 279, 282
Smithsonian Institute	142, 283
SSC	137, 263, 273-274, 276, 284, 286
Stennis Space Center	137, 139
Super Guppy	96, 136-137, 141, 249, 256, 258-279, 282
Systems Tunnel	51, 54
Thermal Control System	60, 213
Thermal Insulation	39-40, 129, 144
Third Stage	12-15, 18-19, 49-51, 54, 60-61, 64-68, 71-76, 78, 81-85, 87, 90, 95-96, 99, 110-111, 121, 136, 140, 188, 207, 211, 249, 256, 260, 262, 265, 267, 270, 272-273, 275, 277-280, 293, 298, 310-311, 316
Thrust Chamber Assembly	39, 78, 81
Thrust Chamber Injector	39, 45, 78, 83
Tullahoma	253
Tulsa (City of)	201, 208, 211-212, 216, 219, 240-241
Turbopump	37, 39, 41-48, 59-60, 69-72, 78, 80-84, 98-99
Ullage Motors	49, 51, 59-61, 234, 239-240, 242-243, 245, 247
VAB	18-19, 93, 101-102, 104-106, 110, 135, 150-153, 162, 164-165, 167, 170, 172, 174, 176, 178, 195, 206-208, 211, 219-220, 222, 224, 226-227, 232, 246, 256, 282, 306-307
Vertical Assembly Building	22, 25-26, 54-55, 93-94, 153, 166, 170, 176, 195, 306-307
Visual Instrumentation	35-36
Von Braun, Wernher	15
West Test Area. MSFC	135, 143, 156-159
Wyle Labs, Huntsville	201-202

DVD-Video/ROM disc

Films Include:

* *Saturn - From Start to Finish.* A film tracing the life of a Saturn V from contractor construction through to the launch of Apollo 11. Includes synchronized stereo audio of the launch.

* Arrival and Installation of S-II-T static firing test stage at Michoud Assembly Facility October 1965

* First test firing of Saturn S-IC-T at Mississippi Test Facility March 1967

* J-2 Engine test in a vacuum at Arnold Engineering Development Center.

* Arrival and test firing of Saturn S-IC-504 at Mississippi Test Facility

* S-IC-501 Test firing at MSFC February 1966

* S-II Battleship test stage firing August 1968

* Arrival and installation of S-IVB Battleship J2 engine at Arnold Engineering Development Center and vacuum engine test.

* J-2 Engine test firing at Santa Susana Field Lab June 1968

* F-1 installation footage in test stand. (silent)

* First test firing of S-IVB-501 at Douglas test stands in Sacramento

Also ROM content

F1 data, History of Saturn V engines, J2 data, Saturn transport records, S-IC manufacturing, S-II Battleship tests, S-II manufacturing, S-IVB Battleship firings.

ROM content can be accessed in the folder marked "Files."

Includes more statistics about the construction of the Saturn V.

For more information and other great space books visit www.apogeebooks.com